COMMERCIAL WIRELESS CIRCUITS and COMPONENTS
HANDBOOK

COMMERCIAL WIRELESS CIRCUITS and COMPONENTS
HANDBOOK

Editor-in-Chief
MIKE GOLIO

CRC PRESS

Boca Raton   London   New York   Washington, D.C.

This material was previously published in *The RF and Microwave Handbook.* © CRC Press LLC 2001

Library of Congress Cataloging-in-Publication Data

Commercial wireless circuits and components handbook / editor-in-chief Mike Golio
    p. cm.
    ISBN 0-8493-1564-6
      1. Radio circuits--Handbooks, manuals, etc. 2. Wireless communication
      systems--Equipment and supplies--Handbooks, manuals, etc. I. Golio, John Michael,
      1954-

TK6560 .C66 2002
621.384'12--dc21

                                                   2002074128

Visit the CRC Press Web site at www.crcpress.com

Preface

The purpose of the *CRC Commercial Wireless Circuits and Components Handbook* is to provide single volume comprehensive coverage of microwave and wireless circuit design. It is intended to be a starting point for any project involving design, development, or acquisition of RF or microwave circuitry. The articles that comprise the handbook provide important information for practicing engineers in industry, government, and academia. The intended audience also includes microwave and other electrical engineers requiring information outside of their area of expertise as well as managers, marketers, and technical support workers who need better understanding of the fields driving their decisions.

The book includes overview articles on the fundamentals of transmitters and receivers, detailed chapters on individual circuit types, including power amplifiers, mixers, oscillators, phased lock loops, filters, switches, low noise amplifiers, and modulation circuitry. Additional chapters cover packaging as well as both large and small signal characterization and high-volume testing techniques for both devices and circuits. Simulation and device modeling for circuit simulation is also included. Finally, all of the articles provide the reader with additional references to related expert literature.

Acknowledgments

This handbook would simply never have been completed if it were not for the efforts of the managing editor, Janet Golio. I am also significantly indebted to the Handbook Editorial Board. This Board contributed to every phase of handbook development. Their efforts are reflected in the organization and outline of the material, selection and recruitment of authors, article contributions, and review of the articles. I am happy to acknowledge their help. I also thank the handbook professionals at CRC Press.

The Editor

Michael Golio is the Director of RF Technology Applications of Thoughtbeam, a Motorola Company. His work focuses on the evaluation and commercialization of emerging compound semiconductor material technologies — especially for RF and microwave applications.

Dr. Golio received his BSEE degree from the University of Illinois in 1976. He worked for 2 years in the Microwave Tunable Devices Organization at Watkins–Johnson before returning to school to complete his MSEE and Ph.D. degrees at North Carolina State University in 1980 and 1983 respectively. His graduate research focused on microwave devices, nonlinear models and carrier transport in compound semiconductors. Upon completion of his graduate work, he served as an Assistant Professor of Electrical Engineering at Arizona State University before joining Motorola Government Electronics Group in 1986. There he directed research on characterization, parameter extraction, and modeling of nonlinear microwave devices. In 1991, he moved to Motorola's Semiconductor Products Sector to develop a GaAs fabrication facility to address commercial products, including chips for cellular phones, digital pagers, and wireless LANs. From 1996 to 2001, Dr. Golio was Director of the RF/Power Design Center at Rockwell Collins in Cedar Rapids, Iowa. The center conducted research and development efforts into RF, microwave, and antenna technologies for commercial and military avionics applications.

Dr. Golio is the author of over 100 publications. He is editor of two successful books: *Microwave MESFETs and HEMTs,* Artech House, 1991, and *RF and Microwave Handbook*, CRC Press, 2000. He has served as organizer for several microwave conferences, workshops, and panel sessions. In 1996 he was elected Fellow of the IEEE. He has served as the Distinguished Microwave Lecturer for the IEEE MTT Society and is currently co-editor of the *IEEE Microwave Magazine*.

Editorial Board

Contributors

Mark Bloom
Motorola, Inc.
Tempe, Arizona

Walter R. Curtice
W.R. Curtice Consulting
Washington Crossing, Pennsylvania

W.R. Deal
Malibu Networks
Calabasas, California

Mike Golio
Motorola, Inc.
Tempe, Arizona

Ron E. Ham
Consulting Engineer
Austin, Texas

Tatsuo Itoh
Electrical Engineering Department
University of California
Los Angeles, California

Christopher Jones
M/A-COM TycoElectronics
Lowell, Massachusetts

J. Stevenson Kenney
School of Electrical and Computer
Engineering
Georgia Institute of Technology
Atlanta, Georgia

Ron Kielmeyer
Motorola, Inc.
Scottsdale, Arizona

Jakub Kucera
Infineon Technologies
Munich, Germany

Jean-Pierre Lanteri
M/A-COM TycoElectronics
Lowell, Massachusetts

Urs Lott
Acter AG
Zurich, Switzerland

John R. Mahon
M/A-COM TycoElectronics
Lowell, Massachusetts

Charles Nelson
Electrical and Electron Engineering Department
California State University
Sacramento, California

Robert Newgard
Rockwell Collins
Cedar Rapids, Iowa

Anthony E. Parker
Department. of Electronics
Macquarie University
Sydney, Australia

Anthony Pavio
Motorola, Inc.
Tempe, Arizona

Jeanne Pavio
Motorola SPS
Phoenix, Arizona

Y. Qian
University of California
Los Angeles, California

Vesna Radisic
HRL Laboratory, LLC
Malibu, California

James Grantley Rathmell
School of Electrical and Information Engineering
The University of Sydney
Sydney, Australia

Alfy Riddle
Macallan Consulting
Milpitas, California

Jonathan B. Scott
Agilent Technologies
Santa Rosa, California

Warren L. Seely
Motorola GSTG, Inc.
Scottsdale, Arizona

John F. Sevic
UltraRF, Inc.
Sunnyvale, California

Joseph Staudinger
Motorola, Inc.
Tempe, Arizona

Michael B. Steer
Electrical and Computer Engineering Department
North Carolina State University
Raleigh, North Carolina

Richard V. Snyder
RS Microwave
Butler, New Jersey

Daniel C. Swanson Jr.
Bartley RF Systems
Amesbury, Massachusetts

R.J. Trew
Virginia Tech University
Blacksburg, Virginia

Contents

1
Receivers

Warren L. Seely
Motorola GSTG, Inc.

1.1 Introduction

An electromagnetic signal picked up by an antenna is fed into a receiver. The ideal receiver rejects all unwanted noise including other signals. It does not add any noise or interference to the desired signal. The signal is converted, regardless of form or format, to fit the characteristics required by the detection scheme in the signal processor, which in turn feeds an intelligible user interface (Fig. 1.1). The unit must require no new processes, materials, or devices not readily available. This ideal receiver adds no weight, size, or cost to the overall system. In addition, it requires no power source and generates no heat. It has an infinite operating lifetime in any environment, and will never be obsolete. It will be flexible, fitting all past, present, and future requirements. It will not require any maintenance, and will be transparent to the user, who will not need to know anything about it in order to use it. It will be fabricated in an "environmentally friendly" manner, visually pleasing to all who see it, and when the user is finally finished with this ideal receiver, he will be able to recycle it in such a way that the environment is improved rather than harmed. Above all else, this ideal receiver must be wanted by consumers in very large quantities, and it must be extremely profitable to produce. Fortunately, nobody really expects to achieve all of these "ideal" characteristics, at least not yet! However, each of these characteristics must be addressed by the engineering design team in order to produce the best product for the application at hand.

1.2 Frequency

Receivers represent a technology with tremendous variety. They include AM, FM, analog, digital, direct conversion, single and multiple conversions, channelized, frequency agile, spread spectrum, chirp, frequency hopping, and others. The applications are left to the imaginations of the people who create them. Radio, telephones, data links, radar, sonar, TV, astronomical, telemetry, and remote control, are just a few of those applications. Regardless of the application, the selection of the operating frequencies is fundamental to obtaining the desired performance.

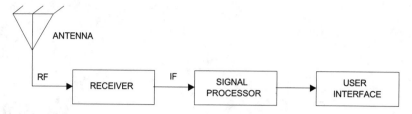

FIGURE 1.1   The receiver.

The actual receiver frequencies are generally beyond the control of the design team, being dictated, controlled, and even licensed by various domestic or foreign government agencies, or by the customer. When a product is targeted for international markets, the allocated frequencies can take on nightmare qualities due to differing allocations, adjacent interfering bands, and neighboring country restrictions or allocations. It will usually prove impossible to get the ideal frequency for any given application, and often the allocated spectrum will be shared with other users and multiple applications. Often the spectrum is available for a price, usually to the highest bidder. Failure to utilize the purchased spectrum within a specified time frame may result in forfeiture of what is now an asset; an expensive mistake. This has opened up the opportunity to speculate and make (or lose) large sums of money by purchasing spectrum to either control a market or resell to other users. For some applications where frequency allocation is up to the user, atmospheric or media absorption, multipathing, and background noise are important factors that must be considered. These effects can be detrimental or used to advantage. An example includes cross links for use with communications satellites, where the cross link is unaffected by absorption since it is above the atmosphere. However, the frequency can be selected to use atmospheric absorption to provide isolation between ground signals and the satellite cross links. Sorting out these problems is time consuming and expensive, but represents a fundamental first step in receiver design.

1.3   Dynamic Range

The receiver should match the dynamic range of the desired signal at the receiver input to the dynamic range of the signal processor. Dynamic range is defined as the range of desirable signal power levels over which the hardware will operate successfully. It is limited by noise, signal compression, and interfering signals and their power levels.

1.3.1   Power and Gain

The power in any signal(s), whether noise, interference or the desired signal, can be measured and expressed in Watts (W), decibels referenced to 1 Watt (dBW), milliwatts (mW) or decibels referenced to one milliwatt (dBm). The power decibel is 10 times the LOG of the dimensionless power ratio. The power gain of a system is the ratio output signal power to the input signal power expressed in decibels (dB). The gain is positive for components in which the output signal is larger than the input, negative if the output signal is smaller. Negative gain is loss, expressed as attenuation (dB). The power gain of a series component chain is found by simple multiplication of the gain ratios, or by summing the decibel gains of the individual components in the chain. All of these relationships are summarized in Fig. 1.2.

1.3.2   Noise

Thermal noise arises from the random movement of charge carriers. The thermal noise power (n_T) is usually expressed in dBm (N_T), and is the product of Boltzman's constant (k), system temperature in degrees Kelvin (T), and a system noise bandwidth in Hertz (b_n). The system noise bandwidth (b_n) is

$$Decibel = 10LOG\left[\frac{p}{p_{ref}}\right]$$

$$g(-) = \frac{p_{out}}{p_{in}}$$

$$P(dBW) = 10LOG\left[\frac{p(W)}{1W}\right]$$

$$G(dB) = 10LOG(g)$$

$$P(dBm) = 10LOG\left[\frac{p(mW)}{1mW}\right]$$

$$g_{total}(-) = g_1 * g_2 *.... * g_N = \frac{p_{Nout}}{p_{1in}}$$

$0dBW = 1Watt$

$0dBm = 1mW$

$1000mW = 1W$

$30dBm = 0dBW$

$$G_{total}(dB) = G_1(dB) + G_2(dB) +...+ G_N(dB) = \Sigma G_i$$
$$Loss(dB) = -G(dB)$$
$$Loss(dB) = Attenuation(dB)$$

FIGURE 1.2 Power and gain relationships.

$$n_T = kTb_n$$

$$b_n(Hz) = \frac{n_{tot-ave}(W)}{n_{pk-ave}(W/Hz)}$$

$$k = 1.38*10^{-23}\frac{W\sec}{K}$$

$$T(K) = T(^0C) + 273.15$$

$$n_T = 1.38*10^{-23}\frac{W\sec}{K}*(25^0C+273.15)K*1Hz = 4.46*10^{-21}W = 4.46*10^{-18}mW$$

$$N_T = 10LOG(n_T) = -204dBW = -174dBm$$

FIGURE 1.3 Noise power relationships.

defined slightly different from system bandwidth. It is determined by measuring or calculating the total system thermal average noise power ($n_{tot-ave}$) over the entire spectrum and dividing it by the system peak average noise power (n_{pk-ave}) in a 1 Hz bandwidth. This has the effect of creating a system noise bandwidth in which the noise is all at one level, that of the peak average noise power. For a 1 Hz system noise bandwidth at the input to a system at room temperature (25°C), the thermal noise power is about −174 dBm. These relationships are summarized in Fig. 1.3.

1.3.3 Receiver Noise

The bottom end of the dynamic range is set by the lowest signal level that can reasonably be expected at the receiver input and by the power level of the smallest acceptably discernible signal as determined at the input to the signal processor. This bottom end is limited by thermal noise at the input, and by the gain distribution and addition of noise as the signal progresses through the receiver. Once a signal is below the minimum discernible signal (MDS) level, it will be lost entirely (except for specialized spread spectrum receivers). The driving requirement is determined by the signal clarity needed at the signal processor. For analog systems, the signal starts to get fuzzy or objectionably noisy at about 10 dB above the noise floor. For digital systems, the allowable bit error rate determines the acceptable margin above the noise floor. Thus the signal with margin sets the threshold minimum desirable signal level.

Noise power at the input to the receiver will be amplified and attenuated like any other signal. Each component in the receiver chain will also add noise. Passive devices such as filters, cables, and attenuators

$$f_n = \frac{{}^{s_i}/_{n_i}}{{}^{s_o}/_{n_o}} = \frac{n_o}{g n_i}$$

$$NF = 10 LOG(f_n)$$

$$T_n = T(f_n - 1) \quad where\, T\ is\ in\ Kelvin$$

$$N_o = NF + G + N_i$$

$$f_t = f_1 + \frac{f_2 - 1}{g_1} + \frac{f_3 - 1}{g_1 * g_2} + \cdots + \frac{f_n - 1}{\Pi g_n}$$

$$\Delta f_{n-bandwidth} = \frac{g_1 f_1 + (f_2 - 1)^{{b_{n2}}/_{b_{n1}}}}{g_1 f_1 + f_2 - 1} \qquad b_{n2} > b_{n1}$$

$$\Delta f_{n-image} = 1 + \frac{l_{ar}}{f_x}$$

$$f_{total} = f_{cascade} * \Delta f_{n-bandwidth} * \Delta f_{n-image}$$

FIGURE 1.4 Receiver noise relationships.

will cause a drop in both signal and noise power alike. These passive devices also contribute a small amount of internally generated thermal noise. Thus the actual noise figure of a passive device is slightly higher than the attenuation of that component. This slight difference is ignored in receiver design since the actual noise figures and losses vary by significantly larger amounts. Passive mixers will generally have a noise figure about 1 dB greater than the conversion loss. Active devices can exhibit loss or gain, and signal and noise power at the input will experience the same effect when transferred to the output. However, the internally generated noise of an active device will be substantial and must be accounted for, requiring reasonably accurate noise figures and gain data on each active component.

The bottom end dynamic range of a receiver component cascade is easily described by the noise equations shown in Fig. 1.4. The first three equations for noise factor (f_n), noise figure (NF), and noise temperature (T_n) are equivalent expressions to quantify noise. The noise factor is a dimensionless ratio of the input signal-to-noise ratio and the output signal-to-noise ratio. Replacing the signal ratio with gain results in the final form shown. Noise figure is the decibel form of noise factor, in units of dB. Noise temperature is the conversion of noise factor to an equivalent input temperature that will produce the output noise power, expressed in Kelvin. Convention dictates using noise temperature when discussing antennas and noise figure for receivers and associated electronics. By taking the decibel equivalent of the noise factor, the expression for noise out (N_o) is obtained, where noise in (N_i) is in dBm and noise figure (NF) and gain (G) are in dB. The cascaded noise factor (f_t) is found from the sum of the added noise due to each cascaded component divided by the total gain preceding that element. Use the cascaded noise factor (f_t) followed by the noise out (N_o) equation to determine the noise level at each point in the receiver.

Noise factor is generally computed for a 1 Hz bandwidth and then adjusted for the narrowest filter in the system, which is usually downstream in the signal processor. Occasionally, it will be necessary to account for noise power added to a cascade when components following the narrowest filter have a relatively broad noise bandwidth. The filter will eliminate noise outside its band up to that filter. Broader band components after the filter will add noise back into the system depending on their noise bandwidth. This additional noise can be accounted for using the equation for $\Delta f_{n-bandwidth}$, where subscript 1 indicates the narrowband component followed by the wideband component (subscript 2). Repeated application of this equation may be necessary if several wideband components are present following the filter. Image noise can be accounted for using the relationship for $\Delta f_{n\text{-image}}$ where l_{ar} is the dimensionless attenuation ratio between the image band and desired signal band and f_x is the noise factor of the system up to the image generator (usually a mixer). Not using an image filter in the system will result in a $\Delta f_{n\text{-image}} = 2$ resulting in a 3 dB increase in noise power. If a filter is used to reject the image by 20 dB, then a substantial reduction in image noise will be achieved. Finally, the corrections for bandwidth and image are easily incorporated using the relationship for the cascaded total noise factor, f_{total}.

A simple single sideband (SSB) receiver example, normalized to a 1 Hz noise bandwidth, is shown in Fig. 1.5. It demonstrates the importance of minimizing the use of lossy components near the receiver front end, as well as the importance of a good LNA. A 10 dB output signal-to-noise margin has been established as part of the design. Using the −174 dBm input thermal noise level and the individual component gains and noise figures, the normalized noise level can be traced through the receiver, resulting in an output noise power of −136.9 dBm. Utilizing each component gain and working backwards from this point with a signal results in the MDS power level in the receiver. Adding the 10 dB signal-to-noise margin to the MDS level results in the signal with margin power level as it progresses through the receiver. The signal and noise levels at the receiver input and output are indicated. The design should minimize the gap between the noise floor and the MDS level. Progressing from the input toward the output, it is readily apparent that the noise floor gets closer to and rapidly converges with the MDS level due to the addition of noise from each component, and that lossy elements near the input hurt system performance. The use of the low noise amplifier as close to the front end of the cascade as possible is critical in order to mask the noise of following components in the cascade and achieve minimum noise figure. The overall cascaded receiver gain is easily determined by the difference in the signal levels from input to output.

The noise floor margins at both the input and output to the receiver are also easily observed, along with the receiver noise figure. Note also that the actual noise power does not drop below the thermal noise floor, which is always the bottom limit. Finally, the actual normalized signal with margin level of −154.9 dBm at the input to the receiver is easily determined.

1.3.4   Intermodulation

Referring to Fig. 1.6, the upper end of the dynamic range is limited by nonlinearity, compression, and the resulting intermodulation in the presence of interfering signals. The in-band two-tone output 3rd order intercept point (3OIP) is a measure of the nonlinearity of a component. This particular product is important because it can easily result in an undesired signal that is close to the desired signal, which would be impossible to eliminate by filtering. By definition, the 3OIP is found by injecting 2 equal amplitude signals (F_1 and F_2) that are not only close to each other in frequency, but are also both within the passband of the component or system. The 3rd order intermodulation products are then given by $\pm nF_1 \pm mF_2$ where $n + m = 3$. For 3rd order products, n and m must be 1 or 2. Since negative frequencies are not real, this will result in two different 3rd order products which are near each other and within the passband. The power in the inter-modulation products is then plotted, and both it and Pout are projected until they intersect, establishing the 3OIP. The desired signal is projected using a 1:1 slope, while the 3rd order products are projected using a 3:1 slope. The output saturation power (P_{sat}) is the maximum power a device will produce. The output 1 dB compression point (1 dB OCP) is the point at which the gain is lowered by 1 dB from small signal conditions due to signal compression on its way to complete saturation. In general, higher values mean better linearity and thus better performance. However, component costs rapidly increase along with these numbers, especially above saturated levels of about +15 dBm, limiting what can be achieved within project constraints. Thus, one generally wants to minimize these parameters in order to produce an affordable receiver. For most components, a beginning assumption of square law operation is reasonable. Under these conditions, the 1 dB OCP is about 3 dB below P_{sat}, and the 3OIP is about 10 dB above the 1 dB OCP. When the input signal is very small (i.e., small signal conditions), Pout increases on a 1:1 slope, and 3rd order products increase on a 3:1 slope. These numbers can vary significantly based on actual component performance and specific loading conditions. This whole process can be reversed, which is where the value of the concept lies. By knowing the small signal gain, Pin or Pout, and the 3OIP,

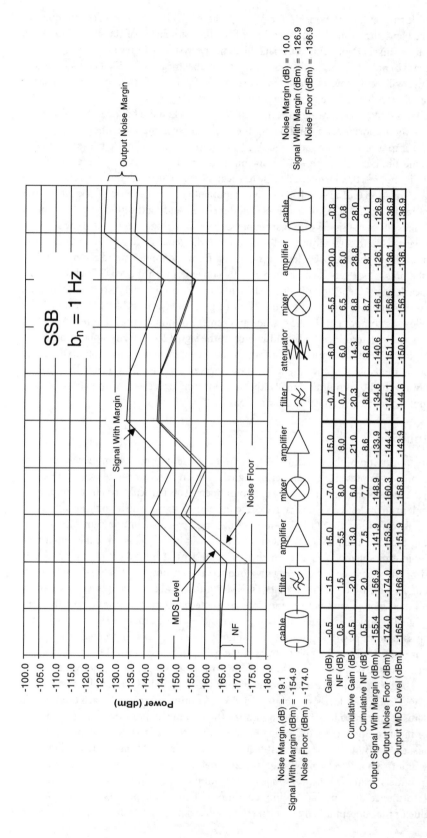

FIGURE 1.5   Example SSB receiver noise and signal cascade normalized to $b_n = 1$.

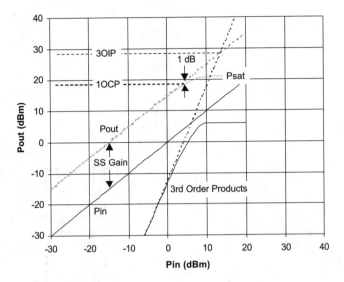

FIGURE 1.6 3OIP, P$_{sat}$, and 1 dB OCP.

all the remaining parameters, including 1OCP, P$_{sat}$, and 3rd order IM levels can be estimated. As components are chosen for specific applications, real data should be utilized where possible. Higher order products may also cause problems, and should be considered also. Finally, any signal can be jammed if the interfering signal is large enough and within the receiver band. The object is to limit the receiver's susceptibility to interference under reasonable conditions.

1.3.5 Receiver Intermodulation

Analog receiver performance will start to suffer when in-band 3rd order products are within 15 dB of the desired signal at the detector. This level determines the maximum signal of interest (MSI). The margin for digital systems will be determined by acceptable bit error rates. The largest signal that the receiver will pass is determined by the saturated power level of the receiver. Saturating the receiver will result in severe performance problems, and will require a finite time period to recover and return to normal performance. Limiting compression to 1 dB will alleviate recovery.

Analyzing the receiver component cascade for 3OIP, P$_{sat}$, 1 dB OCP, and MSI will provide insight into the upper limits of the receiver dynamic range, allowing the designer to select components that will perform together at minimum cost and meet the desired performance (Fig. 1.7). The first equation handles the cascading of known components to determine the cumulated input t 3rdorderinpuintermod

$$\frac{1}{p_{3iip,tot}} = \frac{1}{p_{3iip,1}} + \frac{g_1}{p_{3iip,2}} + \cdots + \frac{\Pi g_n}{p_{3iip,n}}$$

$$P_{3OIP} = P_{3IIP} + G_{ss}$$

$$SFDP = \frac{2}{3}\left(3IIP - MDS_{input}\right) = \frac{2}{3}\left(3OIP - MDS_{output}\right)$$

$$MSI_{out} = NOISEFLOOR_{out} + SPDR$$

FIGURE 1.7 Receiver 3OIP, P$_{sat}$, and 1 dB OCP cascade.

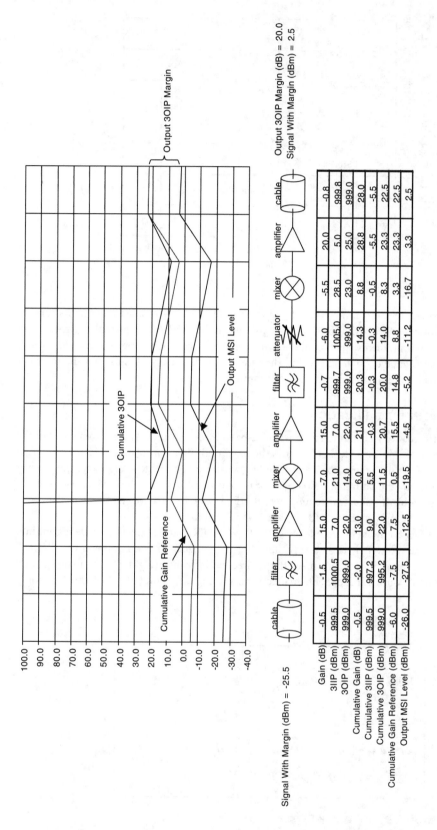

Signal With Margin (dBm) = -25.5

Output 3OIP Margin (dB) = 20.0
Signal With Margin (dBm) = 2.5

	cable	filter	amplifier	mixer	amplifier	filter	attenuator	mixer	amplifier	cable
Gain (dB)	-0.5	-1.5	15.0	-7.0	15.0	-0.7	-6.0	-5.5	20.0	-0.8
3IIP (dBm)	999.5	1000.5	7.0	21.0	7.0	999.7	1005.0	28.5	5.0	999.8
3OIP (dBm)	999.0	999.0	22.0	14.0	22.0	999.0	999.0	23.0	25.0	999.0
Cumulative Gain (dB)	-0.5	-2.0	13.0	6.0	21.0	20.3	14.3	8.8	28.8	28.0
Cumulative 3IIP (dBm)	999.5	997.2	9.0	5.5	-0.3	-0.3	-0.3	-0.5	-5.5	-5.5
Cumulative 3OIP (dBm)	999.0	995.2	22.0	11.5	20.7	20.0	14.0	8.3	23.3	22.5
Cumulative Gain Reference (dBm)	-6.0	-7.5	7.5	0.5	15.5	14.8	8.8	3.3	23.3	22.5
Output MSI Level (dBm)	-26.0	-27.5	-12.5	-19.5	-4.5	-5.2	-11.2	-16.7	3.3	2.5

FIGURE 1.8 Example receiver 3OIP and signal cascade.

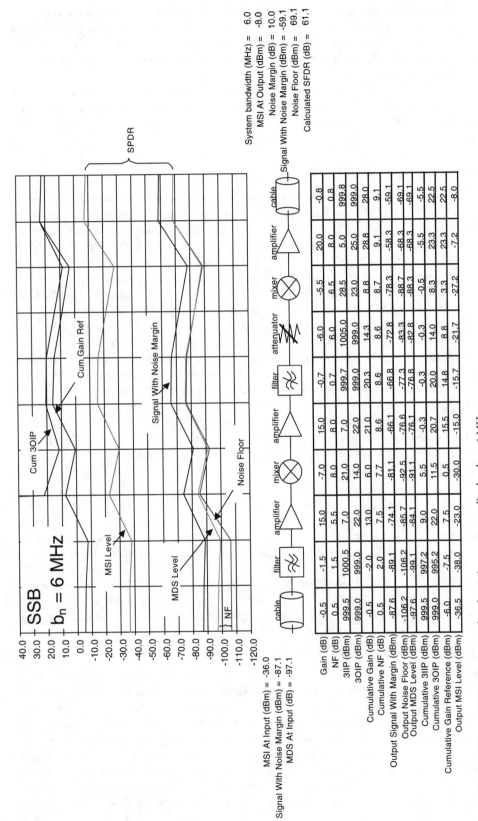

FIGURE 1.9 Example SSB receiver spur free dynamic range normalized to b_n = 6 MHz.

point. After utilizing this equation to determine the cascaded 3OIP up to the component being considered, the second equation can be utilized to determine the associated P_{3OIP}. Successive application will result in completely determining the cascaded performance. The last two equations determine the 3rd order IM spur free dynamic range (SPDR) and the maximum spur-free level or maximum signal of interest (MSI) at the output of the receiver.

An example receiver 3OIP and signal cascade is shown in Fig. 1.8. The results of the cumulative 3OIP are plotted. A gain reference is established by setting and determining the cumulative gain and matching it to the cumulative 3OIP at the output. A design margin of 20 dB is added to set the MSI power level for the cascade.

1.3.6 Receiver Dynamic Range

Combining the results for the noise and intermodulation from the above discussion into one graph results in a graphical representation of the dynamic range of the receiver (Fig. 1.9). Adjusting for the 6 MHz bandwidth moves the noise plots up by 10LOG(6e6) = 67.8 dB. The cumulative 3OIP and gain reference plot remain at the previously determined levels. The SPDR = 2(–5.5 dBm – (–97.1 dBm))/3 = 61.1 dB is calculated and then used to determine the MSI level = –69.1 dBm + 61.1 dB = –8 dBm at the output. The MSI level on the graph is set to this value, backing off to the input by the gain of each component.

The receiver gain, input, and output dynamic ranges, signal levels which can be easily handled, and the appropriate matching signal processing operating range are readily apparent, being between the MSI level and the signal with noise margin. The receiver NF, 3OIP, gain, and SPDR are easily determined from the plot. Weaknesses and choke points, as well as expensive parts are also apparent, and can now be attacked and fixed or improved. In general, components at or near the input to the receiver dominate the noise performance, and thus the lower bounds on dynamic range. Components at or near the output dominate the nonlinear performance, and thus the upper bounds on dynamic range.

The use of 3OIP and noise floors is just one way commonly used to characterize the dynamic range of a receiver. Other methods include determining compression and saturation curves, desensitization, noise power ratios, and intercept analysis for other IM products such as 2OIP up to as high as possibly 15OIP. Specific applications will determine the appropriate analysis required in addition to the main SFDR analysis described above.

1.4   The LO Chain

A reference signal or local oscillator (LO) is generally required in order to up- or downconvert the desired signal for further processing. The design of the LO chain is tied to the receiver components by frequency, power level, and phase noise. The LO signal will have both amplitude and phase noise components, both of which will degrade the desired signal. Often, a frequency agile LO is required. The LO can be generated directly by an oscillator, multiplied or divided from another frequency source, created by mixing several signals, injection or phase locked to a reference source, digitally synthesized, or any combination thereof.

1.4.1   Amplitude and Phase Noise

A pure tone can be represented as a vector of a given amplitude (α) rotating with a fixed angular velocity (ω) as shown in Fig. 1.10. A random noise source can be viewed similarly, but has random phase equally distributed over time about 360°, and random amplitude based on a probability distribution. At any given instant in time the vector will change amplitude and angle. A plot of the noise vector positions for a relatively long period of time would appear solid near the origin and slowly fade away at larger radii from the origin (Fig. 1.10). A plot of the number of hits vs. distance from the origin would result in the probability distribution. This random noise is the same noise present in all electronic systems. Combining the pure tone with the random noise results in the vector addition of both signals (Fig. 1.10). At any given instance in time the combined angular velocity will change by $\Delta\omega$, and the amplitude will change

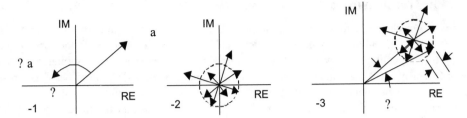

FIGURE 1.10 Phase noise, AM, and PM noise.

by $\Delta\alpha$. The frequency jittering about ω and the amplitude wavering about α result in AM and PM noise components, which distort the signal of interest. Thus, phase noise is a measure of signal stability.

The design of the LO chain usually includes at least one final amplifier stage that is completely saturated. This sets the LO amplitude to a fixed level, minimizes temperature variations, and minimizes or eliminates the AM noise component. Saturation results in a gain reduction of several dB and simultaneously limits the maximum amplitude that can be achieved. Thus as the random noise vector changes the LO amplitude, the saturated amplifier acts to eliminate the output amplitude change. The AM contribution to noise is cleaned up.

The phase noise in the LO chain must be attacked directly at the source. Clean, low phase noise signal generation in oscillators is achieved by the use of very high Q filter components, incorporating bipolar devices as the active oscillator element, maximizing the source power generation, and careful design of the conduction cycle within the oscillator itself. Once a clean signal is created, it must be kept clean.

Frequency multiplying or dividing the signal will also multiply or divide the phase noise by a factor of 20*LOG(N) at any given offset from the base signal. Conversely, if the signal is multiplied or divided, then the spectrum is respectively stretched or contracted by the factor N. The mixing process also mixes the phase noise, but the net result will depend on the mixer types utilized. Injection locking will replicate the injection source modified by the multiplication or division factor N. A phase lock loop exhibits close in noise dependent on the reference source and loop circuitry, but the far out phase noise is set by the source used in the loop itself. Finally, the LO is utilized in the receiver chain to perform frequency conversion. The resulting converted signal will have components of phase noise from the original signal, the LO signal, and from noise in the mixing component.

1.5 The Potential for Trouble

Receivers are designed to work with very small signals, transforming these to much larger signals for handling in the signal processor. This inherent process is open to many pitfalls resulting in the vast majority of problems encountered in creating a viable product. It will only take a very small interfering or spurious signal to wreak havoc. Interfering signals generally can be categorized as externally generated, internally generated, conducted, electromagnetically coupled, piezoelectrically induced, electromechanically induced, and optically coupled or injected. Some will be fixed, others may be intermittent, even environmentally dependent. Most of these problem areas can be directly addressed by simple techniques, precluding their appearance altogether. However, ignoring these potential problem areas usually results in disaster, primarily because they are difficult to pinpoint as to cause and effect, and because eliminating them may be difficult or impossible without making major design changes and fabricating new hardware in order to verify the solution. This can easily turn into a long-term iterative nightmare. Additionally, if multiple problems are present, whether or not they are perceived as multiple problems or as a single problem, the amount of actual time involved in solving them will go up exponentially! Oscillator circuits are generally very susceptible to any and all problems, so special consideration should be given in their design and use. Finally, although the various cause, effect, and insight into curing problems are broken down into component parts in the following discussion, it is often the case that several concepts must be combined to correctly interpret and solve any particular problem at hand.

1.5.1   Electromechanical

Vibrations and mechanical shocks will result in physical relative movement of hardware. Printed circuit boards (PCBs), walls, and lids may bow or flutter. Cables and wire can vibrate. Connectors can move. Solder joints can fracture. PCBs, walls, and lids capacitively load the receiver circuitry, interconnects, and cabling. Movement, even very small deflections, will change this parasitic loading, resulting in small changes in circuit performance. In sensitive areas, such as near oscillators and filters, this movement will induce modulation onto the signals present. In phase-dependent systems, the minute changes in physical makeup and hence phase length of coaxial cable will appear as phase modulation. Connector pins sliding around during vibration can introduce both phase and amplitude noise. These problems are generally addressed by proper mechanical design methods, investigating and eliminating mechanical resonance, and minimizing shock susceptibility. Don't forget that temperature changes will cause expansion and contraction, with similar but slower effects.

1.5.2   Optical Injection

Semiconductor devices are easily affected by electromagnetic energy in the optical region. Photons impinging on the surface of an active semiconductor create extra carriers, which appear as noise. A common occurrence of this happens under fluorescent lighting common in many offices and houses. The 60 Hz "hum" is present in the light given off by these fixtures. The light impinges on the surface of a semiconductor in the receiver, and 60 Hz modulation is introduced into the system. This is easily countered by proper packaging to prevent light from hitting optically sensitive components.

1.5.3   Piezoelectric Effects

Piezoelectric materials are reciprocal, meaning that the application of electric fields or mechanical force changes the electromechanical properties, making devices incorporating these materials highly susceptible to introducing interference. Even properly mounted crystals or SAW devices, such as those utilized in oscillators, will move in frequency or generate modulation sidebands when subjected to mechanical vibration and shock. Special care should therefore be given to any application of these materials in order to minimize these effects. This usually includes working closely with the original equipment manufacturer (OEM) vendors to ensure proper mounting and packaging, followed by extensive testing and evaluation before final part selection and qualification.

1.5.4   Electromagnetic Coupling

Proper design, spacing, shielding, and grounding is essential to eliminate coupled energy between circuits. Improper handling of each can actually be detrimental to achieving performance, adding cost without benefit, or delaying introduction of a product while problems are solved. Proper design techniques will prevent inadvertent detrimental E-M coupling. A simple example is a reject filter intended to minimize LO signal leakage into the receiver, where the filter is capable of the required performance, but the packaging and placement of the filter allow the unwanted LO to bypass the filter and get into the receiver anyway.

It is physically impossible to eliminate all E-M resonant or coupled structures in hardware. A transmission line is created by two or more conductors separated by a dielectric material. A waveguide is created by one or more conductive materials in which a dielectric channel is present, or by two or more nonconductive materials with a large difference in relative dielectric constant. Waveguides do not have to be fully enclosed in order to propagate E-M waves. In order to affect the hardware, the transmission line or waveguide coupling must occur at frequencies that will interfere with operation of the circuits, and a launch into the structure must be provided. Properly sizing the package (a resonant cavity) is only one consideration. Breaking up long, straight edges and introducing interconnecting ground vias on multilayer PCBs can be very effective. Eliminating loops, opens and shorts, sharp bends, and any other "antenna like" structures will help.

E-field coupling usually is associated with high impedance circuits, which allow relatively high E-fields to exist. E-field or capacitive coupling can be eliminated or minimized by any grounded metal shielding. M-field coupling is associated with low impedance circuits in which relatively high currents and the associated magnetic fields are present. M-field or magnetic coupling requires a magnetic shielding material. In either case, the objective is to provide a completely shielded enclosure. Shielding metals must be thick enough to attenuate the interfering signals. This can be determined by E or M skin effect calculations. Alternatively, absorbing materials can also be used. These materials do not eliminate the basic problem, but attempt to mask it, often being very effective, but usually relatively expensive for production environments. Increased spacing of affected circuitry, traces, and wires will reduce coupling. Keeping the E-M fields of necessary but interfering signals orthogonal to each other will add about 20 dB or more to the achieved isolation.

Grounding is a problem that could be considered the "plague" of electronic circuits. Grounding and signal return paths are not always the same, and must be treated accordingly. The subject rates detailed instruction, and indeed entire college level courses are available and recommended for the serious designer. Basically, grounding provides a reference potential, and also prevents an unwanted differential voltage from occurring across either equipment or personnel. In order to accomplish this objective, little or no current must be present. Returns, on the other hand, carry the same current as the circuit, and experience voltage drops accordingly. A return, in order to be effective, must provide the lowest impedance path possible. One way to view this is by considering the area of the circuit loop, and making sure that it is minimized. In addition, the return conductor size should be maximized.

1.6 Summary

A good receiver design will match the maximum dynamic range possible to the signal processor. In order to accomplish this goal, careful attention must be given to the front end noise performance of the receiver and the selection of the low noise amplifier. Equally important in achieving this goal is the linearity of the back end receiver components, which will maximize the SFDR. The basic receiver calculations discussed above can be utilized to estimate the attainable performance. Other methods and parameters may be equally important and should be considered in receiver design. These include phase noise, noise power ratio, higher order intercepts, internal spurious, and desensitization.

Further Reading

Sklar, Bernard, Digital Communications Fundamentals and Applications, Prentice-Hall, Englewood Cliffs, NJ, 1988.

Tsui, Dr. James B., *Microwave Receivers and Related Components*, Avionics Laboratory, Air Force Write Aeronautical Laboratories, 1983.

Watkins-Johnson Company Tech-notes, Receiver Dynamic Range: Part 1, Vol. 14, No. 1.

Watkins-Johnson Company Tech-notes, Receiver Dynamic Range: Part 2, Vol. 14, No. 2.

Steinbrecher, D., Achieving Maximum Dynamic Range in a Modern Receiver, *Microwave Journal*, Sept 1985.

2

Transmitters

Warren L. Seely
Motorola GSTG, Inc.

2.1 Introduction

A signal is generated by the frequency synthesizer and amplified by the transmitter (Fig. 2.1), after which it is fed to the antenna for transmission. Modulation and linearization may be included as part of the synthesized signal, or may be added at some point in the transmitter. The transmitter may include frequency conversion or multiplication to the actual transmit band. An ideal transmitter exhibits many of the traits of the ideal receiver described in the previous section of this handbook. Just as with the receiver, the task of creating a radio transmitter begins with defining the critical requirements, including frequencies, modulations, average and peak powers, efficiencies, adjacent channel power or spillover, and phase noise. Additional transmitter parameters that should be considered include harmonic levels, noise powers, spurious levels, linearity, DC power allocations, and thermal dissipations. Nonelectrical, but equally important considerations include reliability, environmentals such as temperature, humidity, and vibration, mechanicals such as size, weight, and packaging or mounting, interfaces, and even appearance, surface textures, and colors. As with most applications today, cost is becoming a primary driver in the design and production of the finished product. For most RF/microwave transmitters, power amplifier considerations dominate the cost and design concerns.

Safety must also be considered, especially when high voltages or high power levels are involved. To paraphrase a popular educational TV show; "be sure to read and understand all safety related materials

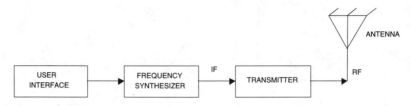

FIGURE 2.1   The transmitter.

that are applicable to your design before you begin. And remember, there is nothing more important than shielding yourself and your coworkers (assuming you like them) from high voltages and high levels of RF/microwave power." Another safety issue for portable products concerns the use of multiple lithium batteries connected in parallel. Special care must be taken to insure that the batteries charge and discharge independent of each other in order to prevent excessive I-R heating, which can cause the batteries to explode. It is your life — spend a little time to become familiar with these important issues.

2.2   ACP, Modulation, Linearity, and Power

The transmitter average and peak output powers are usually determined from link/margin analysis for the overall system. The transmitter linearity requirements are determined from the transmit power levels, phase noise, modulation type, filtering, and allowed adjacent channel power (ACP) spillover. Linearities intimately tied to the transmitter saturated power, which in turn is tied to the 1 dB compression point. The need for high linearity is dependent on the maximum acceptable ACP in adjacent channels or bands. This spillover of power will cause interference in adjacent channels or bands, making it difficult or impossible to use that band. The maximum ACP may be regulated by government agencies, or may be left up to the user. Often the actual transmitted power requirement is less stringent than the linearity requirement in determining the necessary power handling capability or saturated power of the transmitter. In order to achieve required linearity, the transmitter may operate significantly backed off from the saturated power capability, even under peak power operation. Since the cost of a transmitter rapidly increases with its power handling capability, and the linearity requirements are translated into additional power requirements, a great deal of the cost of a transmitter may actually be associated with linearity rather than transmit power level. If the added cost to achieve necessary linearity through additional power capability is significant, linearizing the transmitter can be cost effective.

2.3   Power

The single most important specification affecting the final system cost is often the transmitter saturated power, which is intimately linked to the transmitter linearity requirements. This parameter drives the power amplifier (PA) device size, packaging, thermal paths and related cooling methods, power supply, and DC interconnect cable sizes, weight, and safety, each of which can rapidly drive costs upward. The power level analysis may include losses in the cables and antenna, transmitter and receiver antenna gains, link conditions such as distance, rain, ice, snow, trees, buildings, walls, windows, atmospherics, mountains, waves, water towers, and other issues that might be pertinent to the specific application. The receiver capabilities are crucial in determining the transmitter power requirements. Once a system analysis has been completed indicating satisfactory performance, then the actual power amplifier (PA) requirements are known. The key parameters to the PA design are frequency, bandwidth, peak and average output power, duty cycle, linearity, gain, bias voltage and current, dissipated power, and reliability mean time to failure (MTBF, usually given as maximum junction temperature). Other factors may also be important, such as power added efficiency (PAE), return losses, isolations, stability, load variations, cost, size, weight, serviceability, manufacturability, etc.

2.4 Linearization

Linearity, as previously indicated, is intimately tied to the transmitter power. The need for high linearity is dependent on the maximum acceptable ACP in adjacent channels or bands. This spillover of power will cause interference in those bands, making it difficult or impossible to use that band. The ACP spillover is due to several factors, such as phase noise, modulation type, filtering, and transmit linearity. The basic methods used for linearization include the class A amplifier in back-off, feed forward, Cartesian and polar loops, adaptive predistortion, envelope elimination and recovery (EER), linear amplification using nonlinear components (LINC), combined analog locked-loop universal modulation (CALLUM), I-V trajectory modification, device tailoring, and Dougherty amplification. Each of these methods strives to improve the system linearity while minimizing the overall cost. The methods may be combined for further improvements. Economical use of the methods may require the development of application-specific integrated circuits (ASICs). As demand increases these specialized ICs should become available as building blocks, greatly reducing learning curves, design time and cost.

2.5 Efficiency

Power added efficiency (η_a or *PAE*) is the dimensionless ratio of RF power delivered from a device to the load (p_{out}) minus the input incident RF power ($p_{incident}$) versus the total DC power dissipated in the device (p_{DC}). It is the most commonly used efficiency rating for amplifiers and accounts for both the switching and power gain capabilities of the overall amplifier being considered. High *PAE* is essential to reducing the overall cost of high power transmitter systems, as previously discussed in the power section above. As with power, *PAE* affects the PA device size, packaging, thermal paths and related cooling methods, power supply, and DC interconnect cable sizes, weight, and safety, each of which can rapidly drive up cost.

$$PAE = \eta_a = \frac{p_{load} - p_{incident}}{p_{DC}}$$

2.6 The I-Q Modulator

The I-Q modulator is a basic building block used in numerous applications, and is an essential element of many linearization methods. The basic block diagram is shown in Fig. 2.2, along with the associated symbol that will be used in the following discussions. The modulator consists of two separate mixers that are driven 90° out of phase with each other to generate quadrature (I and Q) signals. The 90° port is usually driven with the high level bias signal, allowing the mixer compression characteristic to minimize amplitude variation from the 90° hybrid. The configuration is reciprocal, allowing either up- or down-conversion.

2.7 Class A Amplifier in Back Off

An amplifier is usually required near the output of any transmitter. The linearity of the amplifier is dependent on the saturated power that the amplifier can produce, the amplifier bias and design, and the characteristics of the active device itself. An estimate of the DC power requirements and dissipation for each stage in the PA chain can be made based on the peak or saturated power, duty cycle, and linearity requirements. The maximum or saturated power (P_{sat} in dBW or dBm, depending on whether power is in W or mW) can be estimated (Fig.2.3) from the product of the RMS voltage and current swings across the RF load, ($V_{sup} - V_{on}$)/2 and I_{on}/2, and from the loss in the output matching circuits (L_{out} in dB). As previously discussed in the receiver section, the 3OIP is about 6 dB above the saturated power for a square law device, but can vary by as much as 4 dB lower to as much as 10 dB higher for a given actual

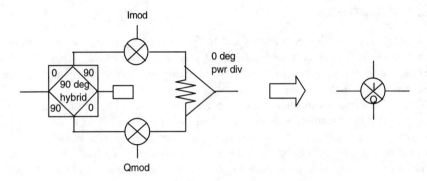

FIGURE 2.2   I-Q modulator block diagram.

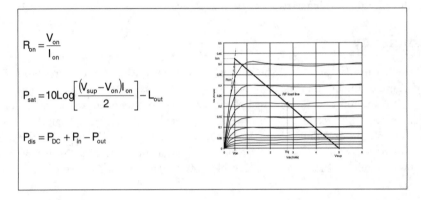

FIGURE 2.3   Class A amplifier saturated power estimation.

device. Thus it is very important to determine the actual 3OIP for a given device, using vendor data, simulation, or measurement. One must take into account the effects of transmitter components both prior to and after the PA, utilizing the same analysis technique used for receiver intermodulation. The ACP output intercept point (AOIP) will be closely correlated to the 3OIP, and will act in much the same way, except for the actual value of the AOIP. The delta between the AOIP and 3OIP will be modulation dependent, and must be determined through simulation or measurement at this time. Once this has been determined, it is a relatively easy matter to determine the required back off from P_{sat} for the output amplifier, often as much as 10 to 15 dB. Under these conditions, amplifiers that require a high intercept but in which the required peak power is much lower that the peak power that is available, linearization can be employed to lower the cost.

2.8   Feed Forward

Although the feed forward amplifier (Fig. 2.4) is a simple concept, it is relatively difficult to implement, especially over temperature and time. The applied signal is amplified to the desired power level by the power amplifier, whose output is sampled. The PA introduces distortion to the system. A portion of the input signal is delayed and then subtracted from the sampled PA signal, nulling out the original signal, leaving only the unwanted distortion created by the PA. This error signal is then adjusted for amplitude and recombined with the distorted signal in the output coupler, canceling out the distortion created by the PA. The resulting linearity improvement is a function of the phase and amplitude balances maintained, especially over temperature and time. The process of generating an error signal will also create nonlinearities, which will limit the ultimate improvements that are attainable, and thus are a critical part of the design.

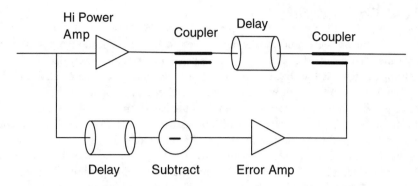

FIGURE 2.4 Feed forward amplifier.

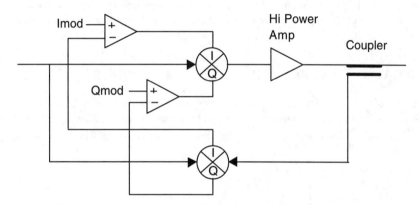

FIGURE 2.5 Cartesian loop.

2.9 Cartesian and Polar Loops

The Cartesian loop (CL) (Fig. 2.5) is capable of both good efficiency and high linearity. The efficiency is primarily determined by the amplifier efficiency. The loop action determines the linearity achieved. A carrier signal is generated and applied to the input of the CL, where it is power divided and applied to two separate I-Q mixers. The high power carrier path is quadrature modulated and then amplified by the output PA, after which the distorted modulated signal is sampled by a coupler. The sample distorted signal is then demodulated by mixing it with the original unmodulated carrier, resulting in distorted modulation in quadrature or I-Q form. These distorted I and Q modulations are then subtracted from the original I and Q modulation to generate error I and Q modulation (hence the name Cartesian), which will continuously correct the nonlinearity of both the power amplifier and the I-Q modulator. Loop gain and phase relationships are critical to the design, and as with any feedback scheme, care must be taken to prevent loop oscillation. The I-Q modulators and the sampling coupler utilize 90-degree power dividers with limited bandwidth over which satisfactory performance can be attained. The loop delay ultimately will limit the attainable bandwidth to about 10%. Even with these difficulties, the CL is a popular choice. Much of the circuitry is required anyway, and can be easily integrated into ASICs, resulting in low production costs.

Whereas the Cartesian loop depends on quadrature I and Q signals, the related polar loop uses amplitude and phase to achieve higher linearity. The method is much more complex since the modulation correction depends on both frequency modulating the carrier as well as amplitude modulating it. Ultimately the performance will be worse than that of the Cartesian loop, as well as being more costly by a considerable margin. For these reasons, it is not used.

2.10 Fixed Predistortion

Fixed predistortion methods are conceptually the simplest form of linearization. A power amplifier will have nonlinearities that distort the original signal. By providing complimentary distortion prior to the PA, the predistorted signal is linearized by the PA. The basic concept can be divided into digital and transfer characteristic methods, both with the same objective. In the digital method (Fig. 2.6), digital signal processing (DSP) is used to provide the required predistortion to the signal. This can be applied at any point in the system, but is usually provided at baseband where it can be cheaply accomplished. The information required for predistortion must be determined and then stored in memory. The DSP then utilizes this information and associated algorithms to predistort the signal, allowing the PA to correct the predistortion, resulting in high linearity. When hardware is used to generate the predistortion, the predistorting transfer characteristic must be determined, and appropriate hardware must be developed. There are no algorithms or methods to accomplish this, so it can be a formidable task. In either case, the improvements in linearity are limited by the lack of any feedback to allow for deviations from the intended operation, and by the ability to actually determine and create the required predistortion. In short, it is cheap, but don't expect dramatic results!

2.11 Adaptive Predistortion

Linearization by adaptive predistortion (Fig. 2.7) is very similar to fixed methods, with the introduction of feedback in the form of an error function that can be actively minimized on a continuous basis. The ability to change under operational conditions requires some form of DSP. The error signal is generated

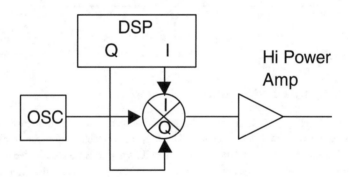

FIGURE 2.6   Fixed digital predistortion.

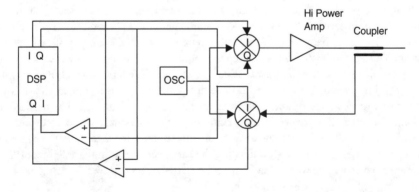

FIGURE 2.7   Adaptive predistortion.

in the same way used for Cartesian loop systems, but is then processed by the DSP, allowing the DSP to minimize the error by modifying or adapting the applied predistortion. The disadvantages in this method center on the speed of the DSP and the inability of the system to react due to loop delay. It must see the error before it can correct for it.

2.12 Envelope Elimination and Recovery (EER)

The highly efficient envelope elimination and recovery amplifier (Fig. 2.8) accepts a fully modulated signal at its input and power divides the signal. One portion of the signal is amplitude detected and filtered to create the low frequency AM component of the original signal. The other portion of the signal is amplitude limited to strip off or eliminate all of the AM envelope, leaving only the FM component or carrier. Each of these components is then separately amplified using high-efficiency techniques. The amplified AM component is then utilized to control the FM amplifier bias, modulating the amplified FM carrier. Thus the original signal is recovered, only amplified. While this process works very well, an alternative is available that utilizes the DSP capabilities to simplify the whole process, cut costs, and improve performance. In a system, the input half of the EER amplifier can be eliminated and the carrier FM modulated directly by DSP-generated tuning of a voltage-controlled oscillator (VCO). The DSP-generated AM is amplified and used to control the FM amplifier bias. The result is the desired modulated carrier.

2.13 Linear Amplification Using Nonlinear Components (LINC)

The LINC transmitter (Fig. 2.9) concept is quite simple. The DSP creates two separate amplitude and phase-modulated signals, each in quadrature (I-Q) format. These signals are upconverted by I-Q modulators to create two separate phase-modulated signals that are separately applied to high-efficiency output power amplifiers. The amplified FM signals are then combined at the output, the signals being such that all of the unwanted distortion is cancelled by combining 180° out of phase, and all of the desired signal components are added by combining in phase. The challenge in this method is in the DSP

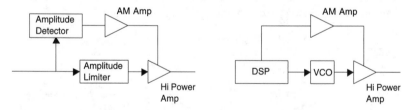

FIGURE 2.8   Envelope elimination and recovery.

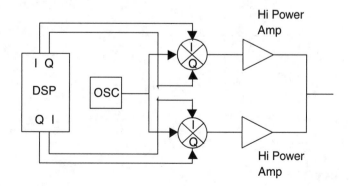

FIGURE 2.9   LINC transmitter.

generation of the original pair of quadrature signals required for the desired cancellation and combination at the output of the transmitter. Another area of concern with this linearization method is the requirement for amplitude and phase matching of the two channels, which must be tightly controlled in order to achieve optimum performance.

2.14 Combined Analog Locked-Loop Universal Modulation (CALLUM)

The CALLUM linearization method is much simpler than it looks at first glance (Fig.2.10). Basically, the top portion of the transmitter is the LINC transmitter discussed above. An output coupler has been added to allow sampling of the output signal, and the bottom half of the diagram delineates the feedback method that generates the two quadrature pairs of error signals in the same way as used in the Cartesian loop or EER methods. This feedback corrects for channel differences in the basic LINC transmitter, substantially improving performance. Since most of the signal processing is performed at the modulation frequencies, the majority of the circuit is available for ASIC implementation.

2.15 I-V Trajectory Modification

In I-V trajectory or cyclic modification, the idea is to create an active I-V characteristic that changes with applied signal level throughout each signal cycle, resulting in improved linear operation (see device tailoring below). A small portion of the signal is tapped off or sampled at the input or output of the amplifier and, based on the continuously sampled signal amplitude, the device bias is continuously modified at each point in the signal cycle. The power range over which high PAE is achieved will be compressed. This method requires a good understanding of the PA device, and excellent modeling. Also, the sampling and bias modification circuitry must be able to react at the same rate or frequencies as the PA itself while providing the required voltage or current to control the PA device. Delay of the sampled signal to the time the bias is modified is critical to obtaining performance. This method is relatively cheap to implement, and can be very effective in improving linearity.

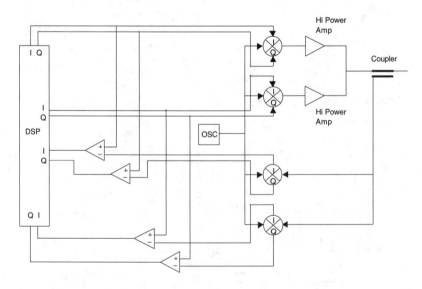

FIGURE 2.10   CALLUM.

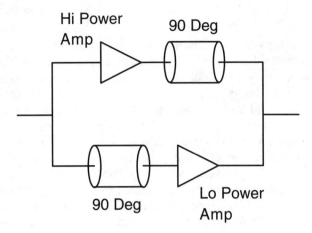

FIGURE 2.11 Dougherty amplifier.

2.16 Dougherty Amplification

The simplified form of the Dougherty amplifier, which maintains high PAE over a much wider power range than a single amplifier, is shown in Fig. 2.11. In the low power path, a 90° phase shifter is used to compensate for the 90° phase shifter/impedance inverter required in the high power path. The low power amplifier is designed to operate efficiently at a given signal level. The class C high power under low power conditions does not turn on, and thus represents high impedance at the input. The high power 90° phase shifter/impedance inverter provides partial matching for the low power amplifier under these conditions. As the signal level increases, the low power amplifier saturates, the class C high power amplifier turns on, and the power of both amplifiers sum at the output. Under these conditions the high power 90° phase shifter/impedance inverter matches the high power amplifier to the load impedance. Although the modulation bandwidths are not a factor in this technique, the bandwidth is limited by the phase and amplitude transfer characteristics of the 90° elements. This concept can be extended by adding more branches, or by replacing the low power amplifier with a complete Dougherty amplifier in itself.

2.17 Device Tailoring

For designers with access to a flexible semiconductor foundry service, linearity can be improved directly at the device level. The most obvious way of accomplishing this is by modifying the semiconductor doping to achieve the desired linearity while maintaining other performance parameters. In the ideal device, both the real and reactive device impedance would remain constant and linear (constant derivatives) as the I-V load line or trajectory is traversed for increasing amplitude signals. This linear operation would continue up to the signal amplitude at which both clipping and pinch-off simultaneously occur (ideal biasing). Thus the ideal device would be perfectly linear for any signal below P_{sat}. The objective should be to come as close to this ideal as possible in order to maximize linearity. At best this is a difficult task involving a great deal of device and process engineering. Another strategy might involve trying to minimize the amplitude-dependent parasitic effects such as leakage currents and capacitive or charge-related problems. A third strategy would be to modify the linearity by paralleling two or more devices of varying design together resulting in the desired performance. This is relatively easy to implement through device layout, with best results achieved when this is accomplished at the lowest level of integration (i.e., multiple or tapered gate lengths within a single gate finger, or a stepped or tapered base layer). The results can be quite dramatic with respect to linearity, and best of all the added recurring cost is minimal. As with trajectory modification, the power range for high efficiency operation is compressed.

2.18 Summary

The key transmitter parameters of ACP, modulation, linearity, and power are all tightly correlated. These parameters must be determined early in transmitter design so that individual component parameters can be determined and flow-down specifications can be made available to component designers. Linear operation is essential to controlling the power spill-over into adjacent channels (ACP). The basic linearization methods commonly used have been described. These include the class A amplifier in back-off, feed forward, Cartesian and polar loops, adaptive predistortion, envelope elimination and recovery (EER), linear amplification using nonlinear components (LINC), combined analog locked-loop universal modulation (CALLUM), I-V trajectory modification, device tailoring, and Dougherty amplification. Combining methods in such a way as to take advantage of multiple aspects of the nonlinear problem can result in very good performance. An example might be the combination of a Cartesian loop with device tailoring. Unfortunately, it is not yet possible to use simple relationships to calculate ACP directly from linearity requirements, or conversely, required linearity given the ACP. The determination of these requirements is highly dependent on the modulation being used. However, simulators are available that have the capability to design and determine the performance that can be expected.

Further Reading

Casadevall, F., The LINC Transmitter, *RF Design*, Feb. 1990.

Boloorian, M. and McGeeham, J., The Frequency-Hopped Cartesian Feedback Linear Transmitter, *IEEE Transactions on Vehicular Technology*, 45, 4, 1996.

Zavosh, F., Runton, D., and Thron, C., Digital Predistortion Linearizes CDMA LDMOS Amps, *Microwaves & RF*, Mar. 2000.

Kenington, P., Methods Linearize RF Transmitters and Power Amps, Part 1, *Microwaves & RF*, Dec. 1998.

Kenington, P., Methods Linearize RF Transmitters and Power Amps, Part 2, *Microwaves & RF*, Jan. 1999.

Correlation Between P1db and ACP in TDMA Power Amplifiers, *Applied Microwave & Wireless*, Mar. 1999.

Bateman, A., Haines, D., and Wilkinson, R., Linear Transceiver Architectures, Communications Research Group, University of Bristol, England.

Sundstrom, L. and Faulkner, M., Quantization Analysis and Design of a Digital Predistortion Linearizer for RF Power Amplifiers, *IEEE Transactions on Vehicular Technology*, 45, 4, 1996.

3

Low Noise Amplifier Design

Jakub Kucera
Infineon Technologies

Urs Lott
Acter AG

3.1 Introduction

Signal amplification is a fundamental function in all wireless communication systems. Amplifiers in the receiving chain that are closest to the antenna receive a weak electric signal. Simultaneously, strong interfering signals may be present. Hence, these low noise amplifiers mainly determine the system noise figure and intermodulation behavior of the overall receiver. The common goals are therefore to minimize the system noise figure, provide enough gain with sufficient linearity, and assure a stable 50 Ω input impedance at a low power consumption.

3.2 Definitions

This section introduces some important definitions used in the design theory of linear RF and microwave amplifiers. Further, it develops some basic principles used in the analysis and design of such amplifiers.

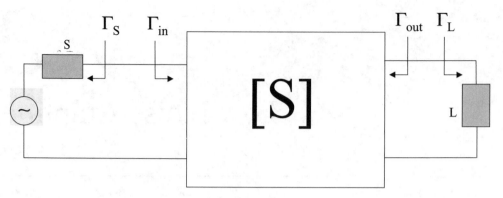

FIGURE 3.1 Amplifier block diagram. Z_S: source impedance, Z_L: load impedance, Γ_S: source reflection coefficient, Γ_{in}: input reflection coefficient, Γ_{out}: output reflection coefficient, Γ_L: load reflection coefficient.

3.2.1 Gain Definitions

Several gain definitions are used in the literature for high-frequency amplifier designs.

The transducer gain G_T is defined as the ratio between the effectively delivered power to the load and the power available from the source. The reflection coefficients are shown in Fig. 3.1.

$$G_T = \frac{1-\left|\Gamma_S\right|^2}{\left|1-\Gamma_S \cdot S_{11}\right|^2} \cdot \left|S_{21}\right|^2 \cdot \frac{1-\left|\Gamma_L\right|^2}{\left|1-\Gamma_L \cdot \Gamma_{OUT}\right|^2}$$

The available gain G_{AV} of a two-port is defined as the ratio of the power available from the output of the two-port and the power available from the source.

$$G_{AV} = \frac{1-\left|\Gamma_S\right|^2}{\left|1-\Gamma_S \cdot S_{11}\right|^2} \cdot \left|S_{21}\right|^2 \cdot \frac{1}{\left|1-\Gamma_{OUT}\right|^2} \quad \text{with } \Gamma_{OUT} = S_{22} + \frac{S_{12} \cdot S_{21} \cdot \Gamma_S}{1-\Gamma_S \cdot S_{11}}$$

The entire available power at one port can be transferred to the load, if the output is terminated with the complex conjugate load.

The available gain G_{AV} is a function of the two-port scattering parameters and of the source reflection coefficient, but independent of the load reflection coefficient Γ_L. The available gain gives a measure for the maximum gain into a conjugately matched load at a given source admittance.

The associated gain G_{ASS} is defined as the available gain under noise matching conditions.

$$G_{ASS} = \frac{1-\left|\Gamma_{opt}\right|^2}{\left|1-\Gamma_{opt} \cdot S_{11}\right|^2} \cdot \left|S_{21}\right|^2 \cdot \frac{1-\left|\Gamma_L\right|^2}{\left|1-\Gamma_L \cdot \Gamma_{OUT}\right|^2}$$

3.2.2   Stability and Stability Circles

The stability of an amplifier is a very important consideration in the amplifier design and can be determined from the scattering parameters of the active device, the matching circuits, and the load terminations (see Fig. 3.1). Two stability conditions can be distinguished: unconditional and conditional stability.

Unconditional stability of a two-port means that the two-port remains stable (i.e., does not start to oscillate) for any passive load at the ports. In terms of the reflection coefficients, the conditions for unconditional stability at a given frequency are given by the following equations

$$\left|\Gamma_{IN}\right| = \left|S_{11} + \frac{S_{12} \cdot S_{21} \cdot \Gamma_L}{1 - \Gamma_L \cdot S_{22}}\right| < 1$$

$$\left|\Gamma_{OUT}\right| = \left|S_{22} + \frac{S_{12} \cdot S_{21} \cdot \Gamma_S}{1 - \Gamma_S \cdot S_{11}}\right| < 1$$

$$\left|\Gamma_S\right| < 1 \ \text{ and } \ \left|\Gamma_L\right| < 1$$

In terms of the scattering parameters of the two-port, unconditional stability is given, when

$$K = \frac{1 - \left|S_{11}\right|^2 - \left|S_{22}\right|^2 + \left|\Delta\right|^2}{2\left|S_{12} \cdot S_{21}\right|} > 1$$

and

$$\left|\Delta\right| < 1$$

with $\Delta = S_{11} \cdot S_{22} - S_{12} \cdot S_{21}$. K is called the stability factor.[1]

If either $\left|S_{11}\right| > 1$ or $\left|S_{22}\right| > 1$, the network cannot be unconditionally stable because the termination $\Gamma_L = 0$ or $\Gamma_S = 0$ will produce or $\left|\Gamma_{IN}\right| > 1$ or $\left|\Gamma_{OUT}\right| > 1$.

The maximum transducer gain is obtained under simultaneous conjugate match conditions $\Gamma_{IN} = \Gamma_S^*$ and $\Gamma_{OUT} = \Gamma_L^*$. Using

$$\Gamma_{IN} = S_{11} + \frac{S_{12} \cdot S_{21} \cdot \Gamma_L}{1 - \Gamma_L \cdot S_{22}} \ \text{ and } \ \Gamma_{OUT} = S_{22} + \frac{S_{12} \cdot S_{21} \cdot \Gamma_S}{1 - \Gamma_S \cdot S_{11}}$$

a closed-form solution for the source and load reflection coefficients Γ_S and Γ_L can be found. However, a simultaneous conjugate match having unconditional stability is not always possible if $K < 1^2$.

Conditional stability of a two-port means that for certain passive loads (represented as $\Gamma_L < 1$ or $\Gamma_S < 1$) oscillation may occur. These values of Γ_L and Γ_S can be determined by drawing the stability circles in a Smith chart. The source and load stability circles are defined as

$$\left|\Gamma_{IN}\right| = 1 \ \text{ and } \ \left|\Gamma_{OUT}\right| = 1$$

On one side of the stability circle boundary, in the Γ_L plane, $\left|\Gamma_{IN}\right| > 1$ and on the other side $\left|\Gamma_{IN}\right| < 1$. Similarly, in the Γ_S plane, $\left|\Gamma_{OUT}\right| > 1$ and on the other side $\left|\Gamma_{OUT}\right| < 1$. The center of the Smith chart ($\Gamma_L = 0$) represents

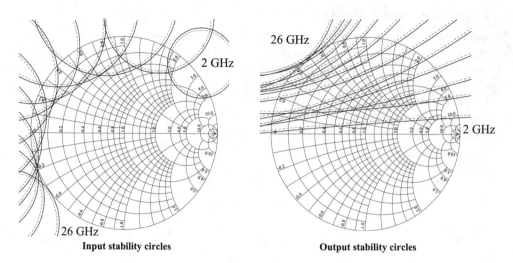

FIGURE 3.2   Source (input) and load (output) stability circles in the Smith chart for the MESFET NE710 over the frequency range from 2 to 26 GHz. Unstable region is indicated by the dotted line.

a stable operating point, if $|S_{11}| < 1$, and an unstable operating point, if $|S_{11}| > 1$ (see Fig. 3.2). Based on these observations, the source and load reflection coefficient region for stable operation can be determined.

With unconditional stability, a complex conjugate match of the two-port is possible. The resulting gain then is called the maximum available gain (MAG) and is expressed as

$$MAG = \left| \frac{S_{21}}{S_{12}} \right| \cdot \left(K - \sqrt{K^2 - 1} \right)$$

The maximum stable gain MSG is defined as the maximum transducer gain for which $K = 1$ holds, namely

$$MSG = \left| \frac{S_{21}}{S_{12}} \right|$$

MSG is often used as a figure of merit for potentially unstable devices (see Fig. 3.3).

It must be mentioned, however, that the stability analysis as presented here in its classical form, is applicable only to a single-stage amplifier. In a multistage environment, the above stability conditions are insufficient, because the input or output planes of an intermediate stage may be terminated with active networks. Thus, taking a multistage amplifier as a single two-port and analyzing its K-factor is helpful, but does not guarantee overall stability. Literature on multistage stability analysis is available.

3.2.3   Representation of Noise in Two-Ports

The LNA can be represented as a noise-free two-port and two partly correlated input noise sources i_n and v_n as shown in Fig. 3.4. The partial correlation between the noise sources i_n and v_n can be described by splitting i_n into a fully correlated part i_c and a noncorrelated part i_u as

$$i_n = i_c + i_u$$

The fully correlated part i_c is defined by the correlation admittance[4] Y_{cor}

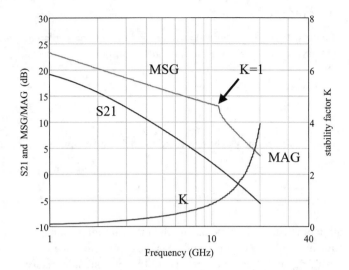

FIGURE 3.3 Maximum stable gain (MSG), maximum available gain (MAG), S_{21}, and stability factor K for a typical MESFET device with 0.6 µm gate length.

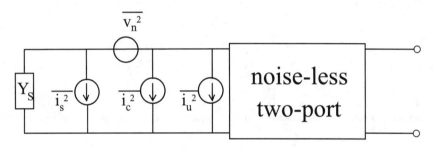

FIGURE 3.4 Representation of noisy two-port as noiseless two-port and two partly correlated noise sources (v_n and $i_c + i_u$) at the input.

$$i_c = Y_{cor} v_n$$

The source impedance $Z_s = R_s + jB_s$ shows thermal noise i_s which depends on the bandwidth as

$$\overline{i_s^2} = 4kTG_s \Delta f$$

Finally, the noise factor F can be expressed in terms of these equivalent input noise generators as

$$F = 1 + \left| Y_s + Y_{cor} \right|^2 \frac{\overline{v_n^2}}{\overline{i_s^2}} + \frac{\overline{i_u^2}}{\overline{i_s^2}}$$

Details on the use of the correlation matrix and the derivation of the noise factor from the noise sources of the two-port can be found in References 4 and 5.

3.2.4   Noise Parameters

The noise factor F of a noisy two-port is defined as the ratio between the available signal-to-noise power ratio at the input to the available signal-to-noise ratio at the output.

$$F = \frac{S_{in}}{N_{in}} \Bigg/ \frac{S_{out}}{N_{out}}$$

The noise factor of the two-port can also be expressed in terms of the source admittance $Y_s = G_s + jB_s$ as

$$F = F_{min} + \frac{R_n}{G_s}\left|Y_s - Y_{opt}\right|^2$$

where F_{min} is the minimum achievable noise factor when the optimum source admittance $Y_{opt} = G_{opt} + jB_{opt}$ is presented to the input of the two-port, and R_n is the equivalent noise resistance of the two-port. Sometimes the values Y_s, Y_{opt}, and R_n are given relative to the reference admittance Y_0.

The noise performance of a two-port is fully characterized at a given frequency by the four noise parameters F_{min}, R_n, and real and imaginary parts of Y_{opt}.

Several other equivalent forms of the above equation exist, one of them describing F as a function of the source reflection coefficient Γ_s.

$$F = F_{min} + \frac{4R_n}{Z_0}\frac{\left|\Gamma_s - \Gamma_{opt}\right|^2}{\left|1 + \Gamma_{opt}\right|^2 \cdot \left(1 - \left|\Gamma_s\right|^2\right)}$$

When measuring noise, the noise factor is often represented in its logarithmic form as the noise figure NF

$$NF = 10\log F$$

Care must be taken not to mix up the linear noise factor and the logarithmic noise figure in noise calculations.

3.2.5   Noise Circles

Noise circles refer to the contours of constant noise figure for a two-port when plotted in the complex plane of the input admittance of the two-port. The minimum noise figure is presented by a dot, while for any given noise figure higher than the minimum, a circle can be drawn. This procedure is adaptable in the source admittance notation as well as in the source reflection coefficient notation. Fig. 3.5 shows the noise circles in the source reflection plane.

Noise circles in combination with gain circles are efficient aids for circuit designers when optimizing the overall amplifier circuit network for low noise with high associated gain.

3.2.6   Friis Formula: Cascading Noisy Two-Ports

When several noisy two-ports are connected in cascade, the overall noise characteristics are described by[6]

$$F_{tot} = F_1 + \frac{F_2 - 1}{G_1} + \frac{F_3 - 1}{G_1 \cdot G_2} + \ldots + \frac{F_i - 1}{G_1 \cdot G_2 \ldots G_{i-1}}$$

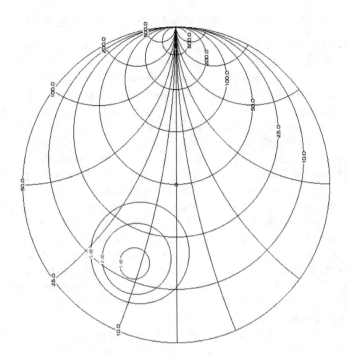

FIGURE 3.5 Noise circles in the input reflection coefficient plane.

where F_i and G_i are noise factor and available gain of the ith two-port. The available gain depends on the output admittance of the previous stage.

3.2.7 Noise Measure M

The overall noise factor of an infinite number of identical cascaded amplifiers is $F = 1 + M$ with

$$M = \frac{F-1}{1-\dfrac{1}{G}}$$

M is here called the noise measure. The noise measure is useful for comparing the noise performance of devices or LNAs with different power gains.

3.3 Design Theory

The apparent structural simplicity of an LNA with its relatively few components is misleading. The design should be easy, but the trade-offs complicate the design. A simultaneous noise and power matching involves a more complicated matching network and the achievable dynamic range is often limited by the given low supply voltage and the maximum allowed current consumption. The LNA must provide enough gain so that the noise contributions from the following components become small. But the maximum tolerable gain is limited by the linearity requirements of the following receiver chain.

The most important design considerations in a high-frequency amplifier are stability, power gain, bandwidth, noise, and DC power consumption requirements.

A systematic mathematical solution, aided by graphical methods, is developed to determine the input and output matching network for a particular noise, gain, stability, and gain criteria. Unconditionally stable designs will not oscillate with any passive termination, while designs with conditional stability require careful analysis of the loading to assure stable operation.

3.3.1   Linear Design Procedure for Single-Stage Amplifiers

1. *Selection of device and circuit topology:* Select the appropriate device based on required gain and noise figure. Also decide on the circuit topology (common-base/gate or common-emitter/source). The most popular circuit topology in the first stage of the LNA is the common-emitter (source) configuration. It is preferred over a common-base (-gate) stage because of its higher power gain and lower noise figure. The common-base (-gate) configuration is a wideband unity-current amplifier with low input impedance ($\approx 1/g_m$) and high predominantly capacitive output impedance. Wideband LNAs requiring good input matching use common-base input stages. At high frequencies the input impedance becomes inductive and can be easily matched.

2. *Sizing and operating point of the active device:* Select a low noise DC operating point and determine scattering and noise parameters of the device. Typically, larger input transistors biased at low current densities are used in low noise designs. At RF frequencies and at a given bias current, unipolar devices such as MOSFET, MESFET, and HEMT are easier to match to 50 Ω when the device width is larger. Both, (hetero-) BJTs and FETs show their lowest intrinsic noise figure when biased at approximately one tenth of the specified maximum current density. Further decreasing the current density will increase the noise figure and reduce the available gain.

3. *Stability and RF feedback:* Evaluate stability of the transistor. If only conditionally stable, either introduce negative feedback (high-resistive DC parallel or inductive series feedback) or draw stability circles to determine loads with stable operation.

4. *Select the source and load impedance:* Based on the available power gain and noise figure circles in the Smith chart, select the load reflection coefficient Γ_L that provides maximum gain, the lowest noise figure (with $\Gamma_S = \Gamma_{opt}$), and good VSWR. In unconditionally stable designs, Γ_L is

$$\Gamma_L = \left(S_{22} + \frac{S_{12} \cdot S_{21} \cdot \Gamma_{opt}}{1 - \Gamma_{opt} \cdot S_{11}} \right)^*$$

   In conditionally stable designs, the optimum reflection coefficient Γ_S may fall into an unstable region in the source reflection coefficient plane. Once Γ_S is selected, Γ_L is selected for the maximum gain $\Gamma_L = \Gamma_{OUT}$, and Γ_L must again be checked to be in the stable region of the load reflection coefficient plane.

5. *Determine the matching circuit:* Based on the required source and load reflection coefficients, the required ideal matching network can be determined. Depending on the center frequency, lumped elements or transmission lines will be applied. In general, there are several different matching circuits available. Based on considerations about insertion loss, simplicity, and reproducibility of each matching circuit, the best selection can be made.

6. *Design the DC bias network:* A suitable DC bias network is crucial for an LNA, which should operate over a wide temperature and supply voltage range and compensate parameter variations of the active device. Further, care must be given that no excessive additional high-frequency noise is injected from the bias network into the signal path, which would degrade the noise figure of the amplifier. High-frequency characteristics including gain, noise figure, and impedance matching are correlated to the device's quiescent current. A resistor bias network is generally avoided because of its poor supply rejection. Active bias networks are capable of compensating temperature effects and rejecting supply voltage variations and are therefore preferred

   For bipolar circuits, a simple grounded emitter DC bias network is shown in Fig. 3.6a. The high-resistive bias network uses series feedback to stabilize the current of the active device against device parameter variations. However, the supply rejection of this network is very poor, which limits its applicability. A bypassed emitter resistor is often used at low frequencies to stabilize the DC bias point (Fig. 3.6b). At RF and microwave frequencies, the bypass capacitor can cause unwanted high-frequency instability and must be applied with care. Furthermore, an emitter

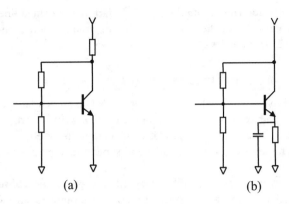

FIGURE 3.6 Passive bias network for bipolar amplifiers.

resistor will degrade the noise figure performance of the amplifier, if the resistor is not fully bypassed at the signal frequency.

More advanced active bias circuits use (temperature compensated) bandgap references and generate a reference current, which is mirrored to the amplifying device through a high value resistor or an RF choke to minimize noise injection. Another popular method is to generate a proportional to absolute temperature (PTAT) current source. The amplifier gain is proportional to the transconductance which itself is proportional to the collector current and inversely proportional to the temperature ($g_m = qI_c/kT$). With the transistor biased with a current proportional to temperature, the gain remains roughly constant over temperature. Combining bandgap circuits with PTAT sources leads to excellent supply and temperature variation suppression.[7]

The implementation of appropriate bias methods for FET amplifiers is generally more involved. The most critical parameter affecting the bias point is the threshold voltage. Stable voltage reference and PTAT current sources are typically based on the Schottky diode barrier height and involve rather sophisticated circuitry.[8]

7. *Optimize entire circuit with lossy matching elements:* The final design optimization of the LNA circuit must include the nonidealities of the matching elements, parasitic components such as bond wire inductance, as well as fabrication tolerance aspects. This last design phase today is usually performed on a computer-aided design (CAD) system.

The dominant features of an LNA (gain, noise, matching properties) can be simulated with excellent accuracy on a linear CAD tool. The active device is characterized by its scattering parameters in the selected bias point, and the four noise parameters. The passive components are described by empirical or equivalent circuit models built into the linear simulation tool. If good models for elements like millimeter-wave transmission lines are not available, these elements must be described by their measured scattering parameters, too.

Alternatively, a nonlinear simulator with a full nonlinear device model allows direct performance analysis over varying bias points. The use of nonlinear CAD is mandatory for compression and intermodulation analysis.

Advanced CAD tools allow for direct numerical optimization of the circuit elements toward user-specified performance goals. However, these optimizers should be used carefully, because it can be very difficult to transform the conflicting design specifications into optimization goals. In many cases, an experienced designer can optimize an LNA faster by using the "tune" tools of a CAD package.

3.4 Practical Design of a Low Noise Amplifier

The last section presented a design procedure to design a stable low noise amplifier based on linear design techniques. In practice, there are nonidealities and constraints on component sizing that typically degrade the amplifier performance and complicate the design. In fact, the presented linear design method does not take power consumption versus linearity explicitly into account. Some guidelines are provided in this section that may facilitate the design.

3.4.1 Hybrid vs. Monolithic Integrated LNA

With the current trend to miniaturized wireless devices, LNAs are often fabricated as monolithic integrated circuits, usually referred to as MMIC (monolithic microwave integrated circuit). High volume applications such as cell phones call for even higher integration in the RF front end. Thus the LNA is integrated together with the mixer, local oscillator, and sometimes even parts of the transmitter or the antenna.[9]

Depending on the IC technology, monolithic integration places several additional constraints on the LNA design. The available range of component values may be limited, in particular the maximum inductance and capacitance values are often smaller than required. Integrated passive components in general have lower quality factors Q because of their small size. In some cases, the first inductor of the matching circuit must be realized as an external component.

The electromagnetic and galvanic coupling between adjacent stages is often high due to the close proximity of the components. Furthermore, the lossy and conducting substrate used in many silicon-based technologies increases coupling. At frequencies below about 10 GHz, transmission lines cannot be used for matching because the required chip area would make the IC too expensive, at least for commercial applications.

Finally, monolithic circuits cannot be tuned in production. On the other hand, monolithic integration also has its advantages. The placement of the components is well controlled and repeatable, and the wiring length between components is short. The number of active devices is almost unlimited and adds very little to the cost of the LNA. Each active device can be sized individually.

For applications with low volume where monolithic integration is not cost effective, LNAs can be built as hybrid circuits, sometimes called MIC (microwave integrated circuit). A packaged transistor is mounted on a ceramic or organic substrates. The matching circuit is realized with transmission lines or lumped elements. Substrates such as alumina allow very high quality transmission line structures to be fabricated. Therefore, in LNAs requiring ultimate performance, e.g., for satellite ground stations, hybrid circuit technology is sometimes used even if monolithic circuits are available.

3.4.2 Multistage Designs

Sometimes a single amplifier stage cannot provide the required gain and multiple gain stages must be provided. Multiple gain stages complicate the design considerably. In particular, the interstage matching must be designed carefully (in particular in narrowband designs) to minimize frequency shifts and ensure stability. The ground lines of the different gain stages must often be isolated from each other to avoid positive feedback, which may cause parasitic oscillations. Moreover, some gain stages may need some resistive feedback to enhance stability.

Probably the most widely used multistage topology is the cascode configuration. A low noise amplifier design that uses a bipolar cascode arrangement as shown in Fig. 3.7 offers performance advantages in wireless applications over other configurations. It consists of a common-emitter stage driving a common-base stage. The cascode derives its excellent high-frequency properties from the fact that the collector load of the common-emitter stage is the very low input impedance of the common-base stage. Consequently, the Miller effect is minimal even for higher load impedances and an excellent reverse isolation is achieved. The cascode has high output impedance, which may become difficult to match to 50 Ω. A careful layout of the cascode amplifier is required to avoid instabilities. They mainly arise from parasitic inductive feedback between the emitter of the lower and the base of the upper transistor. Separating the two ground lines will enhance the high-frequency stability considerably.

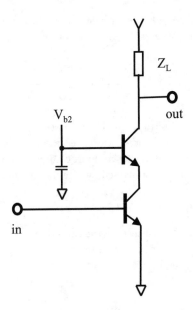

FIGURE 3.7 Cascode amplifier.

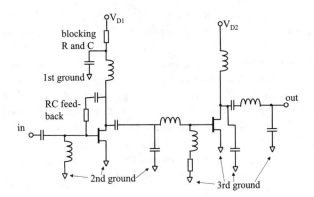

FIGURE 3.8 Design for stability.

3.4.3 Stability Considerations

Figure 3.8 shows a possible strategy for a stable design of a two-stage amplifier. Separated ground and supply lines of the two gain stages minimize positive feedback. RC parallel feedback further enhances in-band stability. Low frequency oscillations caused through unstable bias lines can be attenuated by adding small resistors and blocking capacitors into the supply line.

3.4.4 Feedback

Negative feedback is widely used in amplifier design to stabilize gain against parameter changes in the active device due to supply voltage variations and temperature changes. RF feedback is used in many LNAs to ensure high-frequency stability and make noise and power match coincident. A well-known technique is adding inductance at the emitter (source) of the active device. The inductance L interacts with the base-emitter (gate-source) capacitance C_{IN} and device transconductance g_m to produce a resistive component to the input impedance $g_m \dfrac{L}{C_{IN}}$, while no additional noise source is introduced (except for the parasitic series resistance of the inductor). Neglecting the Miller capacitance, the input impedance of an inductively degenerated FET stage is

$$Z_{IN} = \frac{1}{j\omega C_{IN}} + j\omega L + g_m \frac{L}{C_{IN}}$$

This method of generating a real term to the input impedance is preferable to resistive methods as only negligible additional noise is introduced. Moreover, the inductance has the helpful side effect of shifting the optimum noise match closer to the complex conjugate power match and reducing the signal distortion. However, the benefits are accompanied by a gain reduction.

3.4.5   Impedance Matching

Following the design procedure in the last section, the conditions for a conjugate match at the input and output ports are satisfied at one frequency. Hence, reactive matching inherently leads to a narrowband design. The input bandwidth is given by

$$BW = \frac{f_0}{Q_{IN}}$$

where f_0 is the center frequency and Q_{IN} is the quality factor of the input matching network. The bandwidth can be increased by increasing the capacitance or decreasing the inductance of the matching network.

Using multistage impedance transformators (lumped element filters or tapers) can broaden the bandwidth, but there is a given limit for the reflection coefficient-bandwidth product using reactive elements.[10] In reality, each matching element will contribute some losses, which directly add to the noise figure.

Select an appropriate matching network based on physical size and quality factor (transmission line length, inductance value): long and high-impedance transmission lines show higher insertion loss. Thus, simple matching typically leads to a lower noise figure.

At higher microwave and millimeter-wave frequencies, balanced amplifiers are sometimes used to provide an appropriate noise and power match over a large bandwidth.

3.4.6   Temperature Effects

Typically, LNAs must operate over a wide temperature range. As transistor transconductance is inversely proportional to the absolute temperature, the gain and amplifier stability may change considerably. When designing LNAs with S-parameters at room temperature, a stability margin should be included to avoid unwanted oscillations at low temperatures, as the stability tends to decrease.

3.4.7   Parasitics

Parasitic capacitance, resistance, or inductance can lead to unwanted frequency shifts, instabilities, or degradation in noise figure and gain and rarely can be neglected. Hence, accurate worst-case simulations with the determined parasitics must be made. Depending on the frequency, the parasitics can be estimated based on simple analytical formulas, or must be determined using suitable electromagnetic field-simulators.

3.5   Design Examples

In this section a few design examples of recently implemented low noise amplifiers for frequencies up to 5.8 GHz are presented. They all were manufactured in commercial IC processes.

3.5.1 A Fully Integrated Low Voltage, Low Power LNA at 1.9 GHz[11]

Lowest noise figure can only be achieved when minimizing the number of components contributing to the noise while simultaneously maximizing the gain of the first amplifier stage. Any resistive matching and loading will degrade the noise figure and dynamic behavior and increase power consumption.

In GaAs MESFET processes, the semi-insulating substrate and thick metallization layers allow passive matching components such as spiral inductors and metal-insulator-metal (MIM) capacitors with high quality factors. These lumped passive components are ideally suited for integrated impedance matching at low GHz frequencies. A fully integrated matching network improves the reproducibility and saves board space while it increases expensive chip area.

It is generally known that GaAs MESFETs have excellent minimum noise figures in the lower GHz frequency range. Still few designs achieve noise figures close to the transistor F_{min}. In fact, several factors prevent F_{min} being attained in practice. If a small input device is employed, a large input impedance transformation involving large inductance values is required. Larger MMIC inductors have higher series resistance and consequently introduce more noise. Further a simultaneous noise and power match often needs additional inductive source degeneration, again introducing noise and reducing gain. For a given maximum power dissipation, very large devices, in contrast, must be biased at very low current densities at which the F_{min} and the gain are degraded. Consequently a trade-off must be made for an optimum design.

The employed GaAs technology features three types of active devices: an enhancement and two depletion MESFETs with different threshold voltages. The enhancement device has a higher maximum available gain, a slightly lower minimum noise figure, but somewhat higher distortion compared to the depletion type. Another advantage of the enhancement FET is that a positive gate bias voltage can be used, which greatly simplifies single-supply operation.

Preliminary simulations are performed using a linear simulator based on measured S- and noise parameters of measured active devices at various bias points and using a scalable large signal model within a harmonic balance simulator in order to investigate the influence of the transistor gate width on the RF performance. The current consumption of the transistor is set at 5.5 mA independent of the gate width. The simulations indicate a good compromise between gain, NF, and intermodulation performance at a gate width of 300 μm (Fig. 3.9).

The LNA schematic is shown in Fig. 3.10. The amplifier consists of a single common-source stage, which uses a weak inductive degeneration at the source (the approximately 0.3 nH are realized with several parallel bondwires to ground). The designed amplifier IC is fabricated in a standard 0.6 μm E/D MESFET foundry process. The matching is done on chip using spiral inductors and MIM capacitors. The complete LNA achieves a measured 50 Ω noise figure of 1.1 dB at 1.9 GHz with an associated gain of 16 dB at a very low supply voltage of $V_{dd} = 1$ V and a total current drain of $I_{dd} = 6$ mA. Fig. 3.11 depicts the measured gain and 50 Ω noise figure versus frequency.

The low voltage design with acceptable distortion performance and reasonable power gain can only be achieved using a reactive load with almost no voltage drop.

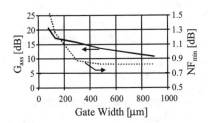 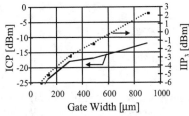

FIGURE 3.9 Simulated performance of an enhancement FET versus device width at a constant power dissipation.

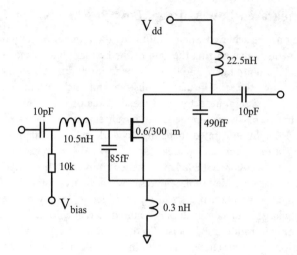

FIGURE 3.10 Schematic diagram of the low voltage GaAs MESFET LNA.

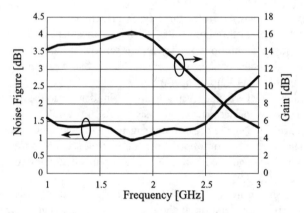

FIGURE 3.11 Measured gain and 50 Ω noise figure vs. frequency.

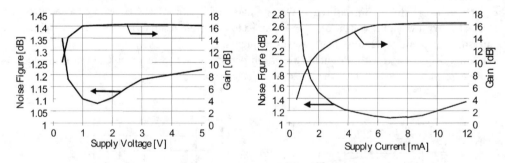

FIGURE 3.12 Measured gain and noise figure vs. supply voltage (I_{dd} = 6 mA) and vs. supply current (V_{dd} = 1 V).

Figure 3.12 shows, respectively, the supply voltage and supply current dependence of the gain and noise figure. As can be seen, the amplifier still achieves 10 dB gain and a 1.35 dB noise figure at a supply voltage of only 0.3 V and a total current consumption of 2.3 mA. Sweeping the supply voltage from 1 to 5 volts, the gain varies less than 0.5 dB and the noise figure less than 0.15 dB, respectively. IIP$_3$ and −1 dB compression point are also insensitive to supply voltage variations as shown in Fig. 3.13.

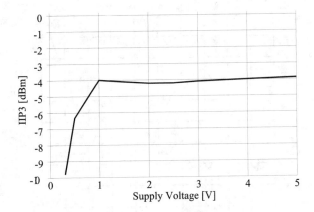

FIGURE 3.13   Measured input IP3 vs. supply voltage.

Below 1 V, however, the active device enters the linear region resulting in a much higher distortion.

Finally, the input and output matchings are measured for the LNA. At the nominal 1 V supply, the input and output return loss are –8 dB and –7 dB, respectively.

3.5.2   A Fully Matched 800 MHz to 5.2 GHz LNA in SiGe HBT Technology

Bipolar technology is particularly well suited for broadband amplifiers because BJTs typically show low input impedances in the vicinity of 50 Ω and hence can be easily matched. A simplified schematic diagram of the monolithic amplifier is shown in Fig. 3.14. For the active devices of the cascode LNA large emitter areas (47 μm^2), biased at low current densities are employed to simplify the simultaneous noise and power match. Input and output matching is consequently obtained simply by the aid of the bondwire inductance at the input and output ports and the chip ground.

The LNA was fabricated with MAXIM's GST-3 SiGe process and subsequently was mounted on a ceramic test package for testing. No additional external components are required for this single-supply LNA.

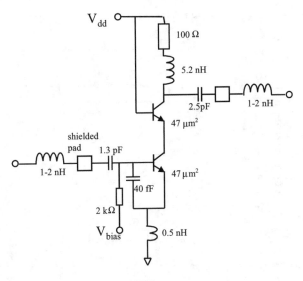

FIGURE 3.14   Schematic diagram of the SiGe HBT LNA.

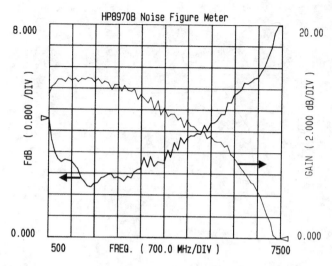

FIGURE 3.15 Measured LNA gain and noise figure vs. frequency (V_{dd} = 3 V, I_{dd} = 8.8 mA).

Figure 3.15 shows the 50 Ω noise figure and associated gain over the frequency range of interest. A relatively flat gain curve is measured from 500 MHz up to 3 GHz. Beyond 3 GHz the gain starts to roll off. The circuit features 14.5 dB of gain along with a 2 dB noise figure at 2 GHz. At 5.2 GHz, the gain is still 10 dB and the noise figure is below 4 dB. The input return loss is less than –10 dB between 2.5 and 6.5 GHz. At 1 GHz it increases to –6 dB.

The distortion performance of the amplifier was measured at the nominal 3 V supply for two frequencies, 2.0 and 5.2 GHz, respectively. At 2 GHz, the –1 dB compression point is +2 dBm at the output. At 5.2 GHz the value degrades to 0 dBm.

The LNA is comprised of two sections: the amplifier core and a PTAT reference. The core is biased with the PTAT to compensate for the gain reduction with increasing temperature. The gain is proportional to the transconductance of the transistor, which itself is proportional to collector current and inversely proportional to temperature. The PTAT biasing increases the collector current with temperature to keep the gain roughly constant over temperature. Simultaneously, the biasing shows a good supply rejection as shown in Fig. 3.16.

A chip photograph of the 0.5 × 0.6 mm² large LNA is shown in Fig. 3.17.

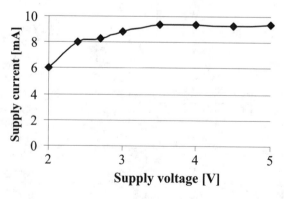

FIGURE 3.16 Supply current vs. supply voltage.

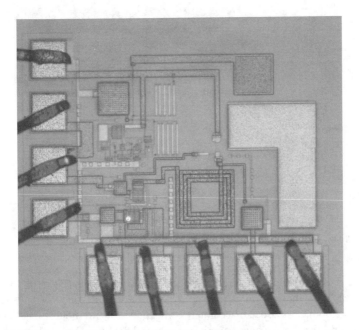

FIGURE 3.17   Chip photograph of the SiGe HBT LNA (0.5×0.6 mm^2).

3.5.3   A Fully Matched Two-Stage Low Power 5.8 GHz LNA[12]

A fully monolithic LNA achieves a noise figure below 2 dB between 4.3 GHz and 5.8 GHz with a gain larger than 15 dB at a DC power consumption of only 6 mW using the enhancement device of a standard 17 GHz f_T 0.6 µm E/D-MESFET process.

A schematic diagram of the integrated LNA core is shown in Fig. 3.18. The circuit consists of two common-source gain stages to provide enough power gain. The first stage uses an on-chip inductive degeneration of the source to achieve a simultaneous noise and power match, and to improve RF stability. Both amplifier stages are biased at the same current. The noise contributions of the biasing resistors are negligible.

The output of each stage is loaded with a band pass LC section to increase the gain at the desired frequency. The load of the first stage, together with the DC block between the stages, is also used for inter-stage matching.

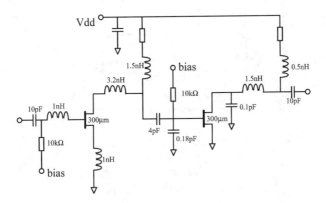

FIGURE 3.18   Schematic diagram of the low noise amplifier.

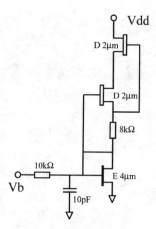

FIGURE 3.19   Schematic diagram of the employed bias circuit.

The DC biasing is done on-chip with a combination of E/D MESFETs (Fig. 3.19). The bias circuit is able to effectively stabilize the bias point for voltages from 1 V to beyond 4 V without any feedback network within the amplifier. It also can accurately compensate for threshold voltage variations.

The correlation of the threshold voltages of enhancement and depletion devices due to simultaneous gate recess etch of both types is used in the bias circuit to reduce the bias current variations over process parameter changes. Figure 3.20 shows the simulated deviation from the nominal current as a function of threshold voltage variations. The device current remains very constant even for extreme threshold voltage shifts.

If the RF input device is small, a large input impedance transformation is required. The third-order intercept point can be degraded and larger inductor values are needed sacrificing chip area and noise figures, due to the additional series resistance of the inductor. If instead a very large device is used, the current consumption is increased, unless the current density is lowered. Below a certain current density the device gain will decrease, the minimum noise figure will increase, and a reliable and reproducible biasing of the device becomes difficult as the device is biased close to the pinch-off voltage. To achieve high quality factors, all inductors are implemented using the two top wiring levels with a total metal thickness of 6 μm. The spiral inductors were analyzed using a 2.5D field simulator in order to accurately determine their equivalent circuit.

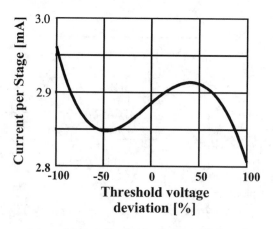

FIGURE 3.20   Simulated current dependence on threshold voltage variations.

FIGURE 3.21 Photograph of the chip mounted in the test package.

Sample test chips were mounted in a ceramic test package (Fig. 3.21) to investigate the influence of the bonding wires and the package parasitics.

In Figs. 3.22 and 3.23 the influence of the bond wires on the input and return loss, gain, and noise figure, respectively, is shown. The optimum input matching is shifted from 5.2 GHz to 5.8 GHz with the bond wire included. In an amplifier stage with moderate feedback one would expect the bond wire to shift the match toward lower frequencies. However, due to the source inductor the inter-stage matching circuit strongly interacts with the input port, causing a frequency shift in the opposite direction.

As expected, the gain curve of the packaged LNA (Fig. 3.23) is flatter and the gain is slightly reduced because of the additional ground inductance arising from the ground bond wires (approx. 40 pH).

At the nominal supply current of 6 mA the measured 50 Ω noise figure is 1.8 dB along with more than 15 dB gain from 5.2 GHz to 5.8 GHz as given in Fig. 3.23. For the packaged LNA the noise figure is slightly degraded due to losses associated with the package and connectors.

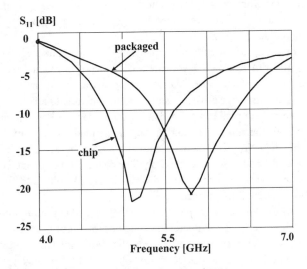

FIGURE 3.22 Input return loss vs. frequency of chip and packaged LNA.

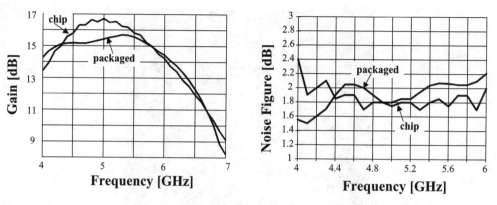

FIGURE 3.23   Gain and noise figure vs. frequency (V_{dd} = 1 V and I_{dd} = 6mA).

At 5.5 GHz the minimum noise figure of the device including the source inductor at the operating bias point is 1.0 dB and the associated gain is 8.5 dB. The minimum noise figure of an amplifier with two identical stages is therefore 1.6 dB. Thus, only a small degradation of the noise figure by the on-chip matching inductor is introduced at the input.

At 5.2 GHz a measured −1 dB compression point of 0 dBm at the output confirms the excellent distortion characteristics of GaAs MESFET devices at very low power consumption. The measured input referenced third order intercept point (IIP3) is -6 dBm.

3.5.4   0.25 μm CMOS LNAs for 900 MHz and 1.9 GHz[13,14]

CMOS technology starts to play a significant role in integrated RF transceivers for the low GHz range with effective gate lengths reaching the deep submicron regions. Competitive circuit performance at low power dissipation is becoming possible even for critical building blocks such as the LNA. In fact, quarter-micron CMOS seems to be the threshold where robust designs can be realized with current consumption competitive to BJT implementations. Further downscaling calls for a reduction in supply voltage which will ultimately limit the distortion performance of CMOS-based designs.

Designing a low noise amplifier in CMOS is complicated by the lossy substrate, which requires a careful layout to avoid noise injection from the substrate. The schematic diagrams of two demonstrated 0.25 μm CMOS LNAs for 900 MHz and 1.9 GHz are shown in Figs. 3.24a and 3.24b, respectively. Both circuits use two stages to realize the desired gain.

The first amplifier consisting of an externally matched cascode input stage and a transimpedance output stage consumes 10.8 mA from a 2.5 V supply. The cascode is formed using two 600-μm wide NMOS devices loaded by a 400 Ω resistor. The inductance of approximately 1.2 nH formed by the bondwire at the source of the first stage is used to simplify the matching of an otherwise purely capacitive input impedance. The directly coupled transimpedance output stage isolates the high-gain cascode and provides a good 50 Ω output matching. A simple biasing is included on the chip. At the nominal power dissipation and 900 MHz, the LNA achieves 16 dB gain and a noise figure of below 2 dB. The input and output return losses are −8 dB and −12 dB, respectively. The distortion performance of the LNA can well be estimated by measuring the input referred third order intercept point and the −1 dB compression point. They are −7 dBm and −20 dBm, respectively.

The 1.9 GHz LNA shown in Fig. 3.24b employs a resistively loaded common-source stage followed by a reactively loaded cascode stage. To use inductors to tune out the output capacitance and to realize the 50 Ω output impedance is a viable alternative to using the transimpedance output stage. The circuit employs a self-biasing method by feeding the DC drain voltage of the first stage to the gates. The supply rejection is consequently poor.

The LNA achieves 21 dB gain and a 3 dB noise figure while drawing 10.8 mA from a 2.7 V supply. In Fig. 3.25 the measured gain and noise figure versus frequency are plotted. At the nominal. bias the input

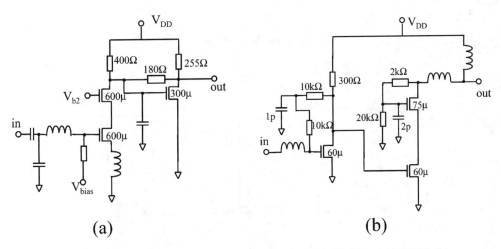

FIGURE 3.24 Schematic diagram of two 0.25 μm CMOS LNAs for 900 MHz (a) and 1900 MHz (b).

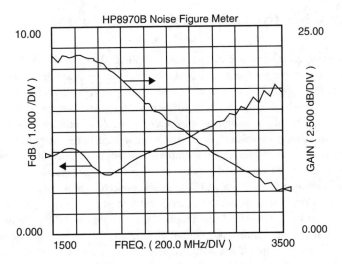

FIGURE 3.25 Measured gain and noise figure of the 1900 MHz CMOS LNA of Figure 3.3(b).

referred −1 dB compression point is −25 dBm, which corresponds to −4 dBm at the output. The input and output return loss are −5 dB and −13 dB, respectively

A comparison between the two amplifiers presented reveals some interesting points:

- The 900 MHz LNA explicitly makes use of the bondwire inductance to reduce the (otherwise purely capacitive) input impedance while the fist stage of the 1.9 GHz amplifier is connected to the chip ground. Both amplifiers use an external inductor for the input matching and both achieve a relatively poor input match.

- No explicit inter-stage matching is employed in either of the amplifiers. The 900 MHz amplifier uses the second stage as an impedance transformer.

- The 900 MHz amplifier employs ten times wider devices biased at lower current densities compared to its 1.9 GHz counterpart. As a consequence, the bias current becomes more sensitive to threshold voltage variations due to fabrication.

- At comparable power consumption the two amplifiers show roughly same distortion performance.

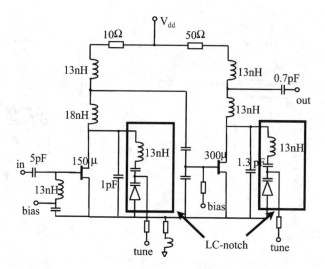

FIGURE 3.26   Schematic diagram of the selective frequency LNA at 2 GHz.

3.5.5   A Highly Selective LNA with Electrically Tunable Image Reject Filter for 2 GHz[15]

LNA designs with purely reactive passive components are inherently narrowband. IC technologies on high resistivity substrates allow reproducible passive components (inductors, capacitors, varactors, transmission lines) with excellent quality factors. They are well suited for designs to include a frequency selectivity which goes beyond a simple matching. In particular, amplifiers with adjustable image rejection can be realized. To show the potential of highly frequency selective LNAs as viable alternative to image reject mixers, an LNA for 1.9 GHz is demonstrated, which allows a tunable suppression of the image frequency. The schematic diagram of the circuit is shown in Fig. 3.26. The amplifier consists of two cascaded common-source stages loaded with LC resonant circuits. Undesired frequencies are suppressed using series notch filters as additional loads. Each of the two notch filters is formed by a series connection of a spiral inductor and a varactor diode. The two notches resonate at the same frequencies and must be isolated by the amplifier stages.

A careful design must be done to avoid unwanted resonances and oscillations. In particular, immunity against variations in the ground inductance and appropriate isolation between the supply lines of the two stages must be included. Only the availability of IC technologies with reproducible high-Q, low-tolerance passive components enables the realization of such highly frequency-selective amplifiers.

The LNA draws 9.5 mA from a 3 V supply. At this power dissipation, the input referred –1 dB compression point is measured at –24 dBm.

The measured input and output reflection is plotted in Fig. 3.27. The tuning voltage is set to 0 V. The excellent input match changes only negligibly with varying tuning voltage. The input matching shows a high-pass characteristic formed by the series C-L combination instead to the commonly used low-pass. So, the inductor can also act as a bias choke and the input matching can contribute to the suppression of lower frequency interferer. Moreover, the employed matching achieves better noise performance than the high-pass matching network.

The power gain vs. frequency for different notch tuning voltages is shown in Fig. 3.28. By varying the tuning voltage from 0.5 V to 1.5 V, the filter center frequency can be adjusted from 1.44 to 1.6 GHz. At all tuning voltages the unwanted signal is suppressed by at least 35 dB

The temperature dependence of gain and noise figure was measured. The temperature coefficients of the gain and noise figure are –0.03 dB/°C and +0.008 dB/°C, respectively. The noise figure of the LNA at different temperatures is plotted in Fig. 3.29.

A chip photograph of the fabricated 1.6 × 1.0 mm² LNA is depicted in Fig.3.30. More than 50% of the chip area is.occupied by the numerous spiral inductors.

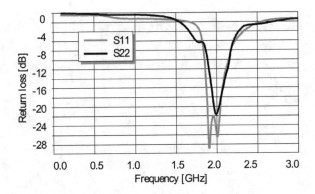

FIGURE 3.27 Measured input and output return loss of the 2 GHz selective LNA.

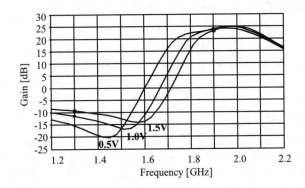

FIGURE 3.28 Selective amplifier gain vs. frequency for different notch filter control voltages.

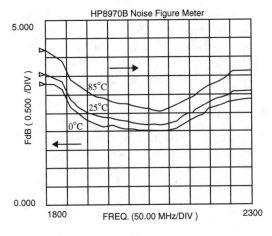

FIGURE 3.29 Amplifier noise figure at various temperatures.

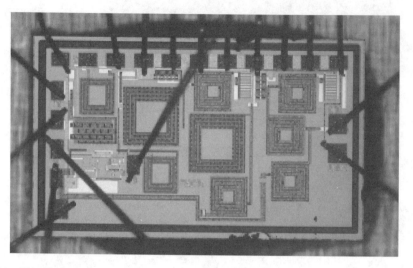

FIGURE 3.30   Chip photograph of the frequency selective LNA.

3.6   Future Trends

RF and microwave functions are increasingly often realized as integrated circuits (ICs) to reduce size and power consumption, enhance reproducibility, minimize costs, and enable mass production.

3.6.1   Design Approach

The classical noise optimization is based on linear methods and does not take power consumption and linearity requirements explicitly into account. Further, these methods offer only little guidance about how to select the active device dimensions. However, LNA circuit design practices are increasingly influenced by the improvements in the device models in terms of accuracy. Powerful optimization tools become available and eases the design procedure. However, a detailed understanding of the basic material will remain necessary for an efficient and robust LNA circuit design.

3.6.2   Device Models

The plurality of bias conditions applied to integrated circuits requires the flexibility of bias-dependent device models. State-of-the-art BJT models (such as Gummel-Poon) already work very well in RF simulations. More recently, sophisticated, semiempirical MOSFET models (such as BSIM3, MM9, or EKV) became suitable for RF simulations. Using accurate models, designs do not need to rely on sample scattering parameters of test devices and tolerance simulations can be implemented.

3.6.3   Circuit Environment

New RF design practices away from the 50 Ω impedance culture will affect the selection of the device size and operation point, but will leave the design procedure basically unchanged. The obstacles in the quest for higher integrated RF radios are the requirements on system noise figure, substrate crosstalk, and parasitic coupling. Trends to alleviate the unwanted coupling involve using fully differential circuit design, which in turn increases the power consumption.

3.6.4   IC Technologies

In recent years, the advances in device shrinking have made silicon devices (BJTs and more recently MOSFETs) become competitive with III-V semiconductors in terms of gain and minimum noise figure at a given power dissipation in the low GHz range.

The introduction of SiGe and SiC layers further enhance the cutoff frequencies and reduce power dissipation of silicon-based transistors. Furthermore, the use of thick (copper) metallization layers allow relatively low-loss passive components such as MIM capacitors and spiral inductors. Silicon-on-insulator (SOI) technologies will further cut substrate losses and parasitic capacitance and reduce bulk crosstalk.

With the scaling toward minimum gate length of below 0.25 μm, the use of CMOS has become a serious option in low-noise amplifier design. In fact, minimum noise figures of 0.5 dB at 2 GHz and cutoff frequencies of above 100 GHz for 0.12 μm devices[16] can easily compete with any other circuit technology. While intrinsic CMOS device F_{min} is becoming excellent for very short gate lengths, there remains the question of how closely amplifier noise figures can approach F_{min} in practice, particularly if there is a constraint on the allowable power consumption.

References

1. J. M. Rollett, Stability and power-gain invariance of linear two ports, *IEEE Trans. on Circuit Theory,* CT-9, 1, 29–32, March 1962, with corrections, CT-10, 1, 107, March 1963.

2. G. Gonzales, *Microwave Transistor Amplifiers Analysis and Design,* 2nd Edition, Prentice Hall, Englewood Cliffs, NJ, 1997.

3. G. Macciarella, et al., Design criteria for multistage microwave amplifiers with match requirements at input and output, *IEEE Trans. Microwave Theory and Techniques,* MTT-41, 1294–98, Aug. 1993.

4. G. D. Vendelin, A. M. Pavio, U. L. Rohde, *Microwave Circuit Design using Linear and Nonlinear Techniques,* 1st Edition, John Wiley & Sons, New York, 1990.

5. H. Hillbrand and P. H. Russer, An efficient method for computer aided noise analysis of linear amplifier networks, *IEEE Trans. on Circuit and Systems,* CAS-23, 4, 235–38, April 1976.

6. H. T. Friis, Noise figure for radio receivers, *Proc. of the IRE,* 419–422, July 1944.

7. H. A. Ainspan, et al., A 5.5 GHz low noise amplifier in SiGe BiCMOS, in *ESSCIRC98 Digest,* 80–83.

8. S. S. Taylor, A GaAs MESFET Schottky diode barrier height reference circuit, *IEEE Journal of Solid-State Circuits,* 32, 12, 2023–29, Dec. 1997.

9. J. J. Kucera, U. Lott, and W. Bächtold, A new antenna switching architecture for mobile handsets, *2000 IEEE Int'l Microwave Symposium Digest,* in press.

10. R. M. Fano, Theoretical limitations on the broad-band matching of arbitrary impedances, *Journal of the Franklin Institute,* 249, 57–83, Jan. 1960, and 139–155, Feb. 1960.

11. J. J. Kucera and W. Bächtold, A 1.9 GHz monolithic 1.1 dB noise figure low power LNA using a standard GaAs MESFET foundry process, *1998 Asia-Pacific Microwave Conference Digest,* 383–386.

12. J. J. Kucera and U. Lott, A 1.8 dB noise figure low DC power MMIC LNA for C-band, *1998 IEEE GaAs IC Symposium Digest,* 221–224.

13. Q. Huang, P. Orsatti and F. Piazza, Broadband, 0.25 μm CMOS LNAs with sub-2dB NF for GSM applications, *IEEE Custom Integrated Circuits Conference,* 67–70, 1998.

14. Ch. Biber, Microwave modeling and circuit design with sub-micron CMOS technologies, PhD thesis, Diss. ETH No. 12505, Zurich, 1998.

15. J. J. Kucera, Highly integrated RF transceivers, PhD thesis, Diss. ETH No. 13361, Zurich, 1999.

16. R. R. J. Vanoppen, et al., RF noise modeling of 0.25μm CMOS and low power LNAs, *1997 IEDM Technical Digest,* 317–320.

4

Microwave Mixer Design

Anthony M. Pavio

Motorola, Inc.

4.1 Introduction

At the beginning of the 20th century, RF detectors were crude, consisting of a semiconductor crystal contacted by a fine wire ("whisker"), which had to be adjusted periodically so that the detector would keep functioning. With the advent of the triode, a significant improvement in receiver sensitivity was obtained by adding amplification in front of and after the detector. A real advance in performance came with the invention by Edwin Armstrong of the super regenerative receiver. Armstrong was also the first to use a vacuum tube as a frequency converter (mixer) to shift the frequency of an incoming signal to an intermediate frequency (IF), where it could be amplified and detected with good selectivity. The superheterodyne receiver, which is the major advance in receiver architecture to date, is still employed in virtually every receiving system.

The mixer, which can consist of any device capable of exhibiting nonlinear performance, is essentially a multiplier or a chopper. That is, if at least two signals are present, their product will be produced at the output of the mixer. This concept is illustrated in Fig. 4.1. The RF signal applied has a carrier frequency of w_s with modulation $M(t)$, and the local oscillator signal (LO or pump) applied has a pure sinusoidal frequency of w_p. From basic trigonometry we know that the product of two sinusoids produces a sum and difference frequency.

The voltage-current relationship for a diode can be described as an infinite power series, where V is the sum of both input signals and I is the total signal current. If the RF signal is substantially smaller than the LO signal and modulation is ignored, the frequency components of the signal are:

$$w_d = nw_p \pm w_s \tag{4.1}$$

As mentioned above, the desired component is usually the difference frequency ($|w_{p+}w_s|$ or $|f_p - f_s|$), but sometimes the sum frequency ($f_s + f_p$) is desired when building an up-converter, or a product related to a harmonic of the LO can be selected.

4-1

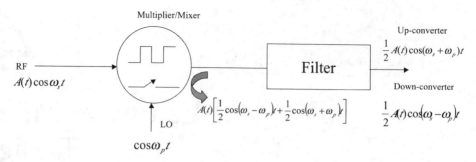

FIGURE 4.1 Ideal mixer model.

A mixer can also be analyzed as a switch that is commutated at a frequency equal to the pump frequency w_p. This is a good first-order approximation of the mixing process for a diode since it is driven from the low-resistance state (forward bias) to the high-resistance state (reverse bias) by a high-level LO signal.

The concept of the switching mixer model can also be applied to field-effect transistors used as voltage-controlled resistors. In this mode, the drain-to-source resistance can be changed from a few ohms to many thousands of ohms simply by changing the gate-to-source potential. At frequencies below 1 GHz, virtually no pump power is required to switch the FET, and since no DC drain bias is required, the resulting FET mixer is passive. However, as the operating frequency is raised above 1 GHz, passive FET mixers require LO drive powers comparable to diode or active FET designs.

Mixers can be divided into several classes: (1) single ended, (2) single balanced, or (3) double balanced. Depending on the application and fabrication constraints, one topology can exhibit advantages over the other types. The simplest topology (Fig. 4.2a) consists of a single diode and filter networks. Although there is no isolation inherent in the structure (balance), if the RF, LO, and IF frequencies are sufficiently separated, the filter (or diplexer) networks can provide the necessary isolation. In addition to simplicity, single diode mixers have several advantages over other configurations. Typically, the best conversion loss is possible with a single device, especially at frequencies where balun or transformer construction is difficult or impractical. Local oscillation requirements are also minimal since only a single diode is employed and DC biasing can easily be accomplished to reduce drive requirements. The disadvantages of the topology are: (1) sensitivity to terminations; (2) no spurious response suppression; (3) minimal tolerance to large signals; and (4) narrow bandwidth due to spacing between the RF filter and mixer diode.The next topology commonly used is the single balanced structure shown in Fig. 4.2b. These structures tend to exhibit slightly higher conversion loss than that of a single-ended design, but since the RF signal is divided between two diodes, the signal power-handling ability is better. More LO power is required, but the structure does provide balance. The double-balanced mixer (Fig. 4.2c) exhibits the best large signal-handling capability, port-to-port isolation, and spurious rejection. Some high-level mixer designs can employ multiple-diode rings with several diodes per leg in order to achieve the ultimate in large-signal performance. Such designs can easily require hundreds of milliwatts of pump power.

4.2   Single-Diode Mixers

The single-diode mixer, although fondly remembered for its use as an AM "crystal" radio or radar detector during World War II, has become less popular due to demanding broadband and high dynamic range requirements encountered at frequencies below 30 GHz. However, there are still many applications at millimeter wave frequencies, as well as consumer applications in the microwave portion of the spectrum, which are adequately served by single-ended designs. The design of single-diode mixers can be approached in the same manner as multi-port network design. The multi-port network contains all mixing product frequencies regardless of whether they are ported to external terminations or terminated internally. With simple mixers, the network's main function is frequency component separation; impedance matching

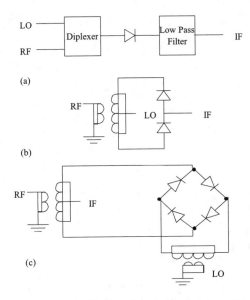

FIGURE 4.2 Typical mixer configurations. (a) Single ended; (b) single balanced; (c) double balanced.

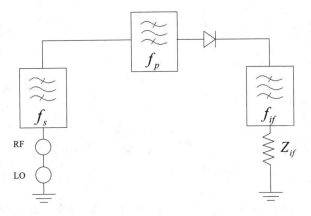

FIGURE 4.3 Filtering requirements for single-diode mixer.

requirements are secondary (Fig. 4.3). Hence, in the simplest approach, the network must be capable of selecting the LO, RF, and IF frequencies (Fig. 4.4).

However, before a network can be designed, the impedance presented to the network by the diode at various frequencies must be determined. Unfortunately, the diode is a nonlinear device; hence, determining its characteristics is more involved than determining an "unknown" impedance with a network analyzer. Since the diode impedance is time varying, it is not readily apparent that a stationary impedance can be found. Stationary impedance values for the RF, LO, and IF frequencies can be measured or determined if sufficient care in analysis or evaluation is taken.

4.3   Single-Balanced Mixers

Balanced mixers offer some unique advantages over single-ended designs such as LO noise suppression and rejection of some spurious products. The dynamic range can also be greater because the input RF signal is divided between several diodes, but this advantage is at the expense of increased pump power. Both the increase in complexity and conversion loss can be attributed to the hybrid or balun, and to the fact that perfect balance and lossless operation cannot be achieved.

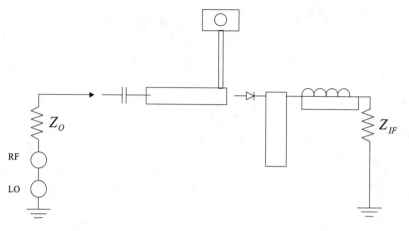

FIGURE 4.4   Typical single-ended mixer.

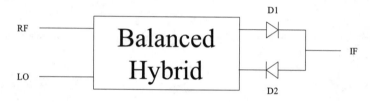

FIGURE 4.5   Single-balanced mixer topology.

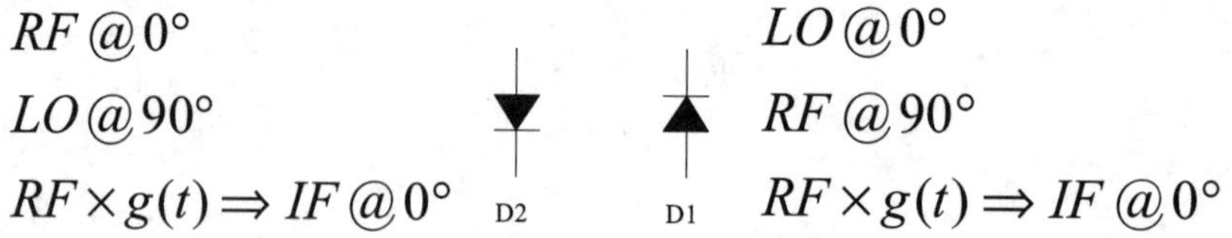

$$RF @ 0°$$
$$LO @ 90°$$
$$RF \times g(t) \Rightarrow IF @ 0°$$ D2

D1 $$LO @ 0°$$
$$RF @ 90°$$
$$RF \times g(t) \Rightarrow IF @ 0°$$

FIGURE 4.6   Signal phase relationships in quadrature coupled hybrid single-balanced mixer.

There are essentially only two design approaches for single-balanced mixers; one employs a 180° hybrid, while the other employs some form of quadrature structure (Fig. 4.5). The numerous variations found in the industry are related to the transmission-line media employed and the ingenuity involved in the design of the hybrid structure. The most common designs for the microwave frequency range employ either a branch-line, Lange, or "rat-race" hybrid structure (Fig. 4.6). At frequencies below about 5 GHz, broadband transformers are very common, while at frequencies above 40 GHz, waveguide and MMIC structures become prevalent.

4.4 Double-Balanced Mixers

The most commonly used mixer today is the double-balanced mixer. It usually consists of four diodes and two baluns or hybrids, although a double-ring or double-star design requires eight diodes and three hybrids. The double-balanced mixer has better isolation and spurious performance than the single-balanced designs described previously, but usually requires greater amounts of LO drive power, are more difficult to assemble, and exhibit somewhat higher conversion loss. However, they are usually the mixer of choice because of their spurious performance and isolation characteristics.

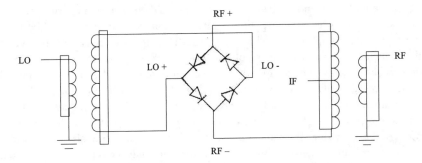

FIGURE 4.7   Transformer coupled double-balanced mixer.

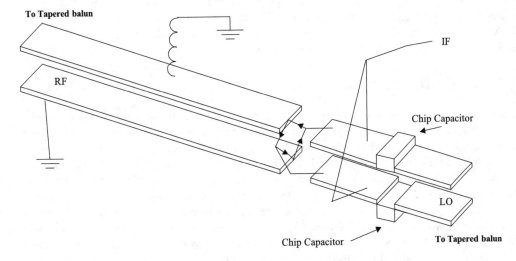

FIGURE 4.8   Double-balanced mixer center section.

A typical single-ring mixer with transformer hybrids is shown in Fig. 4.7. With this configuration the LO voltage is applied across the ring at terminals LO⁻ and LO⁺, and the RF voltage is applied across terminals RF⁻ and RF⁺. As can be seen, if the diodes are identical (matched), nodes RF⁻ and RF⁺ are virtual grounds; thus no LO voltage appears across the secondary of the RF transformer. Similarly, no RF voltage appears across the secondary of the LO balun. Because of the excellent diode matching that can be obtained with diode rings fabricated on a single chip, the L-to-R isolation of microwave mixers can be quite good, typically 30 to 40 dB.

Transmission-line structures which are naturally balanced, such as slotline and finline, can also be used as balanced feed in mixer design. However, all of the structures above, and the more complex transmission-line structures to follow, exhibit one major drawback compared to a transformer hybrid: There is no true RF center tap. As will be seen, this deficiency in transmission-line structures, extensively complicates the design of microwave-balanced mixers.

The lack of a balun center tap does indeed complicate the extraction of IF energy from the structure, but if the IF frequency is low, diplexing can be employed to ease performance degradation. This concept is illustrated in the following example of the center section of a double-balanced 2 to 12 GHz mixer (Fig. 4.8). It will be assumed that because of the soft-substrate transmission-line media and frequency range, a packaged diode ring with known impedances can be used. For Si diodes in this frequency range, the typical LO impedance range (magnitude) is on the order of 75, while the RF impedance is approximately 50. With these values in mind, microstrip-to-parallel plate transmission-line baluns can be fabricated on soft-substrate material.

As can be seen, both the RF and LO baluns terminate at the diode ring and provide the proper phase excitation. But since there is no center tap, the IF must be summed from the top and bottom of either balun. This summing is accomplished with bond wires that have high reactances at microwave frequencies but negligible inductances in the IF passband. Blocking capacitors form the second element in a high-pass filter, preventing the IF energy to be dissipated externally. An IF return path must also be provided at the terminals of the opposite balun. The top conductor side of the balun is grounded with a bond wire, providing a low-impedance path for the IF return and a sufficiently large impedance in shunt with the RF path. The ground-plane side of the balun provides a sufficiently low impedance for the IF return from the bottom side of the diode ring. The balun inductance and blocking capacitor also form a series resonant circuit shunting the IF output; therefore, this resonant frequency must be kept out of the IF passband.

The upper-frequency limit of mixers fabricated using tapered baluns and low parasitic diode packages, along with a lot of care during assembly, can be extended to 40 GHz. Improved "high-end" performance can be obtained by using beam-lead diodes. Although this design technique is very simple, there is little flexibility in obtaining an optimum port VSWR since the baluns are designed to match the magnitude of the diode impedance. The IF frequency response of using this approach is also limited, due to the lack of a balun center tap, to a frequency range below the RF and IF ports.

4.5   FET Mixer Theory

Interest in FET mixers has been very strong due to their excellent conversion gain and intermodulation characteristics. Numerous commercial products employ JFET mixers, but as the frequency of operation approaches 1 GHz, they begin to disappear. At these frequencies and above, the MESFET can easily accomplish the conversion functions that the JFET performs at low frequencies. However, the performance of active FET mixers reported to date by numerous authors has been somewhat disappointing. In short, they have not lived up to expectations, especially concerning noise-figure performance, conversion gain, and circuit-to-circuit repeatability. However, they are simple and low cost, so these sins can be forgiven.

Recently, growing interest is GaAs monolithic circuits is again beginning to heighten interest in active MESFET mixers. This is indeed fortunate, since properly designed FET mixers offer distinct advantages over their passive counterparts. This is especially true in the case of the dual-gate FET mixer; since the additional port allows for some inherent LO-to-RF isolation, it can at times replace single balanced passive approaches. The possibility of conversion gain rather than loss is also an advantage, since the added gain may eliminate the need for excess amplification, thus reducing system complexity.

Unfortunately, there are some drawbacks when designing active mixers. With diode mixers, the design engineer can make excellent first-order performance approximations with linear analysis; also, there is the practical reality that a diode always mixes reasonably well almost independent of the circuit. In active mixer design, these two conditions do not hold. Simulating performance, especially with a dual-gate device, requires some form of nonlinear analysis tool if any circuit information other than small-signal impedance is desired. An analysis of the noise performance is even more difficult.

As we have learned, the dominant nonlinearity of the FET is its transconductance, which is typically (especially with JFETs) a squarelaw function. Hence it makes a very efficient multiplier.

The small-signal circuit [1] shown in Fig. 4.9 denotes the principal elements of the FET that must be considered in the model. The parasitic resistances R_g, R_d, and R_s are small compared to R_{ds} and can be considered constant, but they are important in determining the noise performance of the circuit. The mixing products produced by parametric pumping of the capacitances C_{gs}, C_{dg}, and C_{ds} are typically small and add only second-order effects to the total circuit performance. Time-averaged values of these capacitances can be used in circuit simulation with good results.

This leaves the FET transconductance g_m, which exhibits an extremely strong nonlinear dependence as a function of gate bias. The greatest change is transconductance occurs near pinch off, with the most linear change with respect to gate voltage occurring in the center of the bias range. As the FET is biased toward I_{dss}, the transconductance function again becomes nonlinear. It is in these most nonlinear regions that the FET is most efficient as a mixer.

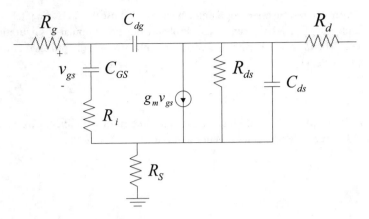

FIGURE 4.9 Typical MESFET model.

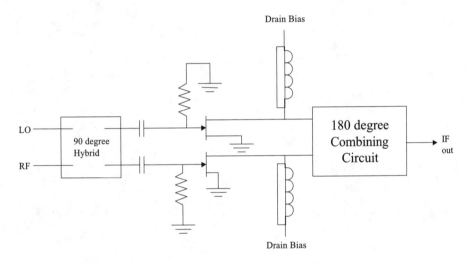

FIGURE 4.10 Typical FET single-balanced mixer.

If we now introduce a second signal, V_c, such that it is substantially smaller than the pump voltage, across the gate-to-source capacitance C_{gs}, the nonlinear action of the transconductance will cause mixing action within the FET producing frequencies $|nw_p \pm w_1|$, where n can be any positive or negative integer. Any practical analysis must include mixing products at both the gate and drain terminal, and at a minimum, allow frequency components in the signal, image, LO, and IF to exist.

Double-balanced FET mixers can also be designed using transformer hybrids [1]. Fig. 4.10 shows a typical balanced FET mixer, which can be designed to operate from VHF to SHF. An additional balun is again required because of the phase relationships of the IF signal. This structure is completely balanced and exhibits spurious rejection performance, similar to diode mixers constructed for the same frequency range. However, the intermodulation and noise-figure performance of such structures is superior to those of simple four-diode designs. For example, third-order intercept points in excess of 33 dBm, with associated gains of 6 dB, are common in such structures. High-level multiple-diode ring mixers, which would require substantially more LO power, would exhibit comparable intermodualtion characteristics, but would never exhibit any gain.

There are a variety of interesting mixer topologies in widespread use that perform vital system functions that cannot be simply classified as balanced mixers. Probably the most popular configuration is the image rejection or single-sideband mixer. However, a variety of subharmonically pumped and self-oscillating mixers are in limited use [1].

Reference

1. G. D. Vendelin, A. M. Pavio, and U. L. Rohde, The Design of Amplifiers Mixers and Oscillators Using the S-Parameter Method, John Wiley and Son, New York, 1990.

5

Modulation and Demodulation Circuitry

Charles Nelson
California State University

5.1 Some Fundamentals: Why Modulate?

Because this chapter uses a building block approach, it may seem to be a long succession of setting up straw men and demolishing them. To some extent, this imitates the development of radio and TV, which has been going on for most of the century just ended. A large number of concepts were developed as the technology advanced; each advance made new demands upon the hardware. At first, many of these advances were made by enthusiastic amateurs who had no fear of failure and viewed radio communication the way Hillary viewed Everest — something to be surmounted "because it was there." Since about World War II, there have been increasing numbers of engineers who understood these principles and could propose problem solutions that might have worked the first or second time they were tried. The author fondly hopes this book will help to grow a new cadre of problem solvers for the 21st century.

What probably first motivated the inventors of radio was the need for ships at sea to make distress calls. It may be interesting to note that the signal to be transmitted was a digital kind of thing called Morse Code. Later, the medium became able to transmit equally crucial analog signals, such as a soldier warning, "Watch out!! The woods to your left are full of the abominable enemy!" Eventually, during a period without widespread military conflict, radio became an entertainment medium, with music, comedy, and news, all made possible by businessmen who were convinced you could be persuaded, by a live voice, to buy soap, and later, detergents, cars, cereals not needing cooking, and so on. The essential low and high frequency content of the signal to be transmitted has been very productive for problems to be solved by radio engineers.

The man on radio, urging you to buy a "pre-owned" Cadillac, puts out most of his sound energy below 1000 Hz. A microphone observes pressure fluctuations corresponding to the sound and generates a corresponding voltage. Knowing that all radio broadcasting is done by feeding a voltage to an antenna, the beginning engineer might be tempted to try sending out the microphone signal directly. A big problem with directly broadcasting such a signal is that an antenna miles long would be required to transmit it efficiently. However, if the frequency of the signal is shifted a good deal higher, effective antennas become much shorter and more feasible to fabricate. This upward translation of the original message spectrum is perhaps the most crucial part of what we have come to call "modulation." However, the necessities of retrieving the original message from the modulated signal may dictate other inclusions in the broadcast signal, such as a small or large voltage at the center, or "carrier" frequency of the modulated signal. The need for a carrier signal is dictated by what scheme is used to transmit the modulated signal, which determines important facts of how the signal can be demodulated.

More perspective on the general problem of modulation is often available by looking at the general form of a modulated signal,

$$f(t) = A(t)\cos\theta(t).$$

If the process of modulation causes the multiplier A(t) out front to vary, it is considered to be some type of "amplitude" modulation. If one is causing the angle to vary, it is said to be "angle" modulation, but there are two basic types of angle modulation. We may write

$$\theta(t) = \omega_c t + \phi(t).$$

If then our modulation process works directly upon $\omega_c = 2\pi f_c$, we say we have performed "frequency" modulation. If, instead, we directly vary the phase factor $\phi(t)$, we say we have performed "phase" modulation. The two kinds of angle modulation are closely related, so that we may do one kind of operation to get the other result, by proper preprocessing of the modulation signal. Specifically, if we put the modulating signal through an integrating circuit before we feed it to a phase modulator, we come out with frequency modulation. This is, in fact, often done. The dual of this operation is possible but is seldom done in practice. Thus, if the modulating signal is fed through a differentiating circuit before it is fed to a frequency modulator, the result will be phase modulation. However, this process offers no advantages to motivate such efforts.

5.2   How to Shift Frequency

Our technique, especially in this chapter, will be to make our proofs as simple as possible; specifically, if trigonometry proves our point, it will be used instead of the convolution theorem of circuit theory. Yet, use of some of the aspects of convolution theory can be enormously enlightening to those who understand. Sometimes, as it will in this first proof, it may also indicate the kind of circuit that will accomplish the task. We will also take liberties with the form of our modulating signal. Sometimes we can be very general, in which case it may be identified as a function m(t). At other times, it may greatly simplify things if we write it very explicitly as a sinusoidal function of time

$$m(t) = \cos\omega_m t.$$

Sometimes, in the theory, this latter option is called "tone modulation," because, if one listened to the modulating signal through a loudspeaker, it could certainly be heard to have a very well-defined "tone" or pitch. We might justify ourselves by saying that theory certainly allows this, because any particular signal we must deal with could, according the theories of Fourier, be represented as a collection, perhaps infinite, of cosine waves of various phases. We might then assess the maximum capabilities of a communication system by choosing the highest value that the modulating signal might have. In AM radio, the highest modulating frequency is typically about $f_m = 5000$ Hz. For FM radio, the highest modulation frequency might be $f_m = 19$ kHz, the frequency of the so-called FM stereo "pilot tone."

In principle, the shifting of a frequency is very simple. This is fairly obvious to those understanding convolution. One theorem of system theory says that multiplication of time functions leads to convolution of the spectra. Let us just multiply the modulating signal by a so-called "carrier" signal. One is allowed to have the mental picture of the carrier signal "carrying" the modulating signal, in the same way that homing pigeons have been used in past wars to carry a light packet containing a message from behind enemy lines to the pigeon's home in friendly territory. So, electronically, for "tone modulation," we need only to accomplish the product

$$\phi(t) = A\cos\omega_m t\cos\omega_c t.$$

Now, we may enjoy the consequences of our assumption of tone modulation by employing trigonometric identities for the sum or difference of two angles:

$$\cos(A+B) = \cos A\cos B - \sin A\sin B \quad \text{and} \quad \cos(A-B) = \cos A\cos B + \sin A\sin B$$

If we add these two expressions and divide by two, we get the identity we need:

$$\cos A\cos B = 0.5\big[\cos(A+B) + \cos(A-B)\big].$$

Stated in words, we might say we got "sum and difference frequencies," but neither of the original frequencies. Let's be just a little more specific and say we started with $f_m = 5000$ Hz and $f_c = 1$ MHz, as would happen if a radio station whose assigned carrier frequency was 1 MHz were simply transmitting a single tone at 5000 Hz. In "real life," this would not be done very often, but the example serves well to illustrate some definitions and principles. The consequence of the mathematical multiplication is that the new signal has two new frequencies at 995 kHz and 1005 kHz. Let's now just add one modulating tone at 3333 Hz. We would have added two frequencies at 9666.667 kHz and 1003.333 kHz. However, if this multiplication was done purely, *there is no carrier frequency term present*. For this reason, we say we have done a type of "suppressed carrier" modulation. Also, furthermore, we have two *new* frequencies for *each* modulating frequency. We define all of those frequencies above the carrier as the "upper sideband" and all the frequencies below the carrier as the "lower sideband." The whole process we have done here is named "double sideband suppressed carrier" modulation, often known by its initials DSB–SC. Communication theory would tell us that the signal spectrum, before and after modulation with a single tone at a frequency f_m, would appear as in Fig. 5.1. Please note that the theory predicts equal positive and negative frequency components. There is no deep philosophical significance to negative frequencies. They simply make the theory symmetrical and a bit more intuitive.

5.3 Analog Multipliers, or "Mixers"

First, there is an unfortunate quirk of terminology; the circuit that multiplies signals together is in communication theory usually called a "mixer." What is unfortunate is that the engineer or technician who produces sound recordings is very apt to feed the outputs of many microphones into potentiometers, the outputs of which are sent in varying amounts to the output of a piece of gear, and *that* component is called a "mixer." Thus, the communication engineer's mixer multiplies and the other adds. Luckily, it will usually be obvious which device one is speaking of.

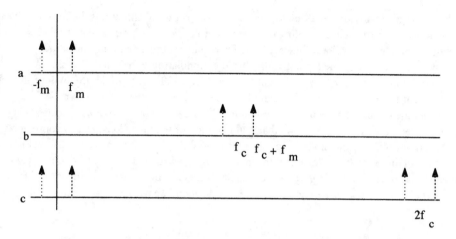

FIGURE 5.1 Unmodulated, modulated, and synchronously demodulated signal spectra. a. Spectrum of tone-modulating signal. b. Spectrum (positive part only) of double sideband suppressed carrier signal. c. Spectrum of synchronously detected DSB–SC signal (except for part near $-2f_c$).

There are available a number of chips (integrated circuits) designed to serve as analog multipliers. The principle is surprisingly simple, although the chip designers have added circuitry which no doubt optimizes the operation and perhaps makes external factors less influential. The reader might remember that the transconductance g_m for a bipolar transistor is proportional to the collector current; its output is proportional to the g_m *and* the input voltage, so in principle one can replace an emitter resistor with the first transistor, which then controls the collector current of the second transistor. If one seeks to fabricate such a circuit out of discrete transistors, one would do well to expect a need to tweak operating conditions considerably before some approximation of analog multiplication occurs. Recommendation: buy the chip. Best satisfaction will probably occur with a "four-quadrant multiplier." The alternative is a "two-quadrant multiplier," which might embarrass one by being easily driven into cut-off.

Another effective analog multiplier is alleged to be the dual-gate FET. The width of the channel in which current flows depends upon the voltage on each of two gates which are insulated from each other. Hence, if different voltages are connected to the two gates, the current that flows is the product of the two voltages. Both devices we have discussed so far have the advantage of having some amplification, so the desired resulting signal has a healthy amplitude. A possible disadvantage may be that spurious signals one does *not* need may also have strong amplitudes.

Actually, the process of multiplication may be the byproduct of any distorting amplifier. One can show this by expressing the output of a distorting amplifier as a Taylor series representing output in terms of input. In principle, such an output would be written

$$V_o = a_0 + a_1(v_1 + v_2) + a_3(v_1 + v_2)^2 + \text{ smaller terms.}$$

One can expand $(v_1 + v_2)^2$ as $v_1^2 + 2v_1v_2 + v_2^2$, so this term yields second harmonic terms of each input plus the product of inputs one was seeking. However, the term $a_1(v_1 + v_2)$ also yielded each input, so the carrier here would not be suppressed. If it is fondly desired to suppress the carrier, one must resort to some sort of "balanced modulator." An "active" (meaning there is amplification provided) form of a balanced modulator may be seen in Fig. 5.2; failure to bias the bases of the transistors should assure that the voltage squared term is large.

One will also find purely passive mixers with diodes connected in the shape of a baseball diamond with one signal fed between first and third base, the other from second to home plate. Such an arrangement has the great advantage of not requiring a power supply; the disadvantage is that the amplitude of the sum or difference frequency may be small.

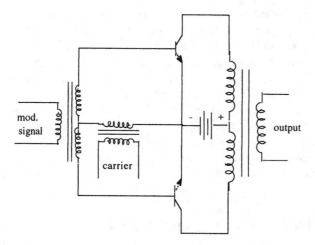

FIGURE 5.2   Balanced modulator.

5.4   Synchronous Detection of Suppressed Carrier Signals

At this point, the reader without experience in radio may be appreciating the mathematical tricks but wondering, if one can accomplish this multiplication, can it be broadcast and the original signal retrieved by a receiver? A straightforward answer might be that multiplying the received signal by another carrier frequency signal such as $\cos \omega_c t$ will shift the signal back exactly to where it started and also up to a center frequency of twice the original carrier. This is depicted in part c of Fig. 5.1. The name of this process is "synchronous detection." (In the days when it was apparently felt that communications enjoyed a touch of class if one used words having Greek roots, they called it "homodyne detection." If the reader reads a wide variety of journals, he/she may still encounter the word.) The good/bad news about synchronous detection is that the signal being used in the detector multiplication must have the *exact frequency and phase* of the original carrier, and such a signal is not easy to supply. One method is to send a "pilot" carrier, which is a small amount of the correct signal. The pilot tone is amplified until it is strong enough to accomplish the detection.

Suppose the pilot signal reaches high enough amplitude but is phase-shifted an amount θ with respect to the original carrier. We would then in our synchronous detector be performing the multiplication:

$$m(t)\cos\omega_c t\cos(\omega_c t+\theta).$$

To understand what we get, let us expand the second cosine using the identity for the sum of two angles,

$$\cos(\omega_c t+\theta)=\cos\omega_c t\cos\theta-\sin\omega_c t\sin\theta.$$

Hence, the output of the synchronous detector may be written as

$$m(t)\cos^2\omega_c t\cos\theta-m(t)\cos\omega_c t\sin\omega_c t\sin\theta=$$
$$(0.5)\big[m(t)\cos\theta(1-\cos2\omega_c t)-m(t)\sin\theta\sin2\omega_c t\big].$$

The latter two terms can be eliminated using a low-pass filter, and one is left with the original modulating signal, m(t), attenuated proportionally to the factor cos θ, so major attenuation does not appear until the phase shift approaches 90°, when the signal would vanish completely. Even this is not totally bad news, as it opens up a new technique called "quadrature amplitude modulation."

The principle of QAM, as it is abbreviated, is that entirely different modulating signals are fed to carrier signals that are 90° out of phase; we could call the carrier signals $\cos \omega_c t$ and $\sin \omega_c t$. The two modulating signals stay perfectly separated if there is no phase shift to the carrier signals fed to the synchronous detectors. The color signals in a color TV system are QAM'ed onto a 3.58 MHz subcarrier to be combined with the black-and-white signals, after they have been demodulated using a carrier generated in synchronism with the "color burst" (several periods of a 3.58 MHz signal), which is cleverly "piggy-backed" onto all the other signals required for driving and synchronizing a color TV receiver.

5.5   Single Sideband Suppressed Carrier

The alert engineering student may have heard the words "single sideband" and be led to wonder if we are proposing sending one more sideband than necessary. Of course it is true, and SSB–SC, as it is abbreviated, is the method of choice for "hams," the amateur radio enthusiasts who love to see night fall, when their low wattage signals can bounce between the earth and a layer of ionized atmospheric gasses 100 or so miles up until they have reached halfway around the world. It turns out that a little phase shift is not a really drastic flaw for voice communications, so the "ham" just adjusts the variable frequency oscillator being used to synchronously demodulate incoming signals until the whistles and squeals become coherent, and then he/she listens

How can one produce single sideband? For many years it was pretty naïve to say, "Well, let's just filter one sideband out!" This would have been very naïve because, of course, one does not have textbook filters with perfectly sharp cut-offs. Recently, however, technology has apparently provided rather good "crystal lattice filters" which are able fairly cleanly to filter the extra sideband. In general, though, the single sideband problem is simplified if the modulating signal does not go to really deep low frequencies; a microphone that does not put out much below 300 Hz might have advantages, as it would leave a transition region of 600 Hz between upper and lower sidebands in which the sideband filter could have its amplitude response "roll off" without letting through much of the sideband to be discarded. Observe Fig. 5.3, showing both sidebands for a baseband signal extending only from 300 Hz to 3.0 kHz.

Another method of producing single sideband, called the "phase-shift method," is suggested if one looks at the mathematical form of just one of the sidebands resulting from tone modulation. Let us just look at a lower sideband. The mathematical form would be

$$v(t) = A\cos\left(\omega_c - \omega_m\right)t = A\cos\omega_c t\cos\omega_m t + A\sin\omega_c t\sin\omega_m t$$

Mathematically, one needs to perform DSB–SC with the original carrier and modulating signals (the cosine terms) and also with the two signals each phase shifted 90°; the resulting two signals are then added to obtain the lower sideband. Obtaining a 90° phase shift is not difficult with the carrier, of which there is only one, but we must be prepared to handle a band of modulating signals, and it is not an elementary task to build a circuit that will produce 90° phase shifts over a range of frequency. However, a reasonable job will be done by the circuit of Fig. 5.4 when the frequency range is limited (e.g., from 300 to 3000 Hz). Note that one does *not* modulate directly with the original modulation signal, but that the network uses each input frequency to generate two signals which are attenuated equal amounts and 90° away from each other. These voltages would be designated in the drawing as V_{xz} and V_{yz}. In calculating such voltages, the reader should note that there are two voltage dividers connected across the modulating voltage, determining V_x and V_y, and that from both of these voltages

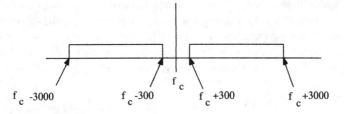

FIGURE 5.3 Double sideband spectrum for modulating signal 300–3000 Hz.

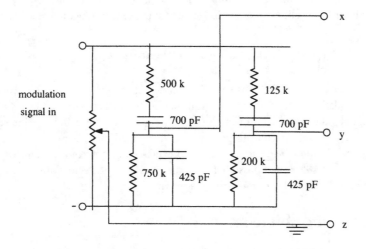

FIGURE 5.4 Audio network for single sideband modulator.

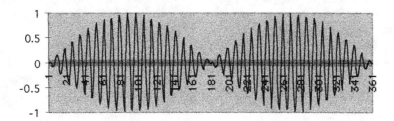

FIGURE 5.5 Double sideband suppressed carrier signal.

is subtracted the voltage from the center-tap to the bottom of the potentiometer. Note also that the resistance of the potentiometer is not relevant as long as it does not load down the source of modulating voltage, and that a good result has been found if the setting of the potentiometer is for 0.224 of the input voltage.

5.6 Amplitude Modulation as Double Sideband with Carrier

The budding engineer must understand that synchronous detectors are more expensive than many people can afford, and that a less expensive detection method is needed. What fills this bill much of the time is called the "envelope detector." Let us examine some waveforms, first for DSB–SC and then for a signal having a large carrier component. Figure 5.5 shows a waveform in which not very different carrier and modulating frequencies were chosen so that a spreadsheet plot would show a few details.

An ideal circuit we call an envelope detector would follow the topmost excursion of the waveform sketched here. Now, the original modulating signal was a sine wave, but the topmost excursion would be a *rectified* sinusoid, thus containing large amounts of harmonic distortion. How can one get a waveform that will be detected without distortion by an envelope detector? What was plotted was $1.0 \cos \omega_c t \cos \omega_m t$. We suspect we must add some amount of carrier $B \cos \omega_c t$. The sum will be

$$\phi_{AM}(t) = B\cos\omega_c t + 1.0\cos\omega_c t\cos\omega_m t = \cos\omega_c t\left[B + 1.0\cos\omega_m t\right].$$

This result is what is commonly called "amplitude modulation." Perhaps the most useful way of writing the time function for an amplitude modulation signal having tone modulation at a frequency f_m follows:

$$\phi_{AM}(t) = A\cos\omega_c t\left[1 + a\cos\omega_m t\right].$$

In this expression, we can say that A is the peak amplitude of the carrier signal that would be present if there were no modulation. The total expression inside the [ ] brackets can be called the "envelope" and the factor "a" can be called the "index of modulation." As we have written it, if the index of modulation were >1, the envelope would attempt to go negative; this would make it necessary, for distortion-free detection, to use synchronous detection. "a" is often expressed as a percentage, and when the index of modulation is less than 100%, it is possible to use the simplest of detectors, the envelope detector. We will look at the envelope detector in more detail a bit later.

5.7   Modulation Efficiency

It is good news that sending a carrier along with two sidebands makes inexpensive detection using an envelope detector possible. The accompanying bad news is that the presence of carrier does not contribute *at all* to useful signal output; the presence of a carrier only leads after detection to DC, which may be filtered out at the earliest opportunity. Sometimes, as in video, the DC is needed to set the brightness level, in which case DC may need to be added back in at an appropriate level.

To express the effectiveness of a communication system in establishing an output signal-to-noise ratio, it is necessary to define a "modulation efficiency," which, in words, is simply the fraction of output power that is put into sidebands. It is easily figured if the modulation is simply one or two purely sinusoidal tones; for real-life modulation signals, one may have to express it in quantities that are less easy to visualize.

For tone modulation, we can calculate modulation efficiency by simply evaluating the carrier power and the power of all sidebands. For tone modulation, we can write:

$$\phi_{AM}(t) = A\cos\omega_c t\left[1 + a\cos\omega_m t\right] =$$

$$A\cos\omega_c t + (aA)/2\left[\cos(\omega_c + \omega_m)t + \cos(\omega_c - \omega_m)t\right].$$

Now, we have all sinusoids, the carrier, and two sidebands of equal amplitudes, so we can write the average power in terms of peak amplitudes as:

$$P = 0.5\left[A^2 + 2\times(aA/2)^2\right] = 0.5A^2\left[1 + a^2/2\right].$$

Then modulation efficiency is the ratio of sideband power to total power, for modulation by a single tone with modulation index "a," is:

$$\eta = \frac{\left(aA/2\right)^2}{0.5A^2\left(1+a^2/2\right)} = \frac{a^2}{2+a^2}.$$

Of course, most practical modulation signals are not so simple as sinusoids. It may be necessary to state how close one is to overmodulating, which is to say, how close to negative modulating signals come to driving the envelope negative. Besides this, what is valuable is a quantity we shall just call "m," which is the ratio of average power to peak power for the modulation function. For some familiar waveforms, if the modulation is sinusoidal, m = 1/2. If modulation were a symmetrical square wave, m = 1.0; any kind of symmetrical triangle wave has m = 1/3. In terms of m, the modulation efficiency is

$$\eta = \frac{ma^2}{1+ma^2}$$

5.8   The Envelope Detector

Much of the detection of modulated signals, whether the signals began life as AM or FM broadcast signals or the sound or the video of TV, is done using envelope detectors. Figure 5.7 shows the basic circuit configuration.

The input signal is of course as shown in Fig. 5.6. It is assumed that the forward resistance of the diode is 100 ohms or less. Thus, the capacitor is small enough that it gets charged up to the peak values of the high frequency signal, but then when input drops from the peak, the diode is reverse-biased so the capacitor can only discharge through R. This discharge voltage is of course given by

$$V\left(0\right)\exp\left(-t/RC\right).$$

Now the problem in AM detection is that we must have the minimum rate of decay of the voltage be at least the maximum decay of the envelope of the modulated wave. We might write the envelope as a function of time:

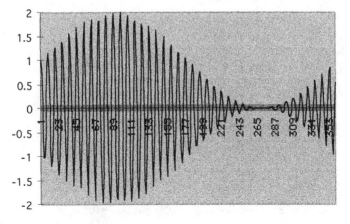

FIGURE 5.6   Amplitude-modulated signal.

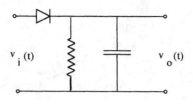

FIGURE 5.7   Simple envelope detector schematic.

$$E(t) = A(1 + a\cos\omega_{mt}t),$$

where A is the amplitude of the carrier before modulation and "a" is the index of modulation, which must be less than one for accurate results with the envelope detector. Then, when we differentiate, we get

$$\frac{dE}{dt} = -\omega_m Aa\sin(\omega_m t).$$

We want this *magnitude* to be less than or equal to the maximum magnitude of the rate of decay of a discharging capacitor, which is E(0)/RC. For what is written as E(0), we will write the instantaneous value of the envelope, and the expression becomes

$$A(1 + a\cos\omega_{mt}t) \geq RC(\omega_m aA\sin(\omega_m t)).$$

The As cancel, and we have

$$RC \leq \frac{1 + a\cos(\omega_m t)}{\omega_m a\sin(\omega_m t)};$$

our major difficulty occurs when the right-hand side has its minimum value.

If we differentiate with respect to $\omega_m t$, we get

$$\frac{\omega_m a\sin\left(\omega_m t \times \left(-\omega_m a\sin(\omega_m t) - \left(1 + a\cos(\omega_m t)(\omega_m)^2 a\cos\omega_m t\right)\right)\right)}{\left(\omega_m a\sin(\omega_m t)\right)^2}.$$

We set the numerator equal to zero to find its maximum. We find we have

$$-(\omega_m a)^2\left[\sin^2(\omega_m t) + \cos^2(\omega_m t)\right] - a(\omega_m)^2\cos\omega_m t$$

$$= -(\omega_m a)^2 - a(\omega_m)^2\cos\omega_m t = 0.$$

Hence, the maximum occurs when $\cos\omega_m t = -a$, and of course by identity, at that time, $\sin\omega_m t = \sqrt{1 - a^2}$.

Inserting these results into our inequality for the time constant RC, we have

$$RC \leq \frac{1-a^2}{\omega_m a \sqrt{1-a^2}} = \frac{\sqrt{1-a^2}}{\omega_m a}.$$

Example 5.1

Suppose we say 2000 Hz is the main problem in our modulation scheme, our modulation index is 0.5, and we choose R = 10k to make it large compared to the diode forward resistance, but not *too* large. What should be the capacitor C?

Solution   We use the equality now and get

$$C = \frac{\sqrt{1-0.5^2}}{0.5 \times 4000\pi \times 10,000} = 13.8 \text{ nF}.$$

5.9   Envelope Detection of SSB Using Injected Carrier

Single sideband, it might be said, is a very forgiving medium. Suppose that one were attempting synchronous detection using a carrier that was off by a Hertz or so, compared to the original carrier. Because synchronous detection works by producing sum and difference frequencies, 1 Hz error in carrier frequency would produce 1 Hz error in the detected frequency. Because SSB is mainly used for speech, it would be challenging indeed to find anything wrong with the reception of a voice one has only ever heard over a static-ridden channel. Similar results would also be felt in the following, where we add a carrier to the sideband and find that we have AM, albeit with a small amount of harmonic distortion.

Example 5.2

Starting with just an upper sideband $B\cos(\omega_c + \omega_m)t$, let us add a carrier term $A\cos\omega_c t$, manipulate the total, and prove that we have an envelope to detect. First we expand the sideband term as

$$\Phi_{SSB}(t) = B\left[\cos\omega_c t \cos\omega_m t - \sin\omega_c t \sin\omega_m t\right].$$

Adding the carrier term $A\cos\omega_c t$ and combining like terms, we have

$$\phi(t) = \cos\omega_c t\left(A + B\cos\omega_m t\right) - B\sin\omega_c t \sin\omega_m t.$$

In the first circuits class, we see that if we want to write a function of one frequency in the form $E(t)\cos(\omega_c t + \text{phase angle})$, the amplitude of the multiplier E is the square root of the squares of the coefficients of $\cos\omega_c t$ and $\sin\omega_c t$. Thus,

$$E(t) = \sqrt{\left(A + B\cos\omega_m t\right)^2 + \left(B\sin\omega_m t\right)^2}$$

$$= \sqrt{A^2 + 2AB\cos\omega_m t + B^2\left(\cos^2\omega_m t + \sin^2\omega_m t\right)}.$$

Now, of course, the coefficient of B^2 is unity for all values of $\omega_m t$. We find that best performance occurs if $B \ll A$. Then we would have our expression for the envelope (and thus it is detectable using an envelope detector):

$$E\big(t\big)=\sqrt{A^2+B^2+2AB\cos\omega_m t}=\sqrt{A^2+B^2}\sqrt{1+\frac{2AB}{A^2+B^2}\cos\omega_m t}\,.$$

Our condition that B ≪ A allows us to say the coefficient of cos ω_mt is really small compared to unity. We use the binomial theorem to approximate the second square root: $(1+x)^n \approx 1 + x/2 + (1/2)^2(-1/2)x^2$ when x ≪ 1. Using our approximation, x ≈ (2B/A) cos ω_mt. In our expansion, the x term is the modulation term we were seeking, the x^2 term contributes second harmonic distortion. Using the various approximations, and stopping after we find the second harmonic (other harmonics *will* be present, of course, but in decreasing amplitudes), we have

$$\text{Detected } f\big(t\big)=B\cos\omega_m t-\big(1/2\big)\big(B^2/A\big)\cos^2\big(\omega_m t\big).$$

When we use trig identities to get the second harmonic, we get another factor of one half; the ratio of detected second harmonic to fundamental is thus (1/4)(B/A). Thus, for example, if B is just 10% of A, second harmonic is only 2.5% of fundamental.

5.10 Direct vs. Indirect Means of Generating FM

Let us first remind ourselves of basics regarding FM. We can write the time function in its simplest form as

$$\phi_{FM}\big(t\big)=A\cos\big(\omega_c t+\beta\sin\omega_m t\big).$$

Now, the alert reader might be saying, "Hold on! That looks a lot like phase modulation. If β = 0, the phase would increase linearly in time, as an unmodulated signal, but gets advanced or retarded a maximum of β." One needs to remember the definition of instantaneous frequency, which is

$$f_i=\frac{1}{2\pi}\frac{d}{dt}\big(\omega_c t+\beta\sin\omega_m t\big)=\frac{1}{2\pi}\big(2\pi f_c+\beta 2\pi f_m\cos\omega_m t\big)=f_c+\beta f_m\cos\omega_m t.$$

Thus, we can say that instantaneous frequency departs from the carrier frequency by a maximum amount βf_m, which is the so-called "frequency deviation." This has been specified as a maximum of 75 kHz for commercial FM radio but 25 kHz for the sound of TV signals.

Now, certainly, the concept of directly generating FM has an intellectual appeal to it. The problems of direct FM are mainly practical; if the very means of putting information onto a high frequency "carrier" is in varying the frequency, it perhaps stands to reason the center value of the operating frequency will not be well nailed down. Direct FM could be accomplished as in Fig. 5.8(a), but Murphy's law would be very dominant and one might expect center frequency to drift continually in one direction all morning and the other way all afternoon, or the like. This system is sometimes stabilized by having an FM detector called a discriminator tuned to the desired center frequency, so that output would be positive if frequency got high and negative for frequency low. Thus, instantaneous output could be used as an error voltage with a long time constant to push the intended center frequency toward the center, whether the average value is above or below.

The best known method of indirect FM gives credit to the man who, more than any other, saw the possibilities of FM and that its apparent defects could be exploited for superior performance, Edwin Armstrong. He started with a crystal-stabilized oscillator around 100 kHz, from which he obtained also a 90° phase-shifted version. A block diagram of just this early part of the Armstrong modulator is shown in Fig. 5.8(a).

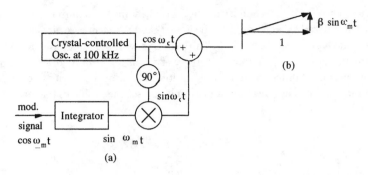

FIGURE 5.8 (a) Crystal-stabilized phase modulator; (b) phasor diagram.

The modulating signal is passed through an integrator before it goes into an analog multiplier, to which is also fed the *phase-shifted* version of the crystal-stabilized signal. Thus, we feed cos $\omega_c t$ and sin $\omega_c t$ sin $\omega_m t$ into a summing amplifier. The phasor diagram shows the two signals with cos ωt as the reference. There is a small phase shift given by $\tan^{-1}(\beta \sin \omega_m t)$ where β here gives the maximum amount of phase shift as a function of time (see Figure 5.8b). To see how good a job we have done, we need to expand $\tan^{-1}(x)$ in a Taylor series. We find that

$$\tan^{-1}(x) \approx x - (x)^3/3 + (x)^5/5.$$

We see that we have a term proportional to the modulating signal (x) and others that must represent odd-order harmonic distortion, if one accounts for the fact that we have resorted to a subterfuge, using a phase modulator to produce frequency modulation. Assuming that our signal finally goes through a frequency detector, we find that the amount of third harmonic as a fraction of the signal output is $\beta^2/4$. Now, in frequency modulation, the maximum amount of modulation which is permitted is in terms of frequency deviation, an amount of 75 kHz. The relation between frequency deviation and maximum phase shift is

$$\Delta f = \beta f_m,$$

where Δf is the frequency deviation, β is maximum phase shift, and f_m is modulation frequency. Since maximum modulation is defined in terms of Δf, the maximum value of β permitted will correspond to *minimum* modulation frequency. Let us do some numbers to illustrate this problem.

Example 5.3

Suppose we have a high fidelity broadcaster wishing to transmit bass down to 50 Hz with maximum third harmonic distortion of 1%. Find the maximum values of β and Δf.

Solution We have $\beta^2/4 = 0.01$. Solving for β, we get $\beta = 0.2$.

Then, $\Delta f = 0.2 \times 50$ Hz $= 10$ Hz.

One can recall that the maximum value of frequency deviation allowed in the FM broadcast band is 75 kHz. Thus, use of the indirect modulator has given us much lower frequency deviation than is allowed, and clearly some kind of desperate measures are required. Such are available, but do complicate the process greatly. Suppose we feed the modulated signal into an amplifier which is not biased for low distortion, that is, its Taylor series looks like

$$a_1 x + a_2 x^2 + a_3 x^3, \text{ etc.}$$

Now the squared term leads to second harmonic, the cubed one gives third harmonic, and so on. The phase-modulated signal looks like A $\cos(\omega_c t + \beta \sin \omega_m t)$ and the term $a_2\, x^2$ *not only doubles the carrier frequency, but also the maximum phase shift* β. Thus, starting with the rather low frequency of 100 kHz, we have a fair amount of multiplying room before we arrive in the FM broadcast band 88 to 108 MHz. Unfortunately, we may need different amounts of multiplication for the carrier frequency than we need for the depth of modulation. Let's carry on our example and see the problems that arise. First, if we wish to go from

$$\Delta f = 10 \text{ Hz to } 75,000 \text{ Hz,}$$

that leads to a total multiplication of 75,000/10 = 7500.

The author likes to say we are limited to frequency doublers and triplers. Let's use as many triplers as possible; we divide the 7500 by 3 until we get close to an even power of 2:

$$7500/3 = 2500;\ 2500/3 = 833,\ 833/3 = 278;\ 278/3 \approx 93,\ 93/3 = 31,$$

which is very close to $32 = (2)^5$.

So, to get our maximum modulation index, we need five each triplers and doublers. However, 7500 × 0.1 MHz = 750 MHz, and we have missed the broadcast band by about 7 times. One more thing we need is a mixer, after a certain amount of multiplication. Let's use all the doublers and one tripler to get a multiplication of 32 × 3 = 96, so the carrier arrives at 9.6 MHz. Suppose our final carrier frequency is 90.9 MHz, and because we have remaining to be used a multiplication of $3^4 = 81$, what comes out of the mixer must be

$$90.9/81 - 1.122 \text{ MHz.}$$

To obtain an output of 1.122 MHz from the mixer, with 9.6 MHz going in, we need a local oscillator of either 10.722 or 8.478 MHz. Note that this local oscillator needs a crystal control also, or the eventual carrier frequency will wander about more than is allowed.

5.11 Quick-and-Dirty FM Slope Detection

A method of FM detection that is barely respectable, but surprisingly effective, is called "slope detection." The principle is to feed an FM signal into a tuned circuit, not right at the resonant frequency but rather somewhat off the peak. Therefore, the frequency variations due to the modulation will drive the signal up and down the resonant curve, producing simultaneous amplitude variations, which then can be detected using an envelope detector. Let us just take a case of FM and a specific tuned circuit and find the degree of AM.

Example 5.4

We have an FM signal centered at 10.7 MHz, with frequency deviation of 75 kHz. We have a purely parallel resonant circuit with a Q = 30, with resonant frequency such that 10.7 MHz is at the lower half-power frequency. Find the output voltage for Δf = +75 kHz and for −75 kHz.

Solution   When we operate close to resonance, adequate accuracy is given by

$$V_o = \frac{V_i}{1 + j2Q\delta'}$$

where δ is the fractional shift of frequency from resonance. If now, 10.7 MHz is the lower half-power point, we can say that $2Q\delta = 1$.

$$\text{Hence, } \delta = 1\big/\big(2 \times 30\big) = \big(f_o - 10.7 \text{ MHz}\big)\big/f_o \, ; \, f_o = 10.881 \text{ MHz}.$$

Now, we evaluate the transfer function at 10.7 MHz ± 75.kHz.
We defined it as 0.7071 at 10.7 MHz. For 10.7 + 0.075 MHz, $\delta = (10.881 - 10.775)/10.881 = 9.774 \times 10^{-3}$, and the magnitude of the transfer function is $|1/(1 + j60\delta)| = 0.8626$.
Because the value was 0.7071 for the unmodulated wave, the modulation index in the positive direction would be

$$\big(0.8624 - 0.7071\big)\big/0.7071 = 0.2196 \text{ or } 21.96\%.$$

For $(10.7 - 0.075)$ MHz, $\delta = (10.881 - 10.625)/10.881 = 0.02356$, and the magnitude of the transfer function is $|1/(1 + j60)| = 0.5775$. The modulation index in the negative direction is $(0.7071 - 0.5775)/0.7071 = 18.32\%$. So, modulation index is not the same for positive as for negative indices. The consequence of such asymmetry is that this process will be subject to harmonic distortion, which is why this process is not quite respectable.

5.12 Lower Distortion FM Detection

We will assume that the reader has been left wanting an FM detector that has much better performance than the slope detector. A number of more complex circuits have a much lower distortion level than the slope detector. One, called the Balanced FM Discriminator, is shown in Fig. 5.9.

Basically, we may consider that the circuit contains two "stagger-tuned" resonant circuits, i.e., they are tuned equidistant on opposite sides of the center frequency, connected back to back. The result is that the nonlinearity of the resonant circuits balance each other out, and the FM detection can be very linear. The engineer designing an FM receiving system has a relatively easy job to access such performance; all that he/she must do is to spend the money to obtain high-quality components.

5.12.1 Phase-Locked Loop

The phase-locked loop is an assembly of circuits or systems that perform a number of functions to accomplish several operations, any one or more of the latter, perhaps being useful and to be capitalized upon. If one looks at a simple block diagram, one will see something like Fig. 5.10.

Thus, one function that will always be found is called a "voltage-controlled oscillator;" the linking of these words means that there is an oscillator which would run freely at some frequency, but that if a non-zero DC voltage is fed into a certain input, the frequency of oscillation will shift to one determined by that input voltage. Another function one will always find (although the nomenclature might vary somewhat) is "phase-comparison." The phase "comparator" will usually be followed by some kind of low-pass filter. Of course, if a comparator is to fulfill its function, it requires two inputs — the phases of which to compare. This operation might be accomplished in various ways; however, one method which might be understood from previous discussions is the analog multiplier. Suppose an analog multiplier receives the inputs $\cos \omega t$ and $\sin (\omega t + \phi)$; their product has a sine and a cosine. Now, a trigonometric identity involving these terms is

$$\sin A \cos B = 0.5 \big[\sin\big(A + B\big) + \sin\big(A - B\big)\big].$$

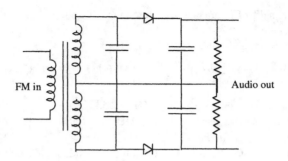

FIGURE 5.9   Balanced FM discriminator.

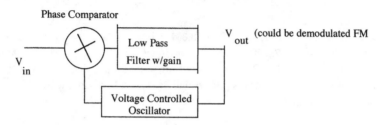

FIGURE 5.10   Basic phase-locked loop.

Thus, the output of a perfect analog multiplier will be $0.5[\sin(2\omega + \phi) + \sin\phi]$. A low-pass filter following the phase comparator is easily arranged; therefore, one is left with a DC term, which, if it is fed to the VCO in such a polarity as to provide negative feedback, will "lock" the VCO to the frequency of the input signal with a fixed phase shift of 90°.

Phase-locked loops (abbreviated PLL) are used in a wide variety of applications. Many of the applications are demodulators of one sort or another, such as synchronous detectors for AM, basic FM, FM–stereo detectors, and in very precise oscillators known as "frequency synthesizers." One of the early uses seemed to be the detection of weak FM signals, where it can be shown that they extend the threshold of usable weak signals a bit.[1] This latter facet of their usefulness seems not to have made a large impact, but the other aspects of PLL usefulness are very commonly seen.

5.13 Digital Means of Modulation

The sections immediately preceding have been concerned with rather traditional analog methods of modulating a carrier. While the beginning engineer can expect to do little or no design in analog communication systems, they serve as an introduction to the digital methods which most certainly will dominate the design work early in the 21st century. Certainly, analog signals will continue to be generated, such as speech, music, and video; however, engineers are finding it so convenient to do digital signal processing that many analog signals are digitized, processed in various performance-enhancing ways, and only restored to analog format shortly before they are fed to a speaker or picture tube. Digital signals can be transmitted in such a way as to use extremely noisy channels. Not long ago, the nightly news brought us video of the Martian landscape. The analog engineer would be appalled to know the number representing traditional signal-to-noise ratio for the Martian signal. The detection problem is greatly simplified because the digital receiver does not need at each instant to try to represent which of an infinite

[1] Taub, H. and Schilling, D.L. *Principles of Communication Circuits,* 2nd Edition, McGraw-Hill, New York, 1986, 426–427.

number of possible analog levels is correct; it simply asks, was the signal sent a one or a zero? *That* is simplicity.

Several methods of digital modulation might be considered extreme examples of some kind of analog modulation. Recall amplitude modulation. The digital rendering of AM is called "amplitude shift keying," abbreviated ASK.

What this might look like on an oscilloscope screen is shown in Fig. 5.11. For example, we might say that the larger amplitude signals represent the logic ones and smaller amplitudes represent logic zeroes. Thus, we have illustrated the modulation of the data stream 10101. If the intensity of modulation were carried to the 100% level, the signal would disappear completely during the intervals corresponding to zeroes. The 100% modulation case is sometimes called on–off keying and abbreviated OOK. The latter case has one advantage if this signal were nearly obscured by large amounts of noise; it is easiest for the digital receiver to distinguish between ones and zeroes if the difference between them is maximized. That is, however, only one aspect of the detection problem. It is also often necessary to know the timing of the bits, and for this one may use the signal to synchronize the oscillator in a phase-locked loop; if, for 50% of the time, there is zero signal by which to be synchronized, the oscillator may drift significantly. In general, a format for digital modulation in which the signal may vanish utterly at intervals is to be adopted with caution and with full cognizance of one's sync problem. Actually, amplitude shift keying is not considered a very high performance means of digital signaling, in much the same was that AM is not greatly valued as a quality means of analog communication. What is mainly used is one or the other of the following methods.

5.13.1 Frequency Shift Keying

Frequency shift keying (abbreviated FSK) can be used in systems having very little to do with high data rate communications; for years it has been the method used in the simple modems one first used to communicate with remote computers. For binary systems, one just sent a pulse of one frequency for a logic one and a second frequency for a logic zero. If one was communicating in a noisy environment, the two signals would be orthogonal, which meant that the two frequencies used were separated by at least the data rate. Now, at first the modem signals were sent over telephone lines which were optimized for voice communications, and were rather limited for data communication. Suppose we consider that for ones we send a 1250 Hz pulse and for zeroes, we send 2250 Hz. In a noisy environment one ought not to attempt sending more than 1000 bits per second (note that 1000 Hz is the exact difference between the two frequencies being used for FSK signaling). Let us instead send at 250 bps. Twelve milliseconds of a 101 bit stream would look as in Figure 5.12.

It is not too difficult to imagine a way to obtain FSK. Assuming one does have access to a VCO, one simply feeds it two different voltage levels for ones and for zeroes. The VCO output is the required output.

5.13.2 Phase Shift Keying

Probably the most commonly used type of digital modulation is some form of phase shift keying. One might simply say there is a carrier frequency f_c and that logic zeroes will be represented by $-\sin 2\pi f_c t$, logic ones by $+\sin 2\pi f_c t$. If the bit rate is 40% of the carrier frequency, the data stream 1010101010 might look as in Fig. 5.13.

In principle, producing binary phase shift keying ought to be fairly straightforward, if one has the polar NRZ (nonreturn to zero, meaning a logic one could be a constant positive voltage for the duration of the bit, zero being an equal negative voltage) bit stream. If then, the bit stream and a carrier signal are fed into an analog multiplier, the output of the multiplier could indeed be considered $\pm\cos \omega_c t$, and the modulation is achieved.

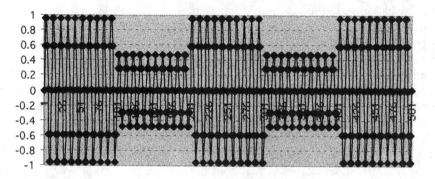

FIGURE 5.11   ASK (amplitude shift keying).

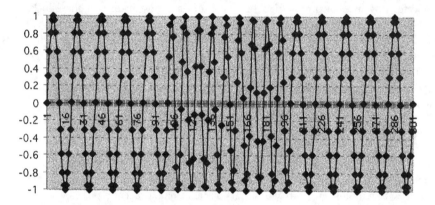

FIGURE 5.12   Frequency shift keying.

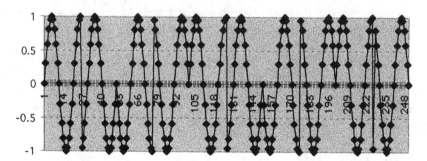

FIGURE 5.13   Phase shift keying.

5.14 Correlation Detection

Many years ago, the communications theorists came up with the idea that if one could build a "matched filter," that is, a special filter designed with the bit waveform in mind, one would startlingly increase the signal-to-noise ratio of the detected signal. Before long, a practically minded communications person had the bright idea that a correlation detector would do the job, at least for rectangular bits. For some reason, as one explains this circuit, one postulates two signals, $s_1(t)$ and $s_2(t)$, which represent, respectively, the signals sent for logic ones and zeroes. The basics of the correlation detector are shown in Fig. 5.14.

Now, a key consideration in the operation of the correlation detector is bit synchronization. It is crucial that the signal $s_1(t)$ be lined up perfectly with the bits being received. Then, the top multiplier "sees" sin $\omega_c t$ coming in one input, and $\pm$sin $\omega_c t$ + noise coming in the other, depending upon whether a one or

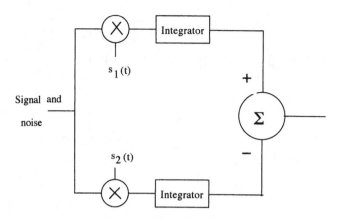

FIGURE 5.14   Correlation detector.

a zero is being received. If it happens that a one is being received, the multiplier is asked to multiply sin $\omega_c t (\sin \omega_c t + \text{noise})$. Of course,

$$\sin^2 \omega_c \left(t \right) = \left(1/2 \right) \left(1 + \cos 2\omega_c t \right).$$

In the integrator, this is integrated over one bit duration, giving a quantity said to be the energy of one bit. The integrator might also be considered to have been asked to integrate $n(t) \sin \omega_c t$, where n is the noise signal. However, the nature of noise is that there is no net area under the curve of its waveform, so considering integration to be a summation, the noise output out of the integrator would simply be the last instantaneous value of the noise voltage at the end of a bit duration, whereas the signal output was bit energy, if the bit synchronization is guaranteed. Meanwhile, the output of the bottom multiplier was the *negative* of the bit energy, so with the signs shown, the output of the summing amplifier is twice the bit energy. Similar reasoning leads to the conclusion that if the instantaneous signal being received were a zero, the summed output would be *minus* twice the bit energy. It takes a rather substantial bit of theory to show that the noise output from the summer is *noise spectral density*. The result may be summarized that the correlation detector can "pull a very noisy signal out of the mud." And, we should assert at this point that the correlation detector can perform wonders for any one of the methods of digital modulation mentioned up to this point.

5.15 Digital QAM

Once the engineer has produced carrier signals that are 90° out of phase with each other, there is no intrinsic specification that the modulation must be analog, as is done for color TV. As a start toward extending the capabilities of PSK, one might consider that one sends bursts of several periods of $\pm\cos \omega_c t$ or $\pm\sin \omega_c t$. This is sometimes called "4-ary" transmission, meaning that there are four different possibilities of what might be sent. Thus, whichever of the possibilities is sent, it may be considered to contain two bits of information. It is a method by which more information may be sent without demanding any more bandwidth, because the duration of the symbol being sent may be no longer or shorter than it was when one was doing binary signaling, sending, for example, simply $\pm\cos \omega_c t$. This idea is sometimes represented in a "constellation," which, for the case we just introduced, would look like part a of Fig. 5.15. However, what is more often done is as shown in Fig. 5.15b, where it could be said that one is sending $\pm\cos (\omega_c t + 45°)$ or $\pm\cos (\omega_c t + 135°)$. It seems as though this may be easier to implement than the case of part a; however, the latter scheme lends itself well to sending 4 bits in a single symbol, as in Fig. 5.15c.

Strictly speaking, one might consider "a" to be the constellation for 4-ary PSK. This leads also to the implication that one could draw a circle with "stars" spaced 45° apart on it and one would have the

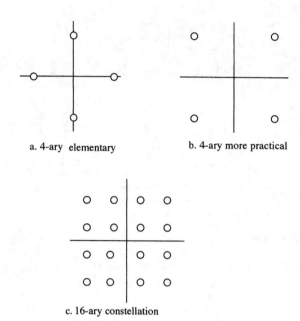

a. 4-ary elementary   b. 4-ary more practical

c. 16-ary constellation

FIGURE 5.15   Constellation showing carrier amplitudes and phase for M'ary signals.

constellation for 8-ary PSK. The perceptive or well-informed reader might have the strong suspicion that crowding more points on the circle makes it possible to have more errors in distinguishing one symbol from adjacent ones, and would be correct in this suspicion. Hence, M'ary communication probably more commonly uses "b" or "c," which should be considered forms of QAM.

6

Power Amplifier Circuits

Mark Bloom
Motorola, Inc.

6.1 Introduction

The Power Amplifier (PA) is typically the last stage in a transmitter system. Its role is to provide the final amplification of signal power to a level that is large enough for microwave propagation through an appropriate antenna. In some systems, the PA is connected directly to an antenna, while in other systems isolators, filters, and switches may follow before the antenna is reached. Often the PA draws 50% or more of the total transmit-current required by a system. If current is an issue (and in most systems it is a key parameter) then the PA design is a critical one to ensure the system current budget is met. Also, in most modern commercial wireless systems, the PA and the associated driver amplifier determine the overall linearity of the transmit chain.

6.2 Design Analysis

As the first step in a PA design, the design must be analyzed and the specification determined in sufficient detail to allow accurate synthesis of the design.

6.2.1 Applications

As the starting point in a design analysis, the application must be carefully considered before the PA specification is defined. Typical considerations are as follows:

- Consumer, high-volume applications: These require *cheap* PAs. To achieve this, customer specifications are barely met. Voltages available to drive the PA are typically low (2.7 to 4.7 V) or limited (i.e., no negative supply to bias a GaAs depletion-MESFET gate), and size is critically important — market pressure is generally forcing the cost and size down, and performance ever upward. Consumer PAs tend to fall into two broad categories of distinction: linear or saturated PAs. Certain process technologies and circuit topologies favor either linear or saturated applications.

- Non-consumer, low-volume applications: These require high performance, and typically require high reliability. The performance often cannot be compromised, which leads to a high cost. High reliability will impact PA design, often with much derating required, which will reduce efficiency.

6.2.2   Modulation Effects

One of the key considerations for a PA is the modulation scheme used. Many PA designs have been single-tone CW, both in simulation and in physical measurement. However, with the widespread implementation of nonconstant envelope modulation schemes for mass wireless markets (i.e., CDMA IS95/98 and NADC IS136), single-tone CW measurements are being used less. More wireless systems are being developed based on spectrally efficient nonconstant envelope modulation, which have a profound impact on PA design.

If a PA is to be used with constant-envelope modulation, then the PA can be operated close to saturation — typically the harmonic content limits the degree of compression permitted in a design. For instance, the widespread GSM wireless standard allows PAs to operate 4 to 5 dB into compression, leading to power-added efficiencies [PAE, see Eq. (6.2)] greater than 60% from a 3 V supply, with 35 dBm of RF output power.

However, if a nonconstant envelope modulation scheme is used, then spectral regrowth, or Adjacent Channel Power (ACP) typically manifests. This distortion can degrade BER for wireless users allocated adjacent (or alternate) frequency bands. Hence most nonconstant envelope schemes have stringent specifications on Adjacent Channel Power Ratio (ACPR). ACPR is defined as the relative difference between the users in-band output power and the users adjacent (or alternate) band output power. For CDMA (IS95/98) a PA can typically operate no more than 1 dB into compression, which limits PAE to around 55% [13]. Methods exist to increase PAE, but these rely on some form of linearization scheme and have proven slow to develop for consumer wireless products.

6.3   Typical PA Specification Parameters

Understanding the specification is critical in choosing the overall amplifier topology and methodology. The key parameters and their impact in design are listed below:

- Small signal gain (S_{21}). Under small-signal operation, a network analyzer can accurately determine small signal gain. Small signal gain is a vector quantity, with magnitude typically expressed in dB, and with phase expressed in degrees. Small signal gain is often the first point in considering how to budget the gain specification for a PA. As a rule of thumb, a power transistor with 20 GHz < Ft < 40 GHz, sized to generate 1 W of RF power will have a small signal gain around 20 dB at 1 GHz when matched for gain using Surface-Mount-Technology (SMT) low-pass matching transformations. Thus gain at 2 GHz (an octave higher) would be 20 dB – 6dB = 14 dB (gain tends to fall off at the rate of –6 dB per octave of frequency).

- Small Signal Return Loss (S_{11} and S_{22}). Again, under small signal conditions, return loss can be calculated with a network analyzer. As will be shown later, output return loss (S_{22}) is often poor in a PA, as the match for good return loss is different for the match for maximum power. Output return loss for a PA can typically be between –5 dB and –10 dB. Input return loss (S_{11}), when matched for maximum gain, can typically be at least –10 dB, and often around –20 dB.

- Output power (P_{out}). The power delivered to a load (typically a 50 Ω termination) can be measured using microwave power meters. For high frequency measurements, units are typically expressed in "dBm," i.e., referred to 1 mW. For instance, 1 W = 30 dBm. Depending on bias and load, power increases linearly with the input power, until the amplifier begins to suffer gain compression — at this point, the gain starts to fall off as a function of input drive. Eventually no more power can be gained out of the PA, leading to the term "saturated output power."

- Efficiency (η). Measured as a percentage between 0% and 100%, two definitions of interest exist. DC-RF efficiency (also referred to a drain or collector efficiency) is simply the ratio of power delivered to a load and the DC power consumed by the PA to deliver that power:

$$\eta_{dc-rf} = \frac{P_{out}}{P_{dc}} \tag{6.1}$$

Power-added efficiency is a more interesting expression, as the input-power to the device is considered, i.e., it is the ratio of the amount of power added by the PA to the DC consumption:

$$\eta_{PowerAdded} = \frac{\left(P_{out} - P_{in}\right)}{P_{dc}} \tag{6.2}$$

By inspection of Eq. (6.2), power-gain effects the value of power-added efficiency calculated.

- Harmonic distortion. Depending on the bias point and class of operation, harmonic content from the PA will generally increase as drive level increases. Generated by clipping of the input signal, odd/even harmonics of the fundamental can become an issue in some systems. Typically harmonics must be at least −30 dBc (referenced to the fundamental). IP_3 and ACPR are phenomena related very closely to harmonic content — the same mechanisms explain all three of these forms of distortion.

6.4 Basic Power Amplifier Concept

The basic problem in designing a power amplifier is in regard to the output match. An inherent trade-off must be made between gain and output power. The problem stems from the fact that the output load a transistor needs for maximum power is *different* from maximum small signal gain.

Transducer power gain, G_T, is defined as the ratio of power delivered to the load to the power available from the source [1–3]. When the small-signal S-parameters and reflection coefficients are normalized to the same reference impedance, then transducer power gain is given as

$$G_T = \frac{P_L}{P_A} = \frac{\left(1-\left|\Gamma_g\right|^2\right)\left|S_{21}\right|^2\left(1-\left|\Gamma_L\right|^2\right)}{\left|\left(1-\Gamma_g S_{11}\right)\left(1-S_{22}'\Gamma_L\right)\right|^2} \tag{6.3}$$

where

$$S_{22}' = S_{22} + \frac{S_{12}S_{21}\Gamma_g}{1-S_{11}\Gamma_g} \tag{6.4}$$

So as to simplify Eq. (6.3), assume the device is unilateral i.e., $S_{12} = 0$. Now unilateral transducer gain, G_{Tu}, is

$$G_{Tu} = \frac{\left(1-\left|\Gamma_g\right|^2\right)\left|S_{21}\right|^2\left(1-\left|\Gamma_L\right|^2\right)}{\left|1-\Gamma_g S_{11}\right|^2\left|1-S_{22}\Gamma_L\right|^2}$$ (6.5)

By inspection of Eq. (6.5), G_{Tu} will be a maximum when $\Gamma_g = S_{11}^*$ and $\Gamma_L = S_{22}^*$. Then, maximum unilateral gain is

$$G_{Tu-max} = \frac{\left|S_{21}\right|^2}{\left(1-\left|S_{11}\right|^2\right)\left(1-\left|S_{22}\right|^2\right)}$$ (6.6)

Hence from inspection of Eqs. (6.5) and (6.7), unilateral transducer gain will be maximized when the input and output terminals are conjugately matched into the load and source. Remember that this derivation is strictly only true for a unilateral device. However, it is a very close approximation with modern device technology.

Now, the actual load required by the output device for maximum gain will be

$$\Gamma_{out} = \left(S_{22} + \frac{S_{21}S_{12}\Gamma_g}{1-S_{11}\Gamma_g}\right)^*$$ (6.7)

Now consider a power amplifier. The load line determines the available output power, as in Fig. 6.1. The output power will be

$$P_{out} = \frac{1}{2}I_{dd}\cdot\left(V_{dd}-V_{knee}\right)$$ (6.8)

So, since $P = IV$ and $V = IR$,

$$R_L = \frac{\left(V_{dd}-V_{knee}\right)^2}{2P_{out}}$$ (6.9)

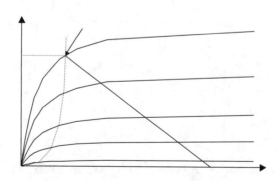

FIGURE 6.1 Load-line for a transistor (this example assumes a FET device).

Limits typically bound the maximum power from a device. For a depletion GaAs FET, I_{dd} is a maximum at a slightly positive gate voltage — too far forward, and the gate will conduct leading to device failure. V_{dd} is either fixed by the available voltage supply, or by the drain-source breakdown characteristics. The saturation voltage determines V_{knee}, which is a function of the size of the FET device and the technology. In practical devices, Eqs. (6.7) and (6.9) yield very different results, i.e.,

$$\left| \left(S_{22} + \frac{S_{21}S_{12}\Gamma_g}{1 - S_{11}\Gamma_g} \right)^* \right| \neq \frac{\left(V_{dd} - V_{knee} \right)^2}{2P_{out}} \tag{6.10}$$

Hence the fundamental problem in trying to simultaneously match for gain and power.

6.5 Analysis of the Specification

The first step in a PA design is to understand the specification, and the customer requirements (which may be slightly different).

6.5.1 Basic Considerations

Initially, available voltages and currents must be considered. Many high-volume commercial PA products require low operating voltages around 3.0 V. This will greatly effect the design — a 3.0 V-supplied PA will have to draw current approximately four times higher than a 12 V-supplied PA. For a typical specification,

$$I_{DD} \approx \frac{P_{OUT}}{\left(V_{DD} - V_{knee} \right) \cdot \left(\dfrac{\eta}{100} \right)} \tag{6.11}$$

where,

I_{DD} is defined in mA
V_{dd} is defined in Volts
V_{knee} is defined in Volts and can be assumed to be zero if V_{dd} is much larger
P_{OUT} is specified in mW
η is between 0 and 100%

Leading on from this, once the DC current requirements have been estimated, then the packaging requirements can be considered.

The key consideration here is that as I_{DD} increases, so does the effect of many circuit parasitics. Series-resistance will play a bigger role in external matching components.

Figure 6.2 shows this effect clearly. As the DC voltage available for the PA falls, the current increases. So, as the Q of the elements (as an example used in a low-pass output-match) lowers, the losses present dissipate more current, leading to a reduction in efficiency.

Thermal design, especially in regard to the active-transistor region and how it is grounded will need to be considered. Thermal design consists of ensuring that the transistor junction temperature, T_j, does not exceed the limit for reliable operation. Each foundry-specific device technology has a maximum junction temperature. Junction temperature can be found from

$$T_j = P_d \cdot \left(\theta_{jc} + \theta_{cs} + \theta_{sa} \right) + T_a \tag{6.12}$$

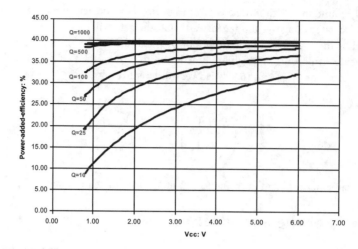

FIGURE 6.2   Effect of Q and Voltage on PAE (for a 1 W PA, with a single L-C low-pass output match with a certain Q).

where,

P_d = power dissipated in junction, which includes DC power and RF power (W)
θ_{jc} = junction-to-case thermal impedance (°C/W)
θ_{cs} = case-to-heat-sink thermal impedance (°C/W)
θ_{sa} = heat-sink-to-ambient thermal impedance (°C/W)
T_a = ambient temperature (°C).

From these simple considerations, an appropriate package can be determined. In the case of a package already specified, then the designer will simply have one less variable to optimize during the design synthesis.

Do not forget that the PA, whether packaged as a ceramic hybrid or a plastic-encapsulated MMIC, is part of an overall system. Consideration must be given to the particular system interface — is the PA package to be soldered to a board, or will epoxy be used? How will grounding of the PA be applied? Will the board ground-plane be a solid shunt of metal, or will board vias be employed to connect with a spatially separated ground-plane? How thick is the board, and of what material? How will it react thermally? Many of these seemingly basic questions are often not answered until the design has been fabricated — of course then it may be too late.

Now the designer has a basic understanding of the environment the PA will operate within. From this, the electrical specification can be analyzed in more detail.

6.5.2   Budgeting

Gain partitioning is the next step in the design. A specification may be for the complete PA to have more small signal gain than a single stage can provide. It is most likely that more than one stage will be required for this to be realized. Hence a gain budget is required to determine the rough gain (and power) levels in each stage.

For instance, consider the simple specification in Table 6.1. The first thing to consider is the maximum output power. Since it is unlikely to find a single transistor with 40 dB power gain, a multistage design is required.

As a rule-of-thumb example, assume a device has 20 dB of small signal gain, but is only conditionally stable. A designer will have to lose some gain (typically 5 dB) to achieve stability. Hence the output stage will have a gain of around 15 dB. At 3 dB gain compression, this implies the output stage will operate with a power gain of 15 dB – 3 dB = 12 dB. However, the previous stages will need to be slightly

TABLE 6.1 A Simple Specification

Parameter	Typical Value	Unit
Frequency	2.0	GHz
Maximum output power	36.0	dBm
Gain at maximum output power	40.0	dB
Amount of gain compression at maximum output power	3.0	dB
Efficiency	65.0	%
Technology of implementation	Bipolar, and V_{cc} = 3.6 V	

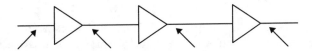

FIGURE 6.3 Gain/power budget.

compressed to obtain good efficiency, so assume the output stage operates at 2 dB compression — now the power gain is 13 dB.

Consider Fig. 6.3. A hypothetical lineup is shown. Generally, gain of a transistor decreases as transistor active-area increases (since the device parasitics increase). Thus stage 1 and stage 2 have more gain than the output stage. Also note that the power gain in is 45 dB. To reduce this (and to help stability) gain in stages 1 and 2 will be lowered to bring the whole design to the 40 dB gain target.

Assume the output stage is running at η = 70%, stage 1 at η = 55%, and stage 2 at η = 40%. Now apply Eq. (6.11). The following collector currents can be calculated:

OP Stage: 1580 mA

Stage 1: 101 mA

Stage 2: 4 mA

Thus total efficiency is 65.6%, which meets the specification. A spreadsheet is useful to perform these basic calculations — the degree of efficiency for each stage can be found very quickly.

6.5.3 Choice of Device

The next step is to estimate the transistor device areas. In this example it is a BiPolar technology, so emitter area is the active parameter. Assume the technology offers reliable operation up to 0.2 mA/μm^2. Then, the following device areas can be calculated from the DC currents above:

OP Stage: 7900 μm^2

Stage 1: 505 μm^2

Stage 2: 20 μm^2

6.5.4 Bias Point and Class of Operation

A PA must be designed to amplify in a certain class of operation. The class is determined by three key factors: Quiescent bias point, matching topology, and transistor configuration. The class determines the maximum potential efficiency η_{max} as well as the relative maximum potential output power P_{relmax}. The classes can also be grouped by their linearity — some classes are highly linear, while others generate a lot of harmonic distortion leading to degraded IP_3 or ACPR. Three main classes exist: A, B, and C. A fourth, class-AB, is a compromise between class-A and class-B. Some other classes exist (D,E,F,S) but these are specialized and generally are not in commercial use (due to their high distortion and bias/matching problems coupled with the complexities of their design). See Reference 1 for a more complete description.

The quiescent bias point of a transistor (V_{dq}, I_{dq} for a FET and V_{cq}, I_{cq} for a bipolar) determines the conduction angle, θ. An amplifier under class-A operation has $\theta = 360°$, a class-AB amplifier obeys $360° > \theta > 180°$ and a class-B amplifier has $\theta = 180°$. A class-C amplifier has $180° > \theta > 0°$.

The first step in determining the class to use is to consider η_{max}. From Reference 1, it can be shown that efficiency is given by

$$\eta = \frac{1}{4}\frac{\theta-\sin\theta}{4\sin\left(\theta/2\right)-\theta/2\cos\left(\theta/2\right)}. \qquad (6.13)$$

From Eq. (6.13), the classical values of η_{max} can be determined: Class-A has $\eta_{max} = 50\%$, class-B has $\eta_{max} = 78.5\%$, and class-C has $\eta_{max} = 100\%$. Note that to obtain η_{max} for a class-C, $\theta = 0°$.

Also, P_{relmax} can be calculated from by Reference 1 by

$$P_{out} \propto \frac{\theta-\sin\theta}{1-\cos\left(\theta/2\right)}. \qquad (6.14)$$

By inspection of Eq. (6.14), P_{out} decreases as conduction angle decreases. Thus to obtain η_{max} for a class-C PA ($\theta = 0°$), $P_{out} = 0$ W. So, practically η_{max} cannot be achieved.

Conduction angle for a PA can be visualized by considering the DC-IV curves. Since V_{dq}, I_{dq} determine quiescent point, V_{dq}, I_{dq} can be superimposed on the DC-IV data.

As can be seen in Fig. 6.4, class-A has the lowest "clipping" of the waveform, and hence the least distortion. As the bias point I_{dq} deepens toward class-C, the harmonic distortion caused by clipping below threshold (on large-signal negative cycles), or from forward gate-conduction (on large-signal positive cycles) increases.

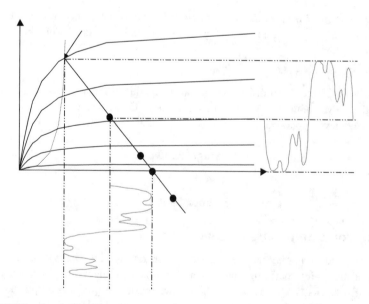

FIGURE 6.4 DC-IV and quiescent bias.

6.6 Topology

Several topologies exist for PAs. Topology refers to the matching techniques to be used within the overall PA lineup.

6.6.1 Reactive Matching

The terminal impedances of the transistor can be matched using reactive elements, such as capacitors and inductors. Reactive elements allow both narrow- and wide-band transformations. The frequency response is critical for an amplifier, especially the impedance presented at the harmonics of the fundamental frequency. The frequency response characteristics of reactive matching may be low pass, high pass, or band pass. These general considerations can be applied when choosing which to use:

- Inductors tend to have a lower Q than capacitors. So, a low pass structure will tend to be more lossy than a high pass. The loss of the matching structure can affect stability, noise figure, and PAE of the amplifier to a great extent, and so this must be considered during the design process. So, input match of a transistor may benefit from being lossy, especially with bipolar designs, which often have base ballasting anyway — the ballasting can be redistributed to be within the input match. However, in a PA, output match generally needs the lowest loss possible to maximize PAE, suggesting a high pass transformation.
- Lossy structures at the input to an amplifier increase noise figure. Some modulation systems (such as IS-95) require very low noise figures, and so the topology may be affected by the noise specifications.
- PAs are often required to have low distortion, which equates to low harmonic content in the output signal. A low pass output match will tend to attenuate the harmonics considerably, leading to improved linearity.
- Inter-stage matches tend to be a compromise of factors — as power levels tend to be relatively high, low pass structures may be avoided due to loss. Also, a high pass structure may allow implicit DC blocking, which most amplifiers require. This leads to a lower passive component count, and reduced cost. To obtain high linearity though, a low pass structure may be beneficial.

As the points above show, choice of matching topology is often a compromise, with each transformer being carefully considered in regard to the specific specifications. It is not possible to generalize, so the whole design must be considered carefully.

6.6.2 Feedback

The use of series or shunt feedback can help ensure stability of a PA. Resistive feedback between drain and gate (typically a few hundred ohms) will increase stability margin, and increase bandwidth [6,7]. See Fig. 6.5 for an example. However, the gain of the transistor will be lower, leading to reduced power-added efficiency or the requirement to add another stage of amplification. Generally gate drain series feedback will not be used on the final amplifier stage due to the degradation in gain

Often a capacitor is required in series with the resistor. This is essential for blocking DC-bias voltages. The R-C time constant of the feedback will have a slight affect on the frequency response of the circuit. A series inductor can also be used with the resistor to adjust the frequency response significantly. Inductance is required in a broadband design where the frequency shaping of the feedback is critical to the gain response.

6.6.3 Balanced Power Amplifier

A power amplifier may be designed to operate in a balanced topology. A balanced amplifier consists of two identical amplifiers (A and B), with two couplers connecting the inputs and the outputs together. Figure 6.6 shows a balanced arrangement.

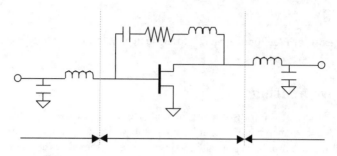

FIGURE 6.5   Series feedback around a simple FET amplifier.

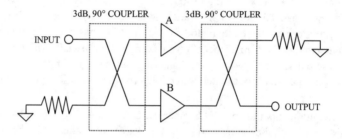

FIGURE 6.6   Diagram of the balanced amplifier.

The advantage of a balanced amplifier is that unwanted reflections are terminated in a load, typically 50 Ω. One amplifier (B) is driven with the signal, while the other (A) is driven by a signal that is 90° out of phase. Any reflected signals are thus 180° out of phase after passing back through the coupler. Hence the unwanted reflections cancel out. The outputs are arranged opposite to the input, so the other output passes through the coupled port, thus recombining the signal in phase.

As shown in [8], transducer gain of a balanced amplifier can be shown to be

$$G_T = \frac{G_1 + G_2 + 2\left(G_1 G_2\right)^{1/2} \cos\left(\varphi_1 - \varphi_2\right)}{4} \tag{6.15}$$

where

$$S_{21A} = G_1^{1/2} \exp\left(j\varphi_1\right) \tag{6.16}$$

$$S_{21B} = G_2^{1/2} \exp\left(j\varphi_2\right) \tag{6.17}$$

So, if the amplifiers A and B are identical, then $G_1 = G_2$ and $\varphi_1 = \varphi_2$. Then Eq. (6.15) resolves to $G_T = G_1$. However, if one amplifier should be turned off or shut down, then G_T falls by a factor of 4 (–6 dB).

The benefit of the balanced power amplifier is that the output ports are very insensitive to mismatch as the return losses are much better than a single amplifier on its own. This results in a very stable PA that is insensitive to mismatch. Note that the output power is the sum of the two amplifiers, but the gain is reduced by twice the coupler loss (assuming identical couplers).

Modern couplers can be purchased in a variety of forms, including lumped ceramic surface mount, which fit the footprint of an 0805 SMT component.

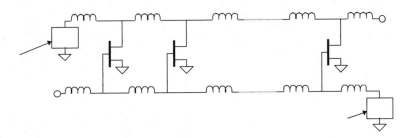

FIGURE 6.7 Distributed power amplifier topology.

6.6.4 Distributed Power Amplifier

Traveling wave, or distributed amplifiers have been around since the 1940s when the first patent was filed [9]. Since then, much work has been performed to optimize and improve the concept, especially in regard to power amplification [10].

The distributed amplifier concept uses the parasitic capacitance/inductance of a transistor as part of two artificial transmission lines. The transmission lines are a lumped equivalent of a distributed line, with some of the lumped elements formed in the transistor. Hence, as the amplifier appears to be a transmission line, it is matched into the required terminations over a very wide bandwidth. That is the benefit of this topology — bandwidths up to many octaves are possible.

Figure 6.7 shows a typical distributed PA topology. A number of FET devices, n, are connected in parallel, with inductances L_g and L_d between the gate and drains.

It has been shown [11] that gain, G, is

$$G = \frac{g_m^2 n^2 Z_o^2}{4}\left(1 - \frac{\alpha_g l_g n}{2}\right)^2 \tag{6.18}$$

where α_g is the effective gate-line attenuation per unit length, and l_g is the length of the gate transmission line per unit cell. From this, is it clear that the gain will increase as more FET stages are added. However, stages cannot be added indefinitely as each FET has parasitic resistance (typically R_i and R_{ds}) which will attenuate the signal — eventually the loss added will exceed the gain benefit of another stage. Note that as the bandwidth generally is very large, the overall gain (even with many FET segments) is fairly low.

Two unusual constraints exist that determine the maximum output power from a distributed power amplifier, in addition to the voltage-swing constraints with all power amplifiers. Firstly, gate periphery cannot be increased without adding more attenuation from the parasitic elements in the FET. Thus gate periphery is a trade-off between gain, bandwidth, and power. Secondly, the drain of each FET must see an impedance that is determined by the characteristic impedance of the drain line. Hence, each FET will not be able to see the impedance that it needs to satisfy maximum power output as in Eq. (6.9). This power mismatch can result in significant reductions in output power.

One final consideration is that as the frequency response is greater than an octave, harmonic terminations will not be correct to maximize efficiency. Thus a traveling wave power amplifier will have lower efficiency than a narrowband reactively matched design.

6.6.5 PA Architecture

The PA architecture may be one of several — Fig. 6.8 shows a tree of the typical types of PA. A MMIC is deemed to have active and passive components on a single substrate. A discrete contains minimal passive components. A plastic package typically only contains an active die, with passive components being external. A module contains an active die and passive discrete components. Table 6.2 details some of the key considerations with each type of PA.s

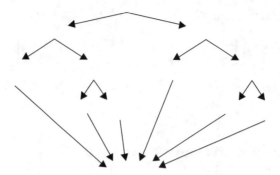

FIGURE 6.8   PA architectures.

6.7   Choice of Active Device Technology

The main active device technologies suitable for PAs are: GaAs MESFET, GaAs HEMT, GaAs HBT, Si MOSFET, Si Bipolar, and SiGe HBT.

It is very difficult to subjectively determine which type of device is most suitable for any particular application. Up to 2 GHz, all can compete. Between 2 GHz and 20 GHz, most Silicon technologies are unsuitable. Beyond 20 GHz only GaAs HEMT and HBT really perform well. Most millimeter-wave devices are HEMT based.

To cloud the issues, research papers may show a particular device technology that is performing beyond what is thought as "normal." This can often be facilitated by hand choosing the best device from a lot of material. Consideration must be made that in production, especially high-volume/low-cost things are much more limited. But remember that what is research today could be production in one or two years.

6.7.1   Gallium Arsenide Solutions

As a rule, GaAs performs better than silicon, but will cost more. High performance GaAs generally require epitaxially grown materials, which can cost around five times the price of a comparable silicon wafer. The majority of the worlds GaAs output is on four-inch material wafers, but GaAs six-inch wafers have been adopted by several foundries. As the demand for GaAs increases (driven primarily by the wireless consumer market) six-inch material will be adopted by more foundries.

6.7.1.1   Heterojunction Bipolar Transistors (HBT)

Generally, HBT technologies have high current density due to their vertical structure — they therefore consume less die area than a FET based technology. However, material structure tends to be more complex than a FET (but more forgiving of certain variations). A typical HBT will consist of seven or eight epitaxially grown layers, whereas an epitaxial FET may only consist of four (plus a super-lattice buffer region). Epitaxial material for HBT devices costs virtually the same as an epitaxial FET. This is because cost of material is dominated by the total thickness of the grown layers. This is similar for both HBT and FET. One consideration is that more care with alignment and registration is required for an HBT (compared to a long gate FET with $L_g = 1$ um).

6.7.1.2   Epitaxial High-Electron Mobility Transistors (HEMT)

HEMT technologies (based on short gate length and epitaxial material) offer best in absolute performance (i.e., noise figure, and PAE). Submicron gate lengths mean cost may be high. Note that most HEMT structures are very sensitive to key layer thickness, which is not such an issue for HBT devices. The sensitivity of layer variations can result in poor yield and hence a higher cost for the customer. Controlling this variation is one of the key issues in choosing a HEMT technology.

TABLE 6.2 PA Architectures

Architecture	Benefit	Disadvantage	Applications
MMIC, Plastic	Low cost	Often will require many external components to get optimal performance Overall size may be large after all the externally required components are added Performance is very dependent on the board it is attached to High volume testing can be difficult	Up to 2–3 GHz Very high volume Requires lots of customer interaction to obtain optimal performance
MMIC, module, ceramic	Can be very small Can operate in many board environments with minimal modifications	High cost Multilayer ceramic technology expensive, which means very complex designs may be large	Up to millimeter wave High volume Often good for a "quick and easy" solution for the customer
MMIC, module, resin laminate	Can be very small Can operate in many board environments with minimal modifications Multilayer easy to implement	Medium cost Performance not as good as ceramic and generally lower E_r means distributed elements are large	Up to 2–3 GHz Very high volume Often good for a "quick and easy" solution for the customer
Discrete, plastic	Very low cost	Will require many external components to get optimal performance Overall size may be large Performance is very dependent on the board it is attached to High volume testing can be difficult	Up to 2–3 GHz High volume Requires huge customer interaction to obtain optimal performance
Discrete, module, ceramic	Can be small Can operate in many board environments with minimal modifications	High cost Multilayer ceramic technology expensive, which means very complex designs may be large Will require many external components to get optimal performance	Up to 20 GHz — beyond this requires very careful passive component selection High volume Often good for a "quick and easy" solution for the customer
Discrete, module, resin laminate	Can be small Can operate in many board environments with minimal modifications Multilayer easy to implement	Medium cost Performance not as good as ceramic and generally lower E_r means distributed elements are large Will require many external components to get optimal performance	Up to 2–3 GHz High volume Often good for a "quick and easy" solution for the customer Good if customer can tolerate a larger module

One clear benefit that a HEMT (or a MESFET) technology has over an HBT is that its on-resistance is a function of gate width (among other parameters). As a consequence, if very low voltage operation is required (sub-3 V) then the on-resistance plays a large part in determining efficiency. As Eq. (6.11) shows, V_{knee} (or saturation voltage) will have an effect determining overall efficiency. V_{knee} can be lowered if the on-resistance is reduced.

One drawback is that most HEMT (or MESFET) technologies require a negative gate voltage. This can prove complex and difficult to generate in a mobile subscriber unit, and most manufacturers would rather avoid the complexity of adding this control. Enhancement-Mode GaAs HEMT and MESFET devices are available, but these can be even more sensitive to material/process variations. Unless a device offers true enhancement-mode operation, a drain-switch is often still required to shut the device down. HBT devices do not have this concern — if the base is held at zero potential, there is no significant collector leakage.

6.7.1.3   Epitaxial Metal Semiconductor Field-Effect Transistors (MESFET)

A MESFET grown with epitaxial layers can perform almost as well as a HEMT for L-band applications. It does not have the same concerns over layer thickness as a HEMT, but it suffers some of the same drawbacks. Yield on this kind of device can be very high, and process variations can be minimized by the use of etch-stop layers.

6.7.1.4   Ion-Implanted MESFET

Ion implantation was used in the first MESFET devices to create the required channel and Ohmic regions. Once the mainstream GaAs device, it has fallen behind in terms of performance compared to epitaxial devices. Ion implantation offers an extremely low-cost device, but suffers in a significant variation in performance from lot to lot. This is due to the intrinsic variation within the ion implantation depth profile. The use of epitaxial material eliminates this issue, but with the added cost that MBE/MOCVD implies.

6.7.2   Silicon Solutions

Silicon offers a low-cost solution for power amplifiers. Even though die are often larger than GaAs, the processed wafer cost of silicon is significantly lower than GaAs. Silicon offers a higher range of integration, coupled with a very well controlled and understood process. On-chip passive elements (predominantly inductors) tend to have more loss than GaAs, but the use of these elements in PA circuits tends to be minimal.

6.7.2.1   CMOS, Bipolar, and BiCMOS

Silicon CMOS has not had much success in power amplifiers due to its comparatively low breakdown and low current density [12]. Bipolar processes can have high figures of merit which allows them to perform comparatively well. However, current densities are lower than GaAs, leading to larger die. Larger die can also lead to stability problems, which a compact design would not suffer from. The use of aluminum interconnect (as opposed to gold used in GaAs) limits the current density significantly. However, with the aluminum/copper interconnect now becoming more common, this issue is less prevalent.

One benefit that silicon bipolar devices have over GaAs Bipolar devices is that the based emitter turn-on voltage, V_{be}, for silicon is around half of a GaAs HBT. Silicon has a V_{be} of around 0.7 V while GaAs V_{be} is around 1.4 V. This allows more flexibility in the circuit design as more devices can be "stacked" between the supply rail and ground. This is a major benefit in wireless systems operated from a battery, which may supply down to 2.8 V (or lower) DC.

6.7.2.2   Laterally Diffused Metal Oxide Semiconductor (LDMOS)

LDMOS is a very cheap silicon process that allows MOS devices to be integrated with passives onto a die. LDMOS uses a slow, well-controlled, and repeatable lateral diffusion process to effectively make the device gate much smaller than its drawn dimension. Since (to a first order) performance is inversely proportional to gate length, LDMOS offers very good RF and microwave performance. However, performance is currently limited to around 3 GHz as an upper maximum.

6.7.2.3   Silicon Germanium (SiGe)

By adding small amounts of germanium into a silicon bipolar process, SiGe devices can be made to perform well at RF/microwave frequencies [14]. The main drawback is that the lattice mismatch between Si and Ge is considerable. Thus devices can only be lightly doped with Ge, which leads to a "weak" heterojunction effect. Since PAs benefit from this heterojunction to achieve higher efficiency, SiGe falls behind GaAs in terms of performance. However, being silicon-based, it is relatively cheap, despite having a fairly complex process. Breakdown voltages in a SiGe process tend to be low, as the devices have often been designed for high-speed digital applications (where SiGe can have major benefits over GaAs). Low breakdown tends to be undesirable for a PA (leading to either device-failure or increased instability, especially under mismatched RF conditions). SiGe devices can be engineered to have higher breakdown, but only at the expense of RF performance. However, some markets such as wireless hand portables are evolving to operate the PA at lower power levels and from lower supply voltages. This could work to favor a SiGe solution over a GaAs solution.

References

1. Ha, T.T., *Solid State Microwave Amplifier Design*, Wiley, New York, 1981.
2. Vendelin, G.D., *Design of Amplifiers and Oscillators by the S-parameter Method*, Wiley, New York, 1982.
3. Bodway, G.E., Two-power power flow analysis using generalized scattering parameters, *Microwave Journal*, 10, 61, 1967.
4. Vendelin, G.D., Pavio, A.M., Rohde, U.L., *Microwave Circuit Design Using Linear and Nonlinear Techniques*, Wiley, New York, 1990.
5. Kraus, H.L., Bostian, C.W., and Raab, F.H., *Solid State Radio Engineering*, McGraw-Hill, New York, 1980.
6. Rigby, P.H., Suffolk, J.R., Peneglly, R.S., Broadband monolithic low-noise feedback amplifiers, *IEEE Microwave and Millimeter Circuits Symposium*, 71, 1983.
7. Jastrzebski, A.J., Bloom, M., Davies, A., Buck, J., and Pennington, D., Design of broad-band MMIC power amplifiers for 6–18 GHz, *IEE Colloquium*, IEE Digest No. 1991/191, 1991.
8. Soares, R. (ed), *GaAs MESFET Circuit Design*, Artech House, Boston, 1988.
9. Percival, W.S., Thermionic valve circuits, British Patent 460562, July 1936.
10. Ayasli, Y., Reynolds, L.D., Mozzi, R.L., Hanes, L.K., 2–20 GHz GaAs traveling-wave power amplifier, *IEEE Trans. MTT*, 32, 290, 1984.
11. Ayasli, Y., Mozzi, R.L., Vorhaus, L.D., Reynolds, L.D., Pucel, R.A., A monolithic GaAs 1–13 GHz GaAs traveling-wave amplifier, *IEEE Trans. MTT*, 30, 976, 1982.
12. Tsai, K-C., Gray, P.R., Techniques in designing CMOS power amplifiers for wireless communications, *JSSC*, July 1999.
13. RF Micro Devices Inc., A Linear, High Efficiency, HBT CDMA Power Amplifier, *Microwave Journal*, January 1997.
14. Crabbe, E.F., Comfort, J.H., Lee, W., Cressler, J.D., Meyerson, B.S., Megdanis, A.C., Sun, J.Y.-C., and Stork, J.M., 73-GHz Self-Aligned SiGe-base bipolar transistor with phosphorus-doped polysilicon emitters, *IEEE Elec. Dev. Lett.*, 16, 1980.

7

Oscillator Circuits

Alfy Riddle

Macallan Consulting

7.1 Introduction

Figure 7.1 shows a variety of styles and packaging options for RF and microwave oscillators. Oscillators serve two purposes: 1) to deliver power within a narrow bandwidth, and 2) to deliver power over a frequency range (i.e., they are tunable). Each purpose has many subcategories and a large range of specifications to define the oscillator. Table 7.1 gives a summary of oscillator specifications.

Fixed oscillators can be used for everything from narrowband power sources to precision clocks. Tunable oscillators are used as swept sources for testing, FM sources in communication systems, and the controlled oscillator in a PLL. Fixed tuned oscillators will have a power supply input and the oscillator output, while tunable sources will have one or more additional inputs to change the oscillator frequency. Some tunable oscillators, particularly those using YIG resonators, will have a second tuning port for small deviations. The theory section will provide the background for understanding all the oscillator specifications.

7.2 Specifications

7.2.1 Power Output

Power output and frequency of oscillation are the most basic oscillator specifications [1]. Oscillators with maximal output power are used in industrial applications and usually have more noise due to their extracting as much power as possible from the resonator and thereby lowering the loaded resonator Q. Power output will vary over temperature, so some designs use a more saturated transistor drive or pass the oscillator signal through a limiter to achieve greater amplitude stability. Both of these actions also increase the oscillator noise and cost. Oscillators optimized for low noise, or jitter, usually have low output power to minimize resonator loading and so these designs rely on post-amplification stages to bring the oscillator power up to useful levels for transmitters and radars. As discussed in the theory section, oscillators create more near-carrier noise than amplifiers, so post-amplifiers usually have a minor

FIGURE 7.1   A picture of various RF and microwave oscillators. Top to bottom and across from the left there is a crystal oscillator and two YIG oscillators, two chip and wire oscillators in TO-8 cans, a microwave IC oscillator, and three discrete PCB VCOs in packages of decreasing size.

TABLE 7.1   Oscillator Specifications

Specification	Characteristic
Power	Minimum output power (over temperature)
	(Flatness over tuning band if tunable)
Frequency	Accuracy (in Hz or ppm)
	Drift over temperature in MHz/degree C
	Aging in ppm/time
	Phase noise in dbc/Hz (or jitter in picoseconds)
	Pulling in Hz (due to load variation)
	Pushing in Hz/V (due to power supply variation)
	Vibration sensitivity in Hz/g acceleration
Tunable	Bandwidth
	Modulation sensitivity in MHz/V
	Modulation sensitivity ratio (max/min sensitivity)
	Tuning range voltage
	Tuning speed in MHz/microsecond
Power Consumption	V, I DC
Package Style	

impact on total oscillator noise. When an oscillator is tunable, the power flatness over the tuning range must also be specified.

7.2.2   Frequency Accuracy and Precision

The frequency accuracy of an oscillator encompasses a large number of sub-specifications because so many things affect an oscillator's frequency. Temperature, internal circuit noise, external vibration, load variations, power supply variations, as well as absolute component tolerance all affect frequency accuracy. We can consider only component tolerances for frequency accuracy and lump all the variations into oscillator precision.

The accuracy of the fundamental frequency of an oscillator is usually specified in ppm or parts per million. So a 2.488 GHz oscillator which is accurate to ±10 ppm will have an output frequency within ±24.88 kHz of 2.488 GHz at the stated temperature, supply voltage, and load impedance. Ambient temperature changes also change the oscillator frequency. The perturbation in a oscillator frequency from temperature is often given in MHz/degree C or ppm/degree C. Manufacturers use several techniques to compensate for temperature changes, such as using an oven to keep the oscillator at a constant temperature such as 70° C, building in a small amount of tuning that is either adjusted digitally or directly from a temperature sensor, and finally resonators can be built with temperature compensating capacitors or cavities [2]. Oscillator components also change with time, which causes a frequency drift due to aging. Aging is usually specified in ppm/year or some other time frame.

Power supply variation affects both the absolute accuracy of an oscillator frequency and the precision with which it maintains that frequency. The sensitivity of an oscillator to power supply variations is called "pushing" and is usually given in MHz/V. Drift in the supply voltage over temperature or with changes in the instrument state affect the accuracy of the frequency while noise on the power supply due to switching circuits will modulate the oscillation frequency through the same pushing mechanism. Time constants in the oscillator bias circuitry will cause the pushing factor to change as the modulation frequency increases, but this is rarely specified. Communication receivers and spectrum analyzers go to great lengths to filter oscillator power supplies.

Load variations also cause changes in oscillation frequency. Because the load on the oscillator output port has some finite coupling to the resonator, changes in the load reactance will change the resonator reactance and so change the oscillation frequency. Typically, a variable length of line is terminated in a standard return loss, such as 12 dB, and the oscillation frequency is measured as the line length is changed. Changing the line length creates a variable load that traces a circle on the Smith Chart. The maximum frequency change is quoted as the "pulling" for the oscillator at the given return loss. Precision oscillators will go through an isolator or a buffer stage to minimize pulling.

Oscillators with cavities and even suspended crystals can be affected by vibration. The vibration sensitivity specification depends on oscillator construction and mounting. Communication systems have been taken down by raindrops hitting the enclosure of an outdoor cavity oscillator. A vibration sensitivity in MHz/g can show sensitivity to vibration, but usually the frequency of the vibration is important as well.

Probably the most common specification of oscillator precision is phase noise or jitter [3]. Phase noise is the frequency domain equivalent of jitter in the time domain. Phase noise will be described in the theory section. Phase noise, FM noise, and jitter are all the same problem with different names. Because an oscillator contains a saturated gain element and a positive feedback loop, it will have very little gain for amplitude noise and a near infinite gain for phase noise. The amplitude and phase variations are with respect to the average oscillation frequency. If an oscillator is measured with a spectrum analyzer with sufficient resolution, the narrow line of the oscillator will appear broadened by noise which falls off at $1/f^3$ or $1/f^2$. The loop gain in an oscillator feedback loop reduces as $1/f^2$ for frequencies other than resonance. The additional $1/f$ factor comes from low frequency modulation within the device or resonator. The phase noise specification is usually given as script $L(f_m) = P_{SSB}(f_m)/Hz/P_C$, which is easily measured with a spectrum analyzer. The original script L definition noted that $P_{SSB}(f_m)/Hz$ was to be the phase noise power in one Hertz of bandwidth, but often people take the spectrum analyzer measurement and call it phase noise because phase noise dominates oscillator noise close to the carrier [4]. In reality, the spectrum analyzer cannot tell the difference between amplitude and phase noise. Whenever the phase noise approaches the noise floor or even a flat noise pedestal, it is likely that a significant amount of amplitude noise is present. In all of the above, f_m denotes the offset frequency from the carrier and corresponds to the frequency modulating the carrier. These same variations are called FM noise when measured with a frequency discriminator.

When digital systems are characterized, the time domain specification of jitter is more common than phase noise. Deviations in expected zero crossing times are measured and accumulated to give peak-to-peak and rms values. These values are given in picoseconds or in UI (unit intervals). UIs are just a fraction of the clock period, so UI = (jitter in picoseconds)/(clock period). Because phase noise shows the phase deviation at each frequency modulating the carrier, we can sum up all of the phase deviations and reach a total phase deviation which, when divided by 360°, also gives the jitter in UIs. This is the same as being able to compute the total power of a signal in frequency or in time. Often communication systems are more sensitive to jitter at certain modulation frequencies than others, so a tolerance plot of phase noise vs. offset frequency is given as a jitter specification [5]. The frequency domain view of jitter also makes it clear why PLLs work as "jitter attenuators." The narrow bandwidth of the PLL feedback loop filters higher modulation frequencies and so reduces the total phase deviations, but only if a significant amount of the jitter is due to frequencies above the loop bandwidth.

7.2.3　Tuning Bandwidth Specifications

For tunable oscillators there is an additional set of specifications. Typically broadband tunable oscillators have their bandwidth specified in terms of minimum and maximum frequency (e.g., f_{Max} and f_{Min}) with the center frequency not mentioned. Narrowband tunable oscillators, those with tuning bandwidths of 10% or less, have their center frequency and bandwidth specified. The tuning range is the voltage range of the tuning port for varactor-tuned oscillators (e.g., $V_{Max} - V_{Min}$) and the current range of the tuning ports for YIG tuned oscillators. Usually the minimum varactor voltage is greater than zero because the varactor diode needs reverse bias to maintain a high Q under the swing of the oscillator signal. The modulation sensitivity is the MHz change per volt at the tuning port. Often the modulation sensitivity is not equal to $(f_{Max} - f_{Min})/(V_{Max} - V_{Min})$ because the tuning sensitivity changes over the tuning range. The modulation sensitivity is usually measured at the center of the tuning range with a small voltage deviation. The modulation sensitivity ratio gives the ratio of maximum to minimum modulation sensitivity over the tuning range. This is especially important for varactor-tuned oscillators used in PLLs because the loop gain will vary by the modulation sensitivity ratio. At low voltages varactors have their maximum capacitance, as shown in Fig. 7.2 [6]. The capacitance rapidly decreases as the tuning voltage increases until a minimum capacitance plateau is reached. The large capacitance change at low voltages means that the oscillator will have a more rapid frequency change at lower tuning voltages than at higher tuning voltages. This also means that the modulation sensitivity is higher at the minimum output frequency than it is at the maximum output frequency. The doping profile of a varactor affects its capacitance vs. voltage curve. The simplest doping profile is an abrupt junction that gives the curve shown in Fig. 7.2. A hyper-abrupt junction C-V curve is also shown in Fig. 7.2. The hyper-abrupt curve will give the oscillator a more linear tuning characteristic, or modulation sensitivity ratio closer to unity. Another approach to linearizing a varactor oscillator is to shape the tuning voltage with an analog diode shaping network or with a digital lookup table [7-8]. YIG oscillators have an inherently linear tuning characteristic as given in Eq. (7.1) [9,56]. In Eq. (7.1), H_O is the magnetic bias field strength in Oersteds (Oe), H_a is the internal anisotropy field, and γ is 2.8 MHz/Oe.

$$f_{YIG} = \gamma\left(H_O \pm H_a\right) \qquad (7.1)$$

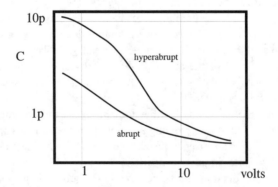

FIGURE 7.2　Varactor C-V plots.

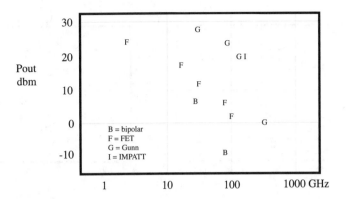

FIGURE 7.3 Device technology and power capability [10–12,57–63].

The last specification is the tuning speed in MHz/second. This parameter determines the maximum modulation rate of an oscillator, or its agility in a frequency hopping application. Varactor-based oscillators can tune much faster than YIG based oscillators. For example, for 10 GHz oscillators, the tuning port of a VCO will typically have a bandwidth of 100 MHz, while the FM coil of a YIG oscillator will have a bandwidth of only 500 kHz.

7.3 Technologies and Capabilities

All of the characteristics discussed above depend on three aspects of oscillator technology: 1) active device; 2) resonator; and 3) packaging. Packaging mainly affects the oscillator's cost, size, temperature stability, susceptibility to mechanical vibration, and susceptibility to interference. Resonator technology mainly affects the oscillators cost, phase noise (jitter), vibration sensitivity, temperature sensitivity, and tuning speed. Device technology mainly affects the oscillator maximum operating frequency, output power, and phase noise (jitter).

Figure 7.3 shows various device technologies and their power vs. frequency capability [10–12,57–63]. While Fig. 7.3 shows fundamental frequency power, in many cases it is most cost effective to use a frequency multiplier to move a lower frequency oscillator up to a higher frequency [13]. While there is always a power loss from frequency multiplication, there is usually very little noise penalty because for equal resonator Qs, oscillator phase noise is proportional to the operating frequency [14]. Frequency multipliers can be simple resistive diode nonlinearities, tuned varactor diode circuits, or PLLs. Resistive nonlinearities are the simplest and broadest band, but have the most loss. Varactor multipliers can be extremely efficient and are used in the highest frequency multipliers [15]. At very high frequencies PLLs are implemented via subharmonic injection locking so that high frequency external frequency dividers and phase comparators are not needed [16,63]. The practical problem with injection locking is that without an external phase detector it is difficult to verify that the oscillator is in lock.

Most small signal oscillators are designed to source 0 to 20 dbm of power. Power oscillators are made to deliver watts of power, but frequency stability suffers due to extracting more power from the resonator and lowering the loaded Q. When stability is a concern, small signal oscillators are built and followed by a carefully designed chain of amplifiers.

7.3.1 Device Technologies

7.3.1.1 Bipolar Transistors

Silicon bipolar transistors are used in most low noise oscillators below 5 GHz. Hetero-junction bipolar transistors (HBTs) are common today and extend the bipolar range to 100 GHz as shown in Fig. 7.3. Bipolar transistors have high gains of over 20 dB at frequencies below 1 GHz and typically have $1/f$ corners

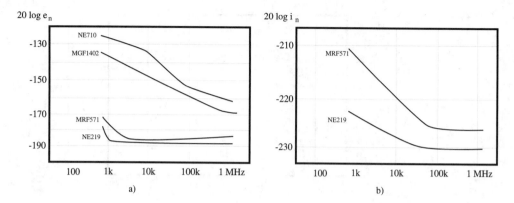

FIGURE 7.4   Device $1/f$ noise comparison for equivalent voltage input noise (a), and equivalent current input noise, (b) [31]. FET input current noise is not shown because it is so small.

in the kHz region. The $1/f$ corner of the oscillator phase noise is less than or equal to the $1/f$ corner of the active device because although the device low frequency noise modulates the device bias and causes phase modulation, it may not be enough to overcome the high frequency noise modulations. In crystal oscillators typically the resonator $1/f$ noise dominates [17,46]. The device $1/f$ noise corner scales with f_{MAX} within a given technology, so smaller devices with higher f_{MAX} will have higher $1/f$ noise corners. A 1 kHz $1/f$ corner is typical for a 10 GHz transistor while a 100 MHz transistor will have a $1/f$ corner in the tens of Hertz. Device technology such as ion implantation will raise the $1/f$ noise corner to 100 kHz or higher by introducing traps in to the device. $1/f$ noise is very sensitive to device construction [18]. The $1/f$ noise in a bipolar transistor is concentrated in the base current, as shown in Fig. 7.4. The $1/f$ noise is due to traps at the base-emitter edge.

7.3.1.2   MOSFETs

CMOS integrated oscillators are becoming more common, although they are limited to the low GHz region [19,20]. Typically two transistors are used in a free running flip-flop configuration. CMOS transistors have lower gain, higher $1/f$ noise corners, and less output power than bipolar transistors, but they offer high integration density and low cost. The higher $1/f$ noise corner of CMOS is mitigated by the reduced modulation sensitivity of the device and by using balanced configurations to reduce noise modulation by symmetry [21].

7.3.1.3   JFETs

JFETs have excellent low frequency noise and limited gain relative to bipolar transistors. While JFETs are found in some extremely low noise discrete oscillators, they are not as common as other devices, especially above 200 MHz [22,54].

7.3.1.4   MESFETs and HEMTs

Above 5 GHz MESFETs and HEMTs are the most common 3-terminal oscillator engine. MESFETs and HEMTs have less gain than bipolar transistors at low frequencies, but have a much higher maximum frequency of operation, or f_{MAX}, as shown in Fig. 7.3. HEMT f_{MAX} is higher than that of MESFETs due to transistor construction that maximizes mobility and provides better channel confinement and control [6]. Both MESFETs and HEMTs have much higher low frequency $1/f$ noise than bipolar transistors, with corner frequencies in the 10 to 100 MHz range being typical for a 600 um device. As with bipolar transistors, the $1/f$ corner scales with f_{MAX} of the device within a given process. Specifically, the $1/f$ level scales with the channel volume beneath the gate structure. GaAs MESFETs and HEMTs have problems with surface traps because of the lack of a native oxide, and there are problems with substrate and channel traps due to the material layering inside the FET. These traps are thermally spread into an approximately continuous $1/f$ distribution at room temperature [23,24]. As shown in Fig. 7.4, the $1/f$ noise for a MESFET or HEMT is mostly in the equivalent gate voltage noise source. The reduced low frequency gain of FETs gives them less modulation sensitivity, so they typically have phase noise levels only 10 dB worse than bipolar oscillators even though the low frequency noise is often 30 dB worse. Figure 7.5 compares noise performance of several other microwave sources.

7.3.1.5 Diodes

Diodes have the highest maximum usable frequencies for solid state devices, as shown in Fig. 7.3. There are many different types of diodes for generating negative resistances and negative conductances. The diode negative immittance cancels the positive loss of the resonator and allows an oscillation to build up from the noise within the device. Traditionally microwave oscillators have been designed as negative immittance devices because it is much easier to measure reflections than to set up feedback loops at microwave frequencies. The distinction between negative resistance devices, such as IMPATT diodes, and negative conductance devices, such as Gunn diodes, is important because the device IV characteristic determines how the device saturates and whether it is stable with a series resonator or shunt resonator.

IMPATT diodes generate negative resistances and so are used at series resonant points in waveguides and planar circuits [25]. The avalanche mode of IMPATT operation creates high power but at the cost of high noise levels. Gunn diodes have an inherently quiet Gunn domain negative conductance that requires a parallel resonant circuit [26]. Gunn diodes are among the quietest high frequency oscillators and exhibit excellent power into the 100s of GHz, as shown in Fig. 7.3. The bulk nature of the Gunn device means that no third terminal metallization is required. This lack of a third terminal and bulk mode of operation reduces the device $1/f$ noise. As with FETs, more advanced material structures, such as using InP rather than GaAs, maximize the high frequency performance of Gunn diodes [12].

7.3.1.6 Multipliers

Frequency multipliers will always be a way of generating the highest frequencies. Reactive multipliers using varactor diodes offer low noise and efficient power generation almost to 1 THz [27]. Resistive multipliers are simple, broadband alternatives to tuned reactive multipliers. Resistive multipliers also suffer significant conversion losses, but are commonly used in broadband instrumentation. All frequency multipliers will increase the phase noise by the same factor that they multiply frequency because frequency and phase are both multiplied, as shown in Eq. (7.2). In dB this would be 20 log N. For example, if the oscillator signal is $V_O(t) = A \, \mathrm{Cos}(\omega_O t + \phi(t))$ then a times two multiplier would generate:

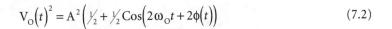

$$V_O(t)^2 = A^2 \left(\tfrac{1}{2} + \tfrac{1}{2} \mathrm{Cos}\left(2\omega_O t + 2\phi(t) \right) \right) \tag{7.2}$$

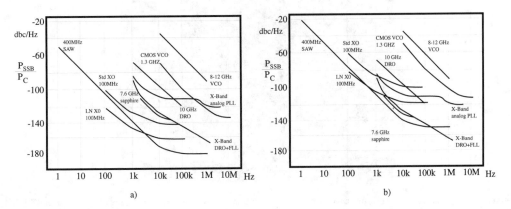

FIGURE 7.5 Oscillator noise performance of some microwave sources: (a) actual; and (b) referred to 10 GHz for comparison (scaled by 20 log[10 GHz/f_{OSC}]) [64–67].

TABLE 7.2 Resonators

Type	Q Range	Range (GHz)	Limitation	Benefit
LC	0.5–200	Hz–100 GHz	Q, lithography	cost
Varactor	0.5–100	Hz–100 GHz	Q, nonlinear, noise	tunable
Stripline	100–1000	MHz–100 GHz	size, lithography	cost, Q
Waveguide	1000–10,000	1–600 GHz	size, cost	Q
YIG	1000	1–50 GHz	cost, magnet, tuning speed	Q, tunable, linear
TL	200–1500	500 MHz–3 GHz	cost	Q, temperature stable
DR	5000–30,000	1–30 GHz	cost, size	Q, temperature stable
Sapphire	50 k	1–10 GHz	cost, size	Q
Quartz	100 k–2.5 M	kHz–500 MHz	frequency	Q, temperature stable
SAW	500 k	1 MHz–2 GHz	frequency, cost	Q

7.3.2 Resonators

Table 7.2 shows an overview of resonator technologies for oscillators. Various abbreviations are used in the above table, with YIG being Yttrium-Iron-Garnet, TL being transmission line, DR being dielectric resonator, and SAW being surface acoustic wave. Resonator choice is a compromise of stability, cost, and size. Generally, Q is proportional to volume, so cost and size tend to increase with Q. Technologies such as quartz, SAW, YIG, and DR allow great reductions in size while achieving high Q by using acoustic, magnetic, and dielectric materials, respectively (see Figure 7.6). Most materials change size with temperature, so temperature-stable cavities have to be made of special materials such as Invar or carbon fiber. Transmission line, dielectric resonator, and quartz resonators can easily have temperature coefficients below 10 ppm. Q changes with frequency for most resonators. Capacitors and dielectric resonators have Qs that decrease with frequency, while inductors and transmission line resonators have Qs that increase with frequency. Quartz resonators are an extremely mature technology with excellent Q, temperature stability, and low cost. Most precision microwave sources use a quartz crystal to control a high frequency tunable oscillator via a PLL. Oscillator noise power, and jitter, is inversely proportional to Q^2, making high resonator Q the most direct way to achieve a low noise oscillator.

FIGURE 7.6 A picture of various resonators. From the left are three transmission line resonators for 500 MHz to 2 GHz operation, two dielectric resonators for 7 and 20 GHz operation, a 10 MHz crystal resonator, and a 300 MHz SAW resonator. The resonators are sitting on top of a 2.5 inch diameter dielectric cylindrical resonator for 850 MHz.

Tunable resonators are very important because they offer the ability to transfer a reference frequency, with or without modulation, through a PLL. Tunable resonators also offer direct modulation and frequency agility for communication and test purposes. Varactor diodes are the most common device for tuning an oscillator. These devices are inexpensive, available in a variety of packages, and can be used at almost any frequency of interest. Varactors also offer rapid tuning for frequency hopping and high speed direct modulation. The only disadvantages of a varactor diode are low Q at high frequencies, low frequency noise, and a nonlinear tuning characteristic [6].

YIG resonators offer the advantages of tuning linearity and high Q. These resonators are excellent for instrumentation and special applications, but suffer from the needing a magnetic bias circuit, which increases the size and cost of the oscillator. Typically YIG resonators have both a broadband and a narrowband tuning port [9, 56]. The narrowband tuning port requires much less inductance and so can be tuned faster than the broadband port, making it more useful for modulation.

7.4 Theory

7.4.1 Introduction

A brief review of oscillator theory will aid in understanding the oscillator specifications mentioned in the first section. First, an overview of oscillator topologies is shown in Fig. 7.7. Traditionally, microwave oscillators have been viewed as one-port circuits with the active device presenting a negative immittance to the resonator, as shown in Fig. 7.7a [28]. The one-port philosophy is easy to measure with a slotted line or a network analyzer. Two-port oscillators are much more common at low frequencies, but probing voltages and currents in a feedback loop will always be difficult at the highest frequencies. Dielectric resonators have made feedback oscillators more common at microwave frequencies [29]. Integrated circuits have made the cross-coupled oscillator configuration, as shown in Fig. 7.7d, popular [30].

One-port analysis does allow confusion over the device acting as a negative impedance or admittance. Knowing the device type is essential for establishing a stable oscillation. For example, a negative conductance Gunn diode has the IV characteristic shown in Fig.7.8. As the oscillation signal grows about the bias point it eventually extends into the positive resistance region. During saturation the load line becomes more horizontal, reducing the negative conductance. The resonator needs to have its minimum conductance at the resonance frequency, so that moving off the resonant frequency would require an increase in the device negative conductance. If a Gunn diode is loaded with a series resonant circuit, which has maximum conductance at resonance, noise will move the oscillation off the series resonance and onto a nearby parasitic parallel resonance where it will stabilize

$$Z_D + Z_R = 0 \tag{7.3}$$

$$Y_D + Y_R = 0 \tag{7.4}$$

$$G H = 1 \tag{7.5}$$

The various oscillation conditions can be defined by the preceding three equations. Eq. (7.3) describes the active device impedance, Z_D, canceling the resonator impedance, Z_R, to support an oscillation. Eq. (7.4) describes the active device admittance, Y_D, canceling the resonator admittance, Y_R, to support oscillation. Eq. (7.3) can be split into real and imaginary parts to give $R_D + R_R = 0$, and $X_D + X_R = 0$. Often X_D is small, so the equations reduce to the active device negative resistance canceling the resonator (and load) resistance while the resonator is just off center frequency enough to cancel the device reactance. From a one-port point of view, when the resonator reactance is zero, no net phase shifts occur from the oscillation signal as it reflects back and forth from the active device to the resonator. Eq. (7.5) is the oscillation equation for a feedback circuit with gain element G and feedback resonator H. Eq. (7.5)

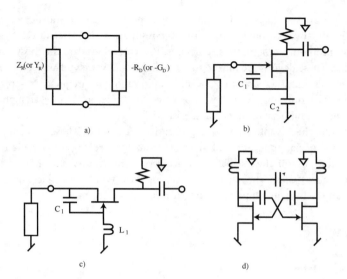

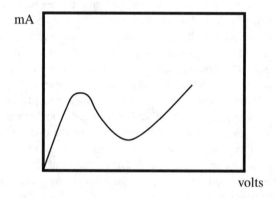

FIGURE 7.7   Oscillator configurations: a) diode; b) source feedback; c) gate feedback; and d) cross-coupled.

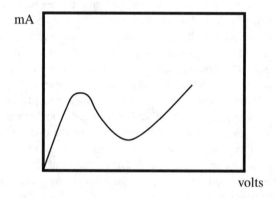

mA

volts

FIGURE 7.8   Gunn diode IV characteristic.

describes a positive feedback situation where the gain cancels the loss in the feedback while the net phase shift around the loop is zero. With zero phase shift around a loop and no loss, a signal will be sustained at the frequency of zero phase. The above equations describe a linear approximation to a stable oscillation. In reality each circuit is nonlinear. Oscillations start from noise in a circuit with positive feedback and grow until the circuit gain element saturates and the above equations are satisfied. Typically oscillators are set up so that the negative immittance, or gain, is 1.5 to 2 times greater than the circuit loss so that the device saturates into a stable oscillation without being driven so hard that its operating point changes excessively [2]. All of the above equations can be brought into a single oscillator theory [31].

At lower frequencies the oscillator output power can be predicted analytically [32]. At microwave frequencies the accuracy of the oscillator frequency and output power is very dependent on the CAD model used. Both harmonic balance and SPICE simulators can be used to predict oscillator output power and frequency, but component and circuit parasitics can make exact frequency predictions difficult. Linear simulators have a long history in oscillator design, as might be predicted by looking at Eqs. (7.3)–(7.5). The operating frequency, tuning range, as well as sensitivities to load, bias, and power supply variations can all be obtained from a linear simulator with bias dependent S-parameters for the active device.

Although many oscillator circuits are used, Figs. 7.7b and 7.7c show two of the most common discrete configurations. Figure 7.7b is used mostly with varactor-tuned inductors and dielectric resonators. Figure 7.7b is also used at lower frequencies where it is known as the Seiler oscillator [33]. Figure 7.7c is used mostly with YIGs, transmission line resonators, and dielectric resonators [34]. Simple analysis using an ideal transconductance for the device shows a negative resistance with capacitance at the gate of Fig. 7.7b and a negative conductance with shunt inductance at the source for Fig. 7.7c. Therefore, Fig. 7.7b operates in the inductive region of a series resonator, and Fig. 7.7c operates in the capacitive region of a parallel resonator. At microwave frequencies C_1 is simply the device capacitance. Close examination of Figs. 7.7b and 7.7c shows that they are both the same circuit. The only real difference is where the tuning takes place. Each circuit uses the device drain to provide load isolation and expects to have a relatively low impedance in the device drain to take power out. Eq. (7.6) shows how the negative input resistance of the source feedback oscillator changes with frequency, and that the residual reactance is capacitive. The input impedance in Fig. 7.7b is that from the resonator looking into the device node. Eq. (7.7) shows how the negative conductance of the gate feedback oscillator varies with frequency and that its residual susceptance is inductive. Note that the conductance is negative above the L_1C_1 resonance frequency and rapidly decreases with frequency. Both of these equations can be used for estimating the tuning bandwidth and oscillation frequency of an oscillator. The resonator used in either case will have its own immittance which can be compared to the device immittance using Eqs. (7.3) or (7.4) to determine if the oscillation condition is satisfied. CAD programs with S-parameter models will provide more accurate device characterization and tuning bandwidth analysis. For example, the configurations of Fig. 7.7 can be analyzed as one-ports for S11, $S11_D$. The resonator S11, $S11_R$, forms a reflection loop so that Eq. (7.5) is valid if $G = S11_D$ and $H = S11_R$.

$$Z_{IN} = -g_m / C_1 C_2 \omega^2 - j(C_1 + C_2) / C_1 C_2 \omega^2 \qquad (7.6)$$

$$Y_{IN} = g_m / (1 - \omega^2 L_1 C_1) + j\omega C_1 / (1 - \omega^2 L_1 C_1) \qquad (7.7)$$

7.4.2   Modulation, Noise, and Temperature

Many things perturb the oscillator frequency. To better study these perturbations consider the idealized oscillator shown in Fig. 7.9. Note that this oscillator is connected in a positive feedback configuration. The active device could be two FETs connected to create positive feedback, or it could be a single negative conductance shunting the resonator. The feedback configuration is chosen because it clarifies that the loaded Q of the oscillator is:

$$Q_L = 1 / (\omega_O L_O G_O). \qquad (7.8)$$

The loaded Q is only determined by circuit losses and external loading. The loaded Q is unaffected by the device g_m or negative conductance. The total losses, G_O, will be made up of internal losses, G_I, and external loading, G_{EXT}, so $G_O = G_I + G_{EXT}$. A little math will show that the total oscillator Q is a parallel combination of the internal Q_I and the external Q_{EXT} as shown in Eq. (7.9).

$$Q_L = (1/Q_I + 1/Q_{EXT})^{-1} \qquad (7.9)$$

The total oscillator Q is important because it determines the sensitivity to perturbations in oscillator frequency. Unfortunately, all we can measure directly is the external Q of an oscillator. Load pulling, as given in the pulling sensitivity specification, can be used to compute the external Q of an oscillator [22]. Load variations affect the oscillation frequency by changing the values of C_O or L_O in the same way G_{EXT} changes the value of G_O. Typically load pulling is measured by placing a 12 dB return loss load on a line

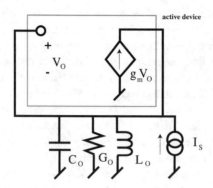

FIGURE 7.9   Idealized oscillator.

stretcher and rotating the load through all phases while recording the maximum frequency deviation. The C_O and L_O of Fig. 7.9 are made up of device capacitance, resonator elements, and load reactance. In low noise oscillators the resonator reactances dominate, but all oscillators have some influence from the other sources of reactance.

Power supply voltage variations change the oscillation frequency by changing the active device reactances and bias. The pushing specification in MHz/V defines the sensitivity to low frequency supply variations. Knowing the spectrum of the power supply variations allows the direct calculation of oscillator FM noise due to power supply variations through the pushing sensitivity. As the frequency of the supply variation is increased, the supply and bias filtering circuitry will reduce its influence on the oscillation frequency, so some measurements may have to be made if the power supply has large variations at high frequencies. Pushing can also be used as a technique to evaluate oscillator Q and minimize phase noise [69].

As discussed in the section on noise, the oscillator phase noise spectrum is the FM noise spectrum divided by f_m^2, where f_m is the modulating, or carrier offset frequency. Oscillator frequency variations are easily measured with a frequency discriminator or other FM detector such as a transmission line discriminator [35]. Spectrum analyzers are also very useful for measuring oscillator phase noise as long as the analyzer noise is much less than that of the measured device. Spectrum analyzers can measure script $L(f_m) = P_{SSB}(f_m)/\text{Hz}/P_C$ where the $P_{SSB}(f_m)/\text{Hz}$ is the phase noise power per Hertz within the $1/f^3$ region close to the carrier. Typically if the noise measurement is very near the carrier frequency, i.e., within the f^{-3} region, the noise is dominated by phase noise. For modulating noise such as power supply pushing the actual phase noise spectrum, $S_\phi(f_m)$, is twice script $L(f_m)$ because the sidebands are correlated [31].

All active devices contain internal noise sources due to resistance (thermal noise), charge crossing an energy barrier (shot noise), and traps (G-R noise) [18]. All of these noise sources perturb the device bias point no matter how well filtered the bias circuit. The most significant source of bias noise is due to traps [31]. These traps are spread in energy distribution by thermal and mechanical processes [23,36,37]. The spreading of the traps in bipolar transistors, FETs, and MOSFETs results in a $1/f$ low frequency noise spectrum in each device. FET based devices have $1/f$ variations in the drain current which converts to an equivalent input noise via the device transconductance. Bipolar devices have $1/f$ variations in the base current due to traps at the edge of the base-emitter junction [38]. The low frequency bias variations change the device reactances and so change the oscillation frequency [31,39–41]. The oscillator sensitivity to bias current or voltage variation is easily simulated and combined with measured device noise to provide a prediction of oscillator noise. Although simulators have estimates of device $1/f$ sources, low frequency noise sources are so dependent on device processing that measuring low frequency noise is the best way to verify noise performance. Once the bias sensitivity and device noise are determined, the oscillator internal phase noise can be computed in a similar manner to power supply pushing. Various schemes using symmetry, low frequency device loading, and even low frequency feedback exist for reducing total oscillator phase noise [42–45]. Even if device noise can be eliminated, resonators such as quartz crystals have their own $1/f$ noise sources that modulate the oscillator spectrum [46].

For voltage- or current-controlled oscillators the noise spectrum contains an additional term due to noise at the modulation port. Even modulating devices, such as varactor diodes, have internal noise that affects the oscillator spectrum. These noise sources can be accounted for just like the pushing and internal contributions. While there are many contributors to oscillator noise, each term is easily accounted for once the modulation sensitivity is known. Some noise contributions, such as power supply switching noise, can come through several paths so the total noise cannot just be a power summation of the individual contributions, but must include any correlation between contributions. In most oscillators one noise source will dominate [31]. For example, in broadband varactor-controlled oscillators, VCOs, the modulation noise usually dominates.

Figure 7.9 also shows a current source, I_S, which adds to the oscillation. This current source is useful for analyzing the oscillator response to high frequency noise and injection locking signals. The modulative low frequency noise sources discussed earlier operate in a multiplicative or nonlinear way, whereas noise sources at the oscillator frequency simply add to the oscillator noise in a linear manner [47]. For additive noise sources, increasing the oscillator power decreases the oscillator phase noise because the additive noise sources are fixed. However, for modulative noise sources increasing the oscillator power does not change the relative noise power because the modulation affects a fraction of the total oscillator power. This seemingly "linear" behavior is the result of any second order nonlinearity, and has caused confusion for many people analyzing oscillator noise. This same fact is why reducing even order nonlinearities through symmetry is effective in reducing oscillator noise.

If I_S only consists of high frequency noise due to thermal and shot noise in the device and circuit, we can produce a simple noise analysis of the oscillator. A more rigorous analysis is given in References 31 and 47. The positive feedback of the oscillator loop will cause the noise at the center of the resonator frequency to be amplified until the amplifier nonlinearity causes device saturation. The amplitude saturation does two things: 1) it reduces the device gain until Eq. (7.5) is satisfied; and 2) it reduces the loop gain to amplitude perturbations. I_S will cause both amplitude and phase variations in the oscillator carrier buildup. Amplitude saturation in the device will effectively strip the amplitude noise off the carrier as it passes through the active device, which means the amplitude noise of V_O will be only I_S passed through the resonator, or

$$S_{V\_AM}(f_m) = \left(S_{Is}(f_O + f_m) + S_{Is}(f_O - f_m)\right) / \left(2V_O^2/2\right) 1 / \left|G_O + j\left(\omega C_O - 1/(\omega L_O)\right)\right|^2, \quad (7.10)$$

which, near to the carrier, simplifies to

$$S_{V\_AM}(f_m) \approx S_{Is}(f_O + f_m) / \left(G_O^2 V_O^2 / 2\left(1 + 4Q^2 f_m^2 / f_O^2\right)\right), \quad (7.11)$$

where $V_O^2/2$ represents the carrier mean square level and $S_{Is}(f_O + f_m)$ is approximately equal to $S_{Is}(f_O - f_m)$.

The noise source, I_S, will perturb the carrier with an equal amount of amplitude and phase noise, which can be seen by envisioning each noise sideband as a phasor rotating about the carrier phasor. The limiting action of the active device does not attenuate the phase modulation of the carrier, so the full loop gain acts on the phase variations causing phase noise to be greatly amplified [48]. The phase noise amplification causes the phase noise to dominate amplitude noise near the carrier. Just as in the amplitude noise case, a low pass equivalent analysis can be performed on the circuit [49]. Solving the low pass equivalent form for phase noise variations is Leeson's model for phase noise [50]. Equation (7.12) gives the phase noise spectral density for the circuit of Fig. 7.9.

$$S_{V\_\phi}(f_m) \approx S_{Is}(f_O + f_m) / \left(G_O^2 V_O^2 / 2 \; 4Q^2 f_m^2 / f_O^2\right), \quad (7.12)$$

The term $S_{Is}(f_O + f_m)/(G_O^2 V_O^2)$ is equal to $kTF/P_C (1 + f_k/f_m)$ in Leeson's analysis, where P_C is the carrier power and F is approximately the device noise figure. The $1 + f_k/f_m$ term accounts for $1/f$ noise modulation with f_k being less than or equal to the device $1/f$ noise corner. Eq. (6.13) is Leeson's equation. Although the preceding analysis assumed uncorrelated high frequency noise sidebands, the inclusion of a $1/f$ term in Leeson's result implies correlated sidebands and a factor of 2 between script L and phase noise spectral density, S_ϕ. Leeson's model for phase noise contains most of the important aspects of oscillator noise in a simple equation. More rigorous oscillator analysis appears with regularity, but Leeson's model continues to be useful and relevant [28,31,39]. Naturally, it is also possible to analyze oscillator noise in the time domain [51,52]. Several important aspects of oscillator noise predicted by Leeson's model are that the noise power decreases as Q^2, that the noise falls off as f_m^{-3} near the carrier, and that the noise power increases as f_O^2. The problems with Leeson's model are the approximations involved. Because oscillators are not linear, noise-matched amplifiers F cannot be used. Because the low frequency noise modulates the carrier, script L does not decrease in proportion to increases in P_C in well-designed, low-noise oscillators. And finally, f_k in the final oscillator depends on all the low frequency noise sources and modulation sensitivities and is usually less than or equal to the device $1/f$ noise corner for the dominant $1/f$ noise source.

$$script\ L(f_m) \approx kTF/P_C(1 + f_k/f_m)/(4Q^2 f_m^2/f_O^2) \tag{7.13}$$

7.4.3   Injection Locking

Very high frequency oscillators are often phase locked by coupling a reference signal directly into the oscillator [28,53,55]. Understanding injection locking can help in understanding oscillator operation in general. In Fig. 7.9 we can let I_S be an injection locking signal, $I_S e^{j\omega t}$. Then $V_O = V_O e^{j\omega t + \phi}$, where the phase shift ϕ occurs if ω is different from the oscillator loop center frequency of ω_O. From linear circuit analysis we can derive Eq. (7,14), and Eq. (7.15) follows from Euler's identity.

$$V_O e^{j\omega t + \phi}\left[1 - \omega^2 L_O C_O + j\omega L_O\left(G_O - g_m\right)\right] = j\omega L_O I_S e^{j\omega t} \tag{7.14}$$

$$V_O\left[1 - \omega^2 L_O C_O + j\omega L_O\left(G_O - g_m\right)\right] = j\omega L_O I_S\left(\cos(\phi) - j\sin(\phi)\right) \tag{7.15}$$

By grouping real and imaginary parts, Eq. (7.15) can be expanded into Eqs. (7.16) and (7.17) which define the locking gain and the bandwidth of locking.

$$I_S = V_O\left(G_O - g_m\right)/\cos(\phi) \tag{7.16}$$

$$\sin(\phi) = \left(1/\omega L_O - \omega C_O\right)V_O/I_S \tag{7.17}$$

Equation (7.16) shows that as the device saturates and g_m approaches G_O, very little current, I_S, is required to maintain lock. However, as the injection frequency shifts away from the resonator frequency, ω_O, ϕ moves away from zero as shown in Eq. (7.17). As ϕ increases from zero to $\pm 90°$, the $\cos(\phi)$ term in Eq. (7.16) decreases to zero and an infinite injection current is required to maintain lock, so ϕ equal to $\pm 90°$ defines the locking bandwidth. Equation (7.17) can be used to translate the $\pm 90°$ limits into frequency limits that define the locking bandwidth. Equation (7.16) shows that very little injection locking signal is required at the band center while more signal is required as the frequency approaches the band edges. Another way of interpreting this is to say there is near infinite injection locking gain at the band center and zero locking gain at the band edges. When I_S is broadband noise this injection gain works to provide the commonly observed high levels of noise that decrease away from the carrier [48]. The relationship between the gain and offset frequency is linked through the loop phase shift. An injection locked oscillator is in fact a first order PLL.

7.5 Summary

Oscillators consist of an active device, a resonator, and a package. These three things determine the frequency, accuracy, available power, and cost of the source. For a simple task such as providing a sinusoid, oscillators require an inordinate number of specifications. The sections on specifications and theory try to provide a background for understanding oscillator requirements, while the technologies and capabilities section tries to show how modern devices and resonators are combined to meet oscillator specifications.

Acknowledgment

Thanks to Mark Shiman of Disman Bakner and Ron Korber of Stellex for providing various samples for the photographs.

References

1. Leier, R.M., and Patston, R.W., Voltage-Controlled Oscillator Evaluation for System Design, *MSN*, 102–125, Nov. 1985.
2. Rogers, R.G., *Low Phase Noise Microwave Oscillator Design*, Artech House, Boston, 1991.
3. Robins, W.P., *Phase Noise in Signal Sources*, Peter Peregrinus Ltd., London, U.K., 1982.
4. Blair, B.E., *Time and Frequency: Theory and Fundamentals*, U.S. Department of Commerce, NBS Monograph 140, 1974.
5. ANSI, *Telecommunications — Synchronous Optical Network (SONET) — Jitter at Network Interfaces*, T1.105.03-1994, 1994.
6. Bahl, I.J., and Bhartia, P.B., *Microwave Solid State Circuit Design*, John Wiley & Sons, New York, 1988.
7. Engineering Staff, *Nonlinear Circuits Handbook*, Analog Devices, Norwood, MA, 1976.
8. Huckleberry, B.E., Design Considerations for a Modern DTO, *Microwave Journal*, 291–295, May 1986.
9. Osbrink, N.K., YIG-Tuned Oscillator Fundamentals, *Microwave System Designer's Handbook*, 207–225, 1983.
10. Heins, M.S., Juneja, T., Fendrich, J.A., Mu, J., Scott, D., Yang, Q., Hattendorf, M., Stillman, G.E., and Feng, M., W-band InGaP/GaAs HBT MMIC Frequency Sources, *IEEE MTT-S Digest*, 239–242, 1999.
11. Dieudonne, J.-M., Adelseck, B., Narozny, P., and Dambkes, H., Advanced MMIC Components for Ka-Band Communication Systems: A Survey, *IEEE MTT-S Digest*, 409–415, 1995.
12. Eisele, H., and Haddad, G.I., Potential and Capabilities of Two-Terminal Devices as Millimeter- and Submillimeter-Wave Fundamental Sources, *IEEE MTT-S Digest*, 933–936, 1999.
13. Faber, M.T., Chramiec, J., and Adamski, M.E., *Microwave and Millimeterwave Diode Frequency Multipliers*, Artech House, Boston, 1995.
14. Scherer, D., Generation of Low Phase Noise Microwave Signals, *RF & Microwave Measurement Symposium*, Hewlett-Packard, Palo Alto, CA, 1983.
15. Bruston, J., Smith, R.P., Martin, S.C., Humphrey, D., Pease, A., and Siegel, P.H., Progress Towards the Realization of MMIC Technology at Submillimeter Wavelengths: A Frequency Multiplier to 320 GHz, *IEEE MTT-S Digest*, 399–402, 1998.

16. Roberts, M.J., Iezekiel, S., and Snowden, C.M., A Compact Subharmonically Pumped MMIC Self-Oscillating Mixer for 77 GHz Applications, *IEEE MTT-S Digest*, 1435–1438, 1998.

17. Bates, P.C., Measure Residual Noise in Quartz Crystals, *Microwaves & RF*, 95–106, Nov. 1999.

18. Ambrozy, A., Electronic Noise, McGraw-Hill, New York, 1982.

19. Banu, M., MOS Oscillators with Multi-Decade Tuning Range and Gigahertz Maximum Speed, *IEEE JSSC*, 1386–1393, Dec. 1988.

20. Svelto, F., Deantoni, S., and Castello, R., A 1.3 GHz Low-Phase Noise Fully Tunable CMOS LC VCO, *IEEE JSSC*, 356–361, Mar. 2000.

21. Aoki, H., and Shimasue, M., Noise Characterization of MOSFET's for RF Oscillator Design, *IEEE MTT-S Digest*, 423–426, 1999.

22. Vendelin, G., Pavio, A.M., and Rohde, U.L., *Microwave Circuit Design*, John Wiley & Sons, 1990.

23. Sodini, D., Touboul, A., Lecoy, G., and Savelli, M., Generation-Recombination Noise in the Channel of GaAs Schottky gate FET, *Electronics Letters*, 42–43, Jan. 22, 1976.

24. Hughes, B., Fernandez, N.G., and Gladstone, J.M., GaAs FETs with a Flicker Noise Corner Below 1 MHz, *WOCSEMMAD*, 20–23, 1986.

25. Goedbloed, J.J., Noise in IMPATT Diode Oscillators, *Philips Research Reports Supplement*, 1–115, 1973.

26. Sze, S.M., *Physics of Semiconductor Devices*, John Wiley & Sons, New York, 1981.

27. Crowe, T.W., Weikle, R.M., and Hesler, J.L., GaAs Devices and Circuits for Terahertz Applications, *IEEE MTT-S Digest*, 929–932, 1999.

28. Kurokawa, K., Noise in Synchronized Oscillators, *IEEE Trans MTT*, 234–240, Apr. 1968.

29. Popovic, N., Review of Some Types of Varactor Tuned DROs, *Applied Microwave and Wireless*, 62–70, Aug. 1999.

30. Abidi, A.A., Radiofrequency CMOS Circuits, *IEEE SCV-MTT Short Course*, Apr. 1997.

31. Riddle, A.N., Oscillator Noise: Theory and Characterization, N.C. State University, Raleigh, NC, PhD Dissertation, 1986.

32. Clarke, K.K., and Hess, D.T., *Communication Circuits: Analysis and Design*, Addison-Wesley, Reading, 1978.

33. Clapp, J.K., Frequency Stable LC Oscillators, *Proc. IRE*, 1295–1300, Aug. 1954.

34. Schiebold, C.F., An Approach to Realizing Multi-Octave Performance in GaAs-FET YIG-Tuned Oscillators, *IEEE MTT-S Digest*, 261–263, 1985.

35. Ondria, J.G., A Microwave System for Measurement of AM and FM Noise Spectra, *IEEE Trans. MTT*, 767–781, Sept. 1968.

36. Rohdin, H., Su, C.-Y., and Stolte, C., A Study of the Relationship Between Low Frequency Noise and Oscillator Phase Noise for GaAs MESFETs, *IEEE MTT-S Digest*, 267–269, 1984.

37. Christenssen, S., Lundstrum, I., and Svensson, C., Low Frequency Noise in MOS Transistors, *Solid-State Electronics*, 797–812, 1968.

38. van der Ziel, A., Noise in Solid State Devices and Lasers, *Proc IEEE*, 1178–1206, Aug. 1970.

39. Siweris, H.V., and Schiek, B., Analysis of Nosie Upconversion in Microwave FET Oscillators, *IEEE Trans. MTT*, 233–242, Mar. 1985.

40. Pucel, R.A, and Curtis, J., Near-Carrier Noise in FET Oscillators, *IEEE MTT-S Digest*, 282–284, 1983.

41. Dallas, P.A., and Everard, J.K.A., Characterization of Flicker Noise in GaAs MESFETs for Oscillator Applications, *IEEE Trans. MTT*, 245–257, Feb. 2000.

42. Chen, H.B., van der Ziel, A., and Amberiadis, K., Oscillators with Odd-Symmetry Characteristics Eliminate Low-Frequency Noise Sidebands, *IEEE Trans. CAS*, 807–809, Sept. 1984.

43. Riddle, A.N., and Trew, R.J., A Novel GaAs FET Oscillator with Low Phase Noise, *IEEE MTT-S Digest*, 257–260, 1985.

44. Tutt, M.N., Pavlidis, D., Khatibzadeh, A., and Bayraktaroglu, B., The Role of Baseband Noise and its Unconversion in HBT Oscillator Phase Noise, *IEEE Trans. MTT*, 1461–1471, July 1995.

45. Prigent, M., and Obregon, J., Phase Noise Reduction in FET Oscillators by Low-Frequency Loading and Feedback Circuitry Optimization, *IEEE Trans. MTT*, 349–352, Mar. 1987.
46. Gagnepain, J.J., Olivier, M., and Walls, F.L., Excess Noise in Quartz Crystal Resonators, *Proc. 36th Symp. Frequency Control*, 218–225, 1983.
47. Edson, W.A., Noise in Oscillators, *Proc. IRE*, 1454–1466, Aug. 1960.
48. Spaelti, A., Der Einfluß des Thermischen Widerstandrauschens und des Schroteffektes auf die Stoermodulation von Oscillatoren, *Bull. Schweiz Elektrotech. Verein*, 419–427, June 1948.
49. Egan, W.F., The Effects of Small Contaminating Signals in Nonlinear Elements used in Frequency Synthesis and Conversion, *Proc. IEEE*, 797–811, July 1981.
50. Leeson, D.B., A Simple Model of Feedback Oscillator Noise Spectrum, *Proc. IEEE*, 329–330, Feb. 1966.
51. Abidi, A.A., and Meyer, R.G., Noise in Relaxation Oscillators, *IEEE JSSC*, 794–802, Dec. 1983.
52. Lee, T.H., and Hajimir, A., Oscillator Phase Noise: A Tutorial, *IEEE JSSC*, 326–336, Mar. 2000.
53. Paciorek, L.J., Injection Locking of Oscillators, *Proc. IEEE*, 1723–1727, Nov. 1965.
54. Rohde, U.L., *Digital PLL Frequency Synthesizers*, Prentice-Hall, Englewood Cliffs, NJ, 1983.
55. Khanna, A.P.S., and Gane, E., A Fast-Locking X-Band Transmission Injection-Locked DRO, *IEEE MTT-S Digest*, 601–604, 1988.
56. Trew, R.J., Design Theory for Broad-Band YIG-Tuned FET Oscillators, *IEEE Trans MTT*, 8–14, Jan. 1979.
57. Prigent, M., Camiiade, M., Dataut, G., Raffet, D., Nebus, J.M., and Obregon, J., High Efficiency Free Running Class F Oscillator, *IEEE MTT-S Digest*, 1317–1324, 1995.
58. Maruhashi, K., Madihian, M., Desclos, L., Onda, K., and Kuzuhama, M., A K-Band monolithic CPW Oscillator Co-Integrated with a Buffer Amplifier, *IEEE MTT-S Digest*, 1321–1324, 1995.
59. Heins, M.S., Barbage, D.W., Fresina, M.T., Ahmari, D.A., Hartmann, Q.J., Stillman, G.E., and Feng, M., Low Phase Noise Ka-Band VCOs Using InGaP/GaAs HBTs and Coplanar Waveguide, *IEEE MTT-S Digest*, 255–258, 1997.
60. Eisele, H., Manns, G.O., and Haddad, G.I., RF Performance Characteristics of InP Millimeter-Wave N+-N--N+ Gunn Devices, *IEEE MTT-S Digest*, 451–454, 1997.
61. Wollitzer, M., Buecher, J., and Luy, J.-F., High Efficiency Planar Oscillator with RF Power of 100 mW Near 40 GHz, *IEEE MTT-S Digest*, 1205–1208, 1997.
62. Siweris, H.J., Werthoff, A., Tischer, H., Schaper, U., Schaefer, A., Verweyen, L., Grave, T., Boeck, G., Schleichtweg, M., and Kellner, W., Low Cost GaAs PHEMT MMICs for Millimeter-Wave Sensor Applications, *IEEE MTT-S Digest*, 227–230, 1998.
63. Kadszus, S., Haydl, W.H., Neumann, M., Bangert, A., and Huelsmann, A., Subharmonically Injection Locked 94 GHz MMIC HEMT Oscillator using Coplanar Technology, *IEEE MTT-S Digest*, 1585–1588, 1998.
64. Galani, Z., Low Noise Microwave Sources for Radar and Missile Systems, Ultra Low Noise Microwave Sources Workshop, 1994.
65. Everard, J.K.A., and Page-Jones, M., Ultra Low Noise Microwave Oscillator with Low Residual Flicker Noise, *IEEE MTT-S Digest*, 693–696, 1995.
66. Vectron, *Frequency Control Products*, Vectron International, 1997.
67. Avantek, *Modular and Oscillator Components*, Avantek, 1989.
68. Feng, Z., Zhang, W., Su, B., Harch, K.F., Gupta, K.C., Bright, V., and Lee, Y.C., Design and Modeling of RF MEMS Tunable Capacitors Using Electro-Thermal Actuators, *IEEE MTT-S Digest*, 1507–1510, 1999.
69. Trans-Tech, Optimize DROs for Low Phase Noise, Application Note No. 1030, *Trans-Tech Temperature Stable Microwave Ceramics*, 1998.

8

Phase Locked Loop Design

Robert Newgard
Rockwell Collins

8.1   Introduction

The objective of this chapter is to present the fundamental considerations that go into phase locked loop (PLL) design. The PLL has been utilized in various systems for many years, but it wasn't until the development of integrated circuits in the 1970s that widespread use as a frequency synthesizer came about. With the expansion in the wireless industry and the ever-increasing demand for higher frequency systems, PLL design has now moved into the microwave realm. This chapter is by no means a complete treatise on PLL design, but rather a synopsis of PLL characteristics and design considerations.

The architecture of the frequency synthesizer is often dependent upon the design of the receiver and exciter, and the PLL design engineer must take this fact and system requirements into account. In addition, many trade-offs must be performed in order to implement a successful frequency synthesizer design; the appropriate injection frequencies must be provided, but these should be generated with consideration given to tuning speed (i.e., settling time), phase noise performance, spurious requirements, and channel spacing to name a few. The following sections provide guidelines related to these trade-offs, and while many details are left to the reader, it should be clear that successful PLL design is no accident

8.2   Roles and Attributes of Phase Locked Loops

In modern wireless communications systems the phase locked loop plays a key role in the performance of the system. The primary function of the PLL is to generate a band of transmit and receive injection frequencies that allow the receiver or transmitter to resolve the required channel spacing. The injections may be synthesized by a combination of PLLs depending upon the system requirements. The spectral purity of the injection signal is a determining factor in the communication systems performance. The PLL generates the injection frequencies from a reference source that is typically a crystal controlled temperature compensated oscillator, but may be any stable reference oscillator (e.g., rubidium, cesium, etc.). The frequency accuracy and temperature stability of the output signal of the PLL is proportional to the frequency accuracy and temperature stability of the crystal oscillator.

The PLL is a negative feedback control system utilizing a phase detector (PD), lowpass filter (LPF), voltage controlled oscillator (VCO), and frequency divider (FD). The basic block diagram of a PLL is shown in Fig. 8.1. The VCO generates an output signal that is dependent upon a DC control voltage at its input. The PD compares the phase of the reference signal to the phase of the divided down VCO signal and generates a correction signal, which is proportional to the phase difference. The LPF's function is to: (1) attenuate the reference sidebands, (2) shape the phase noise, and (3) tailor the PLL's dynamics. Frequency selection of the output of the PLL is accomplished by varying the divisor of the FD. The frequency divider is typically programmable and has enough range to cover the desired amount of frequency tuning bandwidth.

The phase of the PLL's output signal is given by

$$\phi_o = \phi_{ref} N \tag{8.1}$$

where N is the division ratio of the frequency divider and is stepped in integer values. Because frequency is the time derivative of phase, the output frequency of the PLL is given by

$$F_O = \frac{d\phi_o}{dt} = \frac{d\phi_{ref}}{dt} N = F_{ref} N \tag{8.2}$$

which shows that the output frequency is an integer multiple of the reference frequency. The reference frequency is chosen to attain the desired channel spacing, since incrementing N increases the output frequency in multiples of the reference frequency.

The PLL makes a relatively unstable VCO track the phase of the reference signal, which is derived from a stable crystal oscillator. Free running voltage controlled oscillators drift with variation in temperature and power supply noise, as well as noise on the control voltage (a.k.a., tune voltage) line. The action of the feedback loop is to keep the VCO phase locked to the reference oscillator signal.

There are three primary problems that challenge the PLL designer. One is improving the frequency acquisition time or settling time (i.e., when a command occurs to change channels, the PLL takes time

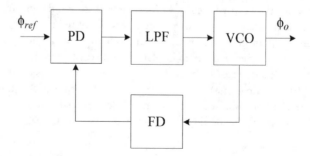

FIGURE 8.1   Phase locked loop.

to move from the old frequency to the new one and acquire lock). The second is reducing sidebands and spurious signals from appearing on the PLL's output. Any discrete frequency components appearing on the VCO control line will modulate the VCO and appear as spurious sidebands on the output of the PLL. The primary discrete spurious frequency source is modulation of the VCO by the error signal, at the comparison frequency, coming from the output of the phase detector. These spurs are referred to as reference sidebands. Other sources of spurious signals are conducted signals on power supplies (e.g., VCO power supply, phase detector power supply, etc.), radiated signals (e.g., induced on the VCO tank coil and loop filter coils), and isolation from other signal sources (e.g., reverse isolation from the programmable divider to the PLL output). In addition to these spurious signal sources, mechanical vibration of the synthesizer assembly may induce unwanted sidebands on the VCO by physically modulating the printed wiring board and/or the VCO's tank coil. The spurious signals are reduced through good design of the PLL's dynamics, by using good RF shielding techniques, and by providing mechanical support to the assemblies. The third problem that challenges the PLL designer is phase noise performance. As mentioned before, the long-term frequency stability of the output signal of the PLL is determined by the frequency standard used, but the phase noise performance, and thus the short-term stability, is dependent upon the design of the PLL. In the receive path, in order to downconvert the modulated radio frequency (RF) signal, the output signal of the PLL (a.k.a., local oscillator or LO) is mixed with the RF signal. The phase noise of the LO is superimposed onto the intermediate frequency (IF) or baseband signal and thereby affects the receiver's selectivity. In the transmit path, because the LO is mixed with the IF signal or baseband signal to generate the modulated RF signal and again the phase noise of the LO is superimposed, the transmit noise floor or signal-to-noise ratio (SNR) is a function of the LO's phase noise. Also, the performance of the receiver in the presence of a strong adjacent channel signal is affected by the phase noise performance of the LO. The adjacent channel signals mix with the LO's phase noise and produce noise signals at the IF, thereby decreasing the receiver's selectivity. In order to better understand these design considerations, the designer needs to develop a clear understanding of the mathematical models that are used to characterize PLL behavior.

8.3   Transfer Function of the Basic PLL

By making the assumption that the PLL is continuous in time, basic feedback control theory utilizing Laplace Transforms can be utilized to determine the loop's behavior, provided that the loop bandwidth is much, much less than the reference frequency. While in practice it is true that the phase detector and frequency dividers are not continuous in time, it is necessary to make this assumption in order to model the stability of the PLL using the Laplace transform. When wide loop bandwidth synthesizers are designed, the sampling nature of the frequency divider and phase detector cannot be ignored. The time delay of these devices will introduce phase shift (i.e., reduction in phase margin), thereby affecting the dynamic performance of the PLL. Another assumption is that the PLL has reached steady state (i.e., it has reached a phase locked condition).

The characteristics described in this section do not address the acquisition of phase lock. The block diagram of a PLL and the gain of each of the functional blocks is shown in Fig. 8.2. The phase detector is shown as an adder and gain block in order to clarify the understanding of the functionality of the phase detector. The forward gain, G(s), is used represent the product of the transfer function of each individual block within the forward path of the PLL. Likewise, the feedback gain, H(s), represents the product of each individual transfer function within the feedback path of the PLL. The equations describing the PLL shown in Fig. 8.2, in terms of the transform variables, are

$$\phi_O(s) = \phi_e(s)G(s) \qquad (8.3)$$

$$\phi_b(s) = H(s)\phi_o(s) \qquad (8.4)$$

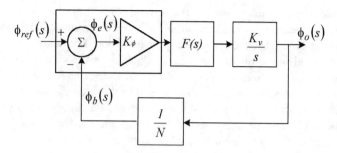

FIGURE 8.2 PLL gain block diagram.

$$\phi_e\big(s\big)=\phi_{ref}\big(s\big)-\phi_b\big(s\big) \tag{8.5}$$

where classical control theory notation is used. The overall closed-loop transfer function is found by solving the above equations.[1]

$$A_{CL}\big(s\big)=\frac{\phi_o\big(s\big)}{\phi_{ref}\big(s\big)}=\frac{G\big(s\big)}{1+G\big(s\big)H\big(s\big)} \tag{8.6}$$

The denominator of the closed-loop response is defined as the characteristic equation. The forward and feedback gain of the PLL shown in Fig. 8.2 are

$$G\big(s\big)=K_\phi F\big(s\big)\frac{K_v}{s} \tag{8.7}$$

$$H\big(s\big)=\frac{1}{N} \tag{8.8}$$

Therefore, the transfer function of the PLL shown in Fig. 8.2 is

$$A_{CL}\big(s\big)=\frac{K_\phi F\big(s\big)\dfrac{K_v}{s}}{1+K_\phi F\big(s\big)\dfrac{K_v}{s}\dfrac{1}{N}} \tag{8.9}$$

The transfer function given in Eq. (8.9) is referred to as the closed-loop response. The open-loop transfer function is defined as the ratio of the output of the feedback path $\phi_b(s)$ to the system error signal $\phi_e(s)$. The open-loop transfer function is used in the analysis of the PLL's stability.

$$A_{OL}\big(s\big)=\frac{\phi_b}{\phi_e}=G\big(s\big)H\big(s\big)=K_\phi F\big(s\big)\frac{K_v}{s}\frac{1}{N}=M\angle\alpha \tag{8.10}$$

The closed-loop and open-loop response, Eqs. (8.6) and (8.10) respectively, yield a phasor quantity for each unique complex parameter s. For the open-loop response, the magnitude is M and the phase angle is α. As can be seen from Eq. (8.10), the open-loop response appears in the denominator of the closed-loop response. The frequency at which the magnitude of the open-loop response equals one is used to determine the stability of the PLL. As described in the following section, the phase of the open-loop response at this point is critical in determining the loop stability.

8.4   Stability

There are many ways to evaluate the stability of a PLL, but a very popular method is to analyze the stability by plotting the open-loop gain and phase margin as a function of frequency. A feedback control system will become unstable if the magnitude of the open-loop response of the system exceeds unity at the frequency for which the open-loop phase shift is equal to ±180°. The magnitude of the open-loop response at this point is referred to as the gain margin. For a stable PLL the gain margin should be greater than 10 dB. Also, as a measure of relative stability, the phase margin of the PLL is 180° plus the phase angle where the magnitude of the open-loop response is equal to unity (i.e., 0 dB). The.frequency at which this occurs is referred to as the open-loop bandwidth. In other words, the phase margin is the amount of phase shift at the loop bandwidth that would produce instability. For a stable PLL the phase margin should be greater than 30°. It is common practice to plot the log magnitude and phase margin of the open-loop transfer function to analyze stability. In designing the PLL, it is imperative that enough phase margin is allowed such that the loop's closed-loop gain response will not have peaking. The plot shown in Fig. 8.3 shows the closed-loop gain of a PLL with three values of phase margin (10°, 45°, and 60°). When adequate phase margin is not provided, the loop will be unstable. The PLL output signal can be observed on a spectrum analyzer. Any discrete spurs separated from the desired output signal by the loop bandwidth and harmonics thereof, indicates inadequate phase margin. The designer should make sure that variations in loop bandwidth, that occur as the PLL output signal is tuned, do not cause a loss in phase margin and thereby have an adverse affect on loop stability. Loop bandwidth variations are caused by changes in VCO gain, phase detector gain, component temperature coefficients, and loop division ratio.

8.5   Type and Order

It is imperative in modeling the transient and steady-state response of PLLs to develop an understanding of how the PLL will respond to various inputs. Most common PLL designs fall into two categories, Type

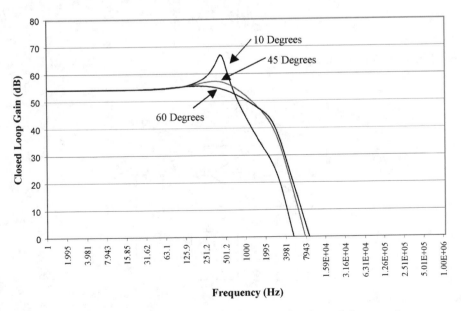

FIGURE 8.3   Closed-loop magnitude response of a PLL with 10°, 45°, and 60° of phase margin.

I and Type II, albeit the PLL type is not limited. The type of system refers to the number of poles in the open-loop gain located at the origin (i.e., the number of perfect integrators in the PLL). The order of the system refers to the degree of the characteristic equation or the denominator of the closed-loop transfer function. As shown in Fig. 8.2, there are two blocks that are a function of frequency, the loop filter and VCO. Therefore, the filter block, F(s), is the factor that determines the type and order of the PLL. The control system examples that follow will further the reader's understanding of PLL design.

8.5.1   Type I First-Order Loop

The first PLL introduced is a type I, first-order; although it is not practical due to the fact that the sidebands caused by the error signal, $\phi_e(s)$, are in most cases too high without a loop filter. This is dependent upon many of the system parameters (e.g., reference signal frequency, VCO gain, division ratio, etc.). It is presented here as a basis for furthering the reader's understanding of the analysis of phase locked loops. A simplified block diagram is shown in Fig. 8.4. As can be seen from the block diagram, the only integrator is the VCO. The assumption is made at this point that the phase detector doesn't have a pole at the origin, but such is not always the case. The closed-loop transfer function is given by

$$A_{CL}(s) = \frac{K_\phi K_v}{s + \frac{K_\phi K_v}{N}} \qquad (8.11)$$

The uncompensated loop bandwidth is commonly defined as

$$\omega_n = \frac{K_\phi K_v}{N} \qquad (812)$$

which is, in this case type I first-order, the loop's bandwidth, since there is no compensation by a loop filter. Substituting eq. (8.12) into Eq. (8.13), the closed-loop transfer function becomes

$$A_{CL}(s) = \frac{N\omega_n}{s + \omega_n} \qquad (8.13)$$

The open-loop transfer function is given by

$$A_{OL}(s) = \frac{K_\phi K_v}{sN} \qquad (8.14)$$

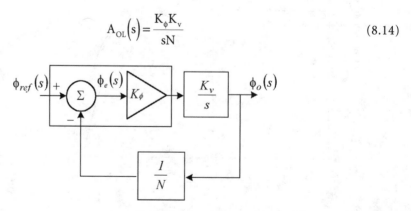

FIGURE 8.4   Type I first-order PLL.

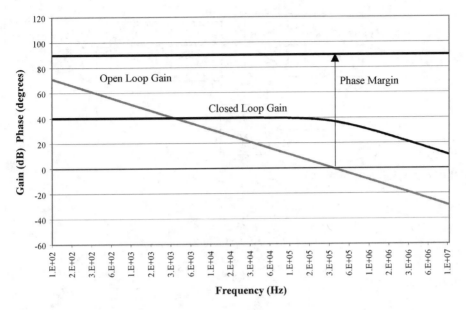

FIGURE 8.5 Type I first-order transfer functions.

The plot in Fig. 8.5 shows the open-loop gain and margin phase along with the closed-loop gain. It can be seen from the closed-loop gain that this example has 40 dB of gain inside the loop bandwidth. The loop bandwidth is defined, with respect to frequency, as the point where the open-loop gain equals one. The phase margin is equal to 90°, which is more than adequate for stability. The problem here is the reference sideband spurs will not be attenuated by the loop. Hence, it is imperative that the PLL designer adds a loop filter to the design and thereby the order of the PLL is increased. By examining the closed-loop transfer function, Eq. (8.13), it can be seen that it is a lowpass filter response with gain inside of the loop bandwidth. If the bandwidth of loop is narrow enough and the reference frequency is high enough the loop will provide attenuation, albeit the slope of the attenuation is 20 dB/decade. The limitations on the design due to system requirements, of vibration and settling time, typically force the designer to add additional filtering to the forward path of the PLL.

8.5.2 Type I Second-Order Loop

A lowpass filter is added in the forward path of the PLL in order to attenuate the reference sideband spurs. In this example, a single pole filter has been added and hence the PLL becomes a type I, second order loop. A simplified block diagram is shown in Fig. 8.6. The transfer function of the loop filter, F(s), is given by

$$F\left(s\right) = \frac{1}{RCs + 1} \tag{8.15}$$

The closed-loop transfer function of the PLL is given by

$$A_{CL}\left(s\right) = \frac{K_\phi K_v F\left(s\right)}{s + \dfrac{K_\phi K_v F\left(s\right)}{N}} \tag{8.16}$$

By substituting Eq. (8.15) into Eq. (8.16) and simplifying, the order of the PLL is plainly seen

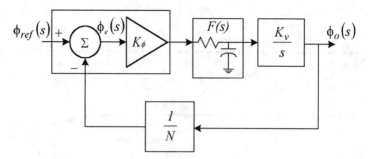

FIGURE 8.6   Type I second-order PLL.

$$A_{CL}(s) = \frac{K_\phi K_v \dfrac{1}{RCs+1}}{s + \dfrac{K_\phi K_v \dfrac{1}{RCs+1}}{N}} = \frac{K_\phi K_v}{RCs^2 + s + \dfrac{K_\phi K_v}{N}} \tag{8.17}$$

The open-loop transfer function is given by

$$A_{OL}(s) = K_\phi \frac{K_v}{s} \frac{1}{RCs+1} \frac{1}{N} = \frac{K_\phi K_v}{s(NRCs+N)} \tag{8.18}$$

Figure 8.7 shows an example of the closed-loop gain, the open-loop gain and phase margin of a type I, second-order loop. The addition of the filter has added phase shift to the open-loop response, but at 75°, the phase margin is adequate for stability. The loop filter, however, still doesn't offer much filtering for reference signal spurs. Higher order filters are typically added to the PLL in order to provide the appropriate attenuation, but as can be seen in Fig. 8.7, the additional filtering adds phase shift. The goal here is to maximize the filter's attenuation while realizing minimum phase shift. An elliptic filter is often used due to the fact that they have higher selectivity (i.e., the passband is closest to the stopband) compared to other filters. The higher selectivity results in a minimization of phase shift.

8.5.3   Phase Errors for Type I and Type II PLL

Dependent upon the system requirements, the PLL will have to respond to various kinds of inputs (i.e., phase of reference signal, change in division ratio, etc.). The designer is required to know how the PLL will respond to these inputs when the loop has reached steady state. In classical control theory a system is characterized by its response to step changes in position, velocity, and acceleration. In PLL design these changes correspond to step changes in phase, frequency, and time-varying frequency. The steady state response is determined by using the Laplace final value theorem, which is

$$\underset{t\to\infty}{\text{Lim}}\Big[\phi_e(t)\Big] = \underset{s\to 0}{\text{Lim}}\Big[s\phi_e(s)\Big] \tag{8.19}$$

For the Type I PLL, shown in Fig. 8.4, the function $\phi_e(s)$ is the phase error signal generated within the phase detector and is referred to as the system error.

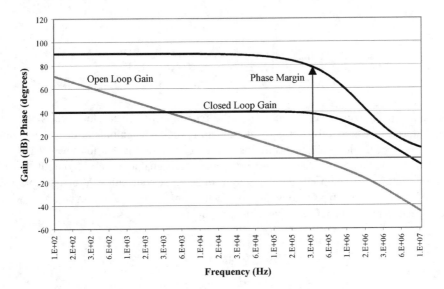

FIGURE 8.7   Type I second-order transfer functions.

$$\phi_e\!\left(s\right) = \frac{1}{1 + \dfrac{K_\phi K_v}{sN}}\,\phi_{ref}\!\left(s\right) \tag{8.20}$$

A phase unit step function, u(t), is applied to the input and the Laplace transform gives

$$\phi_{ref}\!\left(s\right) = \frac{A}{s} \tag{8.21}$$

where A is the magnitude of the phase step in radians. This would represent the input signal shifting phase of A radians. By substituting Eqs. (8.20) and (8.21) into the Laplace final value theorem

$$\underset{t\to\infty}{\text{Lim}}\!\left[\phi_e\!\left(t\right)\right] = \underset{s\to 0}{\text{Lim}}\!\left[s\,\frac{1}{1 + \dfrac{K_\phi K_v}{sN}}\,\frac{A}{s} \right] = 0 \tag{8.22}$$

Thus, when a step phase change is applied to the Type I PLL, the final value of the system error is zero, which better be the case or we don't have a phase locked loop. Next a unit step function of frequency is applied to the PLL. Phase is the integral of frequency, therefore the reference signal becomes

$$\phi_{ref}\!\left(s\right) = \frac{A}{s^2} \tag{8.23}$$

Once again by substituting Eq. (8.20) and (8.23) into the Laplace final value theorem

$$\underset{t\to\infty}{\text{Lim}}\left[\phi_e\left(t\right)\right] = \underset{s\to0}{\text{Lim}}\left[s\,\dfrac{1}{1+\dfrac{K_\phi K_v}{sN}}\,\dfrac{A}{s^2}\right] = \dfrac{AN}{K_\phi K_v} \qquad (8.24)$$

Thus, when a step frequency change is applied to the Type I PLL, the final value of the system error is a constant, but as can be seen, this constant is dependent upon the magnitude of the change, along with the division ratio of the loop. What this means to the PLL designer is that for any given N or output frequency, there will be a phase error between the reference signal and the PLL output. If the system cannot tolerate this error and needs the PLL output to be phase coherent with the reference signal, the designer will have to use a Type II loop. Next we will examine the case of a Type I loop with a time varying frequency input. The reference signal is given by

$$\phi_{ref}\left(s\right) = \dfrac{A}{s^3} \qquad (8.25)$$

Substituting into the Laplace Final Value Theorem

$$\underset{t\to\infty}{\text{Lim}}\left[\phi_e\left(t\right)\right] = \underset{s\to0}{\text{Lim}}\left[s\,\dfrac{1}{1+\dfrac{K_\phi K_v}{sN}}\,\dfrac{A}{s^3}\right] = \infty \qquad (8.26)$$

What this indicates is that the phase error signal is continually increasing. The Laplace Final Value Theorem can be applied to any Type of PLL and Table 8.1 is a quick reference to the system phase error within Type I and Type II loops.

TABLE 8.1   System Phase Error

Input Signal ϕ_{ref}	Type I	Type II
Phase	0	0
Frequency	Constant	0
Time varying frequency	Continually increasing	Constant

8.5.4   Type II Third-Order Loop

With the introduction of an integrator and a single pole RC as the lowpass filter, F(s), the PLL becomes a Type II third order system. A popular configuration of the loop filter is shown in Fig. 8.8, which may be used with a proportional or pseudo-differential phase detector. These types of phase detectors will be discussed later in this section. In order to prevent slew rate limiting in the amplifier, capacitor C1 is commonly added to realize a single pole in front of the amplifier.

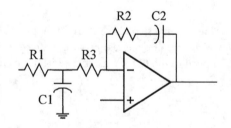

FIGURE 8.8   Integrator loop filter design.

The transfer function for the loop filter given in Fig. 8.8 is given by

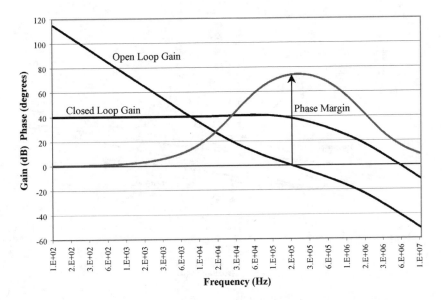

FIGURE 8.9 Type II second-order transfer functions.

$$F(s) = \left[\frac{sR_2C_2 + 1}{s(R_1 + R_3)C_2} \right] \left[\frac{1}{s\left[\frac{R_1R_3}{R_1 + R_3} \right]C_1 + 1} \right] \qquad (8.27)$$

The zero is added to the filter transfer function to pull the phase margin up toward 90°. By substituting Eq. (8.27) into Eq. (8.16), the closed-loop transfer function becomes

$$A_{CL}(s) = \frac{K_\phi K_v (R_2C_2s + 1)}{(R_1 + R_3)C_1C_2s^3 + (R_1 + R_3)C_2s^2 + \dfrac{K_\phi K_v R_2 C_2}{N}s + \dfrac{K_\phi K_v}{N}} \qquad (8.28)$$

Figure 8.9 shows the closed-loop gain, the open-loop gain and phase margin of a type II, third-order loop. The additional integrator causes the phase response to start from 0°. As mentioned earlier, the zero was added to the filter to move the phase response toward 90°. Due to the additional integrator, the usable loop bandwidth is narrower than the Type I PLL. The benefit of using the integrator cannot be seen in the closed-loop response but will be shown later in the section on filter design. Typically to achieve the necessary attenuation of the reference sidebands, the PLL needs to be designed with a higher order.

8.5.5 Higher Order Loops

While the above examples serve well to further the understanding of PLLs, practical requirements often drive the designer to higher order loops. To make the proper trade-off between settling time and spurious signals at the PLL output, higher order filters are often necessary to minimize the amount of phase shift and maximize the amount of reference spur attenuation. Higher order filters have a steeper attenuation characteristic thereby achieving less phase shift. Figure 8.10 illustrates this by plotting filter attenuation

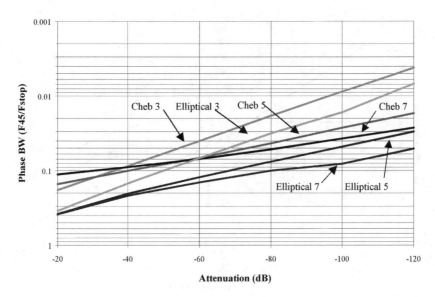

FIGURE 8.10   Lowpass filter phase shift comparisons.

vs. the phase bandwidth (i.e., filter's frequency at which the phase response is equal to 45° divided by the frequency of the stop-band attenuation).

The system specifications, in some applications, require that the noise from the PLL meet a certain shape factor. The noise sources from within the loop can be tailored by the design of the loop's lowpass filter (e.g., dual stop-band filter). The noise shaping requirements typically forces the design to use high order filters.

8.6   Phase Noise

The phase noise performance of a PLL is a critical parameter in the design of any system. The phase noise model developed in this chapter is directed toward the identification of the major contributors to the overall phase noise in a PLL, and evaluating the relative contributions of the significant sources that contribute to the output power spectral density. In certain cases, the source of noise in the loop can be pinpointed, but often it is difficult to characterize the noise precisely enough to make the necessary trade-offs. Most PLL designers are familiar with the different sources of noise that exist; in particular these include the frequency standard, VCO, frequency divider, phase detector, and integrator/low pass filter (i.e., active filter).

The spectral characteristics of the oscillators (i.e., VCO and frequency standard) have been modeled in the past and are relatively well understood. Mathematically, an ideal sinewave can be described by the following equation

$$V(t) = V_o \sin(\omega t) \qquad (8.29)$$

where V_o is the nominal amplitude and ω is the carrier frequency expressed in radians/second. In the real world, the sinewaves have error components related to both the phase and amplitude. A real sinewave signal is better modeled by

$$V(t) = \left[V_o + \varepsilon(t)\right]\sin\left[\omega t + \Delta\phi(t)\right] \qquad (8.30)$$

where $\varepsilon(t)$ is the amplitude fluctuation and $\Delta\phi(t)$ is the randomly fluctuating phase noise term. Both of these terms, $\varepsilon(t)$ and $\Delta\phi(t)$, are stationary random processes and are narrowband with respect to ω. For the purpose of this discussion, the amplitude spectral density will be ignored since it is of negligible significance compared to the phase perturbations.

There are two types of fluctuating phase terms. The first is the discrete signal components, which appear in the spectral density plot. These are commonly referred to as spurious signals. The second type of phase instability is random in nature and is commonly called phase noise. There are many sources of random phase perturbations in any electronic system, such as thermal, shot, and flicker noise. One description of phase noise is the spectral density of phase fluctuations on a per-Hertz basis. The term spectral density describes the energy distribution as a continuous function, expressed in units of phase variance per unit bandwidth. The spectral density is described by the following equation

$$S_\phi\left(f_m\right) = \frac{\Delta\phi^2_{rms}\left(f_m\right)}{\text{measurementBW}} \tag{8.31}$$

The units of spectral density are rad^2/Hz. The U.S. National Bureau of Standards has defined the single sideband spectral density as

$$L\left(f_m\right) = \frac{P_{ssb}}{P_s} \tag{8.32}$$

where P_{ssb} is the power in one hertz of bandwidth at one phase modulation sideband and P_s is the total signal power. The single sideband spectral density, $L(f_m)$, is directly related to the spectral density, $S_\phi(f_m)$, by

$$L\left(f_m\right) \cong \frac{1}{2}S_\phi\left(f_m\right) \tag{8.33}$$

This holds true only if the modulation sideband, P_{ssb}, is such that the total phase deviation is much less than 1 radian. $L(f_m)$ is expressed in dBc/Hz or dB relative to the carrier on a per hertz basis.

For the purpose of evaluating the noise performance of the PLL, each of the functional blocks is considered noiseless and a noise signal is added into the PLL at a summing junction in from of each of the functional blocks. In Fig. 8.11, the noise sources within the PLL are shown along with the gains of the various blocks. In evaluating the contribution of each of the noise sources to the overall noise at the output of the PLL, each one will be considered alone. Since these noise sources are independent, super-position may be used to determine the phase noise at the output of the PLL. The transfer function for each of the noise sources is easily written. Once again, the transfer functions are derived using classic control theory. Two additional gain blocks have been added to the basic block diagram. The first one is the additional filter after the loop integrator as discussed in the section on higher order PLLs. The second is the inclusion of the VCO's modulation bandwidth. Dependent upon the design of the VCO's input circuitry, the VCO's tune voltage input will have a finite bandwidth within which it will respond to an input signal. The 3 dB point of this response is defined as the modulation bandwidth. If the forward gain is defined as $G(s)$ and the feedback is defined as $H(s)$, then the closed-loop gain is given by

$$A_{CL}\left(s\right) = \frac{G\left(s\right)}{1 + H\left(s\right)G\left(s\right)} \tag{8.34}$$

By applying Eq. (8.34) to the PLL, the closed-loop gain for each of the noise sources is determined. This is done to characterize the loop's overall phase noise performance. By plotting the PLL's response to the

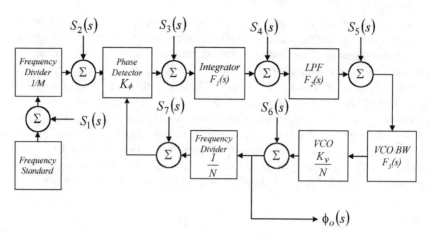

FIGURE 8.11 Phase noise sources in a PLL.

individual noise sources, the proper trade-off for the optimization of the loop's performance (i.e., phase noise and settling time) can be made. The transfer function for each of the noise sources is given in Table 8.2

The complete equation for the output phase noise of the PLL as a function of frequency is given by

$$S_\phi\left(f_m\right) = \left|A_1\right|^2 S_1 + \left|A_2\right|^2 S_2 + \left|A_3\right|^2 S_3 + \left|A_4\right|^2 S_4 + \left|A_5\right|^2 S_5 + \left|A_6\right|^2 S_6 + \left|A_7\right|^2 S_7 \qquad (8.35)$$

where the S_x's are power spectral densities that are a function of the offset frequency, f_m, from the carrier and the A_x's are a function of the complex variable s. By substituting simplified equations from Table 8.2 into Eq. (8.35), some interesting conclusions can be drawn.

$$\Phi(s) = \frac{K(s)}{s + K(s)}\left[\frac{N}{M}S_1 + N\left(S_2 + S_7\right) + \frac{N}{K_\phi}S_3 + \frac{N}{K_\phi F_1}S_4 + \frac{N}{K_\phi F_1 F_2}S_5\right] + \frac{s}{s + K(s)}S_6 \qquad (8.36)$$

First, the frequency standard noise S_1, reference frequency divider S_2, phase detector noise S_3, integrator noise S_4, loop low pass filter noise S_5, and feedback divider noise S_7 are all acted upon by a common PLL function. All these noise sources are passed through a low pass filter. The actual cutoff frequency is determined by the designer in the choosing of the various parameters that establish the PLL compensated loop bandwidth, $K(s)$. The next noise component in Eq. (8.36) to be considered is that of the VCO. The loop acts upon the VCO phase noise as if it was passed through a high pass filter. At offset frequencies that are much less than the compensated loop bandwidth, the dominant noise sources are the digital noise and frequency standard noise. At offset frequencies that are much greater than the compensated loop bandwidth, the dominant noise source is the VCO noise. As the loop's compensated bandwidth is approached, the PLL's output noise is a summation of all the noise sources. Care needs to be taken in the designing of the loop parameters, such that peaking of the noise at the loop's bandwidth doesn't occur. A general rule of thumb is that the designer would like to set the loop bandwidth at the point where the VCO phase noise crosses the digital noise. In doing so, the optimum noise performance of the overall PLL can be achieved, but this is not always possible due to settling time requirements. Next, it looks very advantageous to increase M and thereby decrease the contribution of the frequency standard phase noise. But if increasing M lowers the reference frequency, then N must increase if the overall multiplication factor to the output of the PLL is to remain the same. This actually decreases the frequency standard noise, but increases the multiplication of the phase noise contribution of the integrator, phase

Table 8.2 Phase Noise Sources Transfer Functions

Source	Transfer Function	Simplification $K(s) = \dfrac{K_\phi F_1(s)F_2(s)F_3(s)K_v}{N}$
Frequency Standard	$A_1(s) = \left[\dfrac{K_\phi F_1(s)F_2(s)F_3(s)K_v}{s + \dfrac{K_\phi F_1(s)F_2(s)F_3(s)K_v}{N}}\right]\dfrac{1}{M}$	$A_1(s) = \left[\dfrac{K(s)}{s + K(s)}\right]\dfrac{N}{M}$
Reference Divider	$A_2(s) = \dfrac{K_\phi F_1(s)F_2(s)F_3(s)K_v}{s + \dfrac{K_\phi F_1(s)F_2(s)F_3(s)K_v}{N}}$	$A_2(s) = \left[\dfrac{K(s)}{s + K(s)}\right]N$
Phase Detector	$A_3(s) = \dfrac{F_1(s)F_2(s)F_3(s)K_v}{s + \dfrac{K_\phi F_1(s)F_2(s)F_3(s)K_v}{N}}$	$A_3(s) = \left[\dfrac{K(s)}{s + K(s)}\right]\dfrac{N}{K_\phi}$
Integrator	$A_4(s) = \dfrac{F_2(s)F_3(s)K_v}{s + \dfrac{K_\phi F_1(s)F_2(s)F_3(s)K_v}{N}}$	$A_4(s) = \left[\dfrac{K(s)}{s + K(s)}\right]\dfrac{N}{K_\phi F_1(s)}$
Lowpass Filter	$A_5(s) = \dfrac{F_3(s)K_v}{s + \dfrac{K_\phi F_1(s)F_2(s)F_3(s)K_v}{N}}$	$A_5(s) = \left[\dfrac{K(s)}{s + K(s)}\right]\dfrac{N}{K_\phi F_1(s)F_2(s)}$
VCO	$A_6(s) = \dfrac{s}{s + \dfrac{K_\phi F_1(s)F_2(s)F_3(s)K_v}{N}}$	$A_6(s) = \left[\dfrac{s}{s + K(s)}\right]$
Feedback Divider	$A_7(s) = \dfrac{K_\phi F_1(s)F_2(s)F_3(s)K_v}{s + \dfrac{K_\phi F_1(s)F_2(s)F_3(s)K_v}{N}}$	$A_7(s) = \left[\dfrac{K(s)}{s + K(s)}\right]N$

detector, loop filter, and loop divider. Therefore, from Eq. (8.36) the designer would want to keep N as low as possible to keep the noise inside the loop bandwidth as low as possible.

The evaluation of the phase noise performance of a PLL can be an arduous task, but a simplified approach follows. The phase noise inside the loop bandwidth is multiplied by reference noise and digital noise. The noise outside of the loop bandwidth is primarily that of the VCO. While this certainly is an oversimplification, it is helpful in understanding the noise performance of a PLL. Writing the transfer function for each of the noise sources within the loop will help to clarify how the loop acts upon each noise source as well as what trade-off can be made in filter design for phase noise performance and settling time. Also, by looking at the individual contributions of each of the phase noise sources, the designer can determine where to focus their energy in reducing the overall phase noise performance of the PLL.

8.7   Phase Detector Design

The phase detector produces an output signal that is proportional to the phase difference between the reference input, ϕ_{ref}, and the phase of the divided down VCO signal, ϕ_o/N. The most commonly used phase detector is a phase-frequency detector. In an out-of-lock condition, the output of the phase-frequency detector latches (i.e., the AC component is removed), thereby the error signal goes to the low or high rail, depending upon the direction of phase error. The phase detector having the ability to perform frequency discrimination has greatly simplified the complexity of the PLL circuitry. In much of the literature written on PLL design, the analog phase detector is addressed. With the advent of VLSI design, the digital phase detector is most favored. There are three basic types of digital phase-frequency detectors: (1) the charge pump, (2) the proportional or pulse width modulated, and (3) the pseudo-differential.

There are many factors to consider when determining which kind of phase detector the designer should use (e.g., PLL's tuning bandwidth, settling time, power consumption, phase error, etc.), but the most prevalent phase-frequency detector has a charge pump output. The three kinds of digital phase-frequency detectors mentioned above will be discussed.

8.7.1   Charge Pump Phase-Frequency Detector

The charge pump phase-frequency detector is used predominately within the commercial PLL ASIC industry. The charge pump phase detector output is a current that has an average value equal to the system phase error.[2] In Fig. 8.12 the configuration of the PLL utilizing a charge pump phase-frequency detector is shown. One advantage of the charge pump phase-frequency detector is the reduction in complexity of the PLL. A second advantage is the ability to program the current, thereby being able to adjust the gain of the phase detector for optimum loop performance. A third advantage is that for narrow tuning bandwidth or fixed injection PLLs, only a passive filter on the charge pump output is required, thereby reducing the cost and size of the PLL. If the tuning bandwidth of the PLL is wide (e.g., octave) an op amp will usually have to be added to increase the dynamic range of the tuning voltage supplied to the VCO. A disadvantage of the charge pump phase frequency detector is the leakage current. All attempts must be made to reduce the leakage current on the charge pump's output. As the level of the leakage current increases, so will the amount of loop filtering needed to suppress the phase error spurious signals on the VCO's output.

The charge pump phase-frequency detector has three states: (1) sourcing current, (2) sinking current, and (3) high impedance or tri-state. The amount of current being sourced or sunk is defined to be I_ϕ. Since the phase detector operates over a 2π range, the gain of the phase detector is therefore $I_\phi/2\pi$. If ϕ_o/N is leading ϕ_{ref}, then the charge pump phase detector is sinking I_ϕ current, which is defined as the pull down current. If ϕ_o/N is lagging ϕ_{ref}, then the charge pump phase detector is sourcing I_ϕ current,

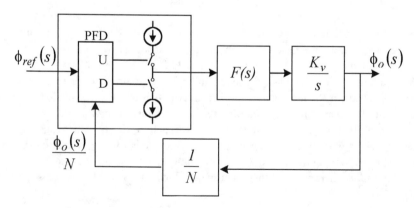

FIGURE 8.12   Charge pump phase detector PLL.

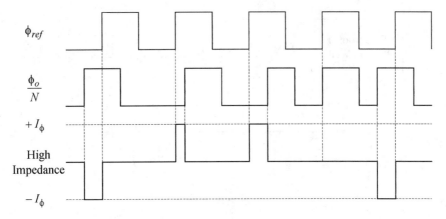

FIGURE 8.13 Charge pump phase detector output waveform.

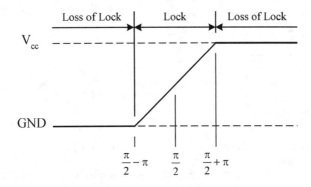

FIGURE 8.14 Proportional phase detector average voltage output.

which is defined as the pull up current. When the two input waveforms have nearly identical phase, the charge pump phase detector is tri-stated. This is illustrated in Fig. 8.13. The pull up and pull down currents must be equal for the gain of the phase detector to be linear. The charge pump current needs to be constant over the operating temperature of the system and operating voltage of the phase detector, due to the fact that the phase detector gain, and thereby the loop gain changes, are proportional to the charge pump current.

8.7.2   Proportional Phase-Frequency Detector

The proportional phase-frequency detector output is a variable pulse width rectangular wave, which has a duty cycle proportional to the phase difference between the two inputs. The phase range is 2π radians, with positive latching at either end. The gain curve of the proportional phase-frequency detector is shown in Fig. 8.14, which shows the average value of the phase-frequency detector output. The slope of the gain curve is approximately equal to V_{CC} divided by 2π, and is defined as K_ϕ. If the phase difference exceeds the $\frac{\pi}{2} \pm \pi$ range, thereby causing a loss-of-lock condition, the frequency discrimination capability of the phase-frequency detector causes the output to rail at VCC or GND, depending upon the sense of the phase error. An advantage of this type of phase-frequency detector is the gain linearity. A disadvantage of using the proportional phase-frequency detector is the amount of filtering needed to attenuate the reference sidebands, which will be discussed in further detail in the section on loop filter design. Another disadvantage is the phase error associated with a Type I loop. Of course an integrator may be added as the loop filter, thereby making the loop a Type II and removing the phase error associated with a change in frequency.

There are two common methods to extend the frequency tuning bandwidth of the PLL using a proportional phase detector, which are shown in Fig. 8.15. The first method is the use of an operational

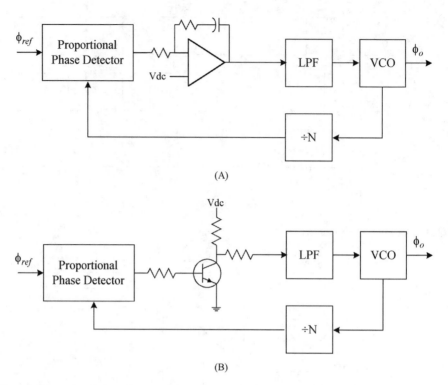

FIGURE 8.15 Proportional phase detector (A) Type II PLL and (B) Type I PLL.

amplifier as an integrator. The second is to add a discrete amplifier, which extends the tuning range with less added noise than an operational amplifier. The amplifiers effectively step up the voltage that is input to the VCO, thereby reducing the amount of VCO gain necessary to cover a given tuning bandwidth.

8.7.3   Pseudo-Differential

The pseudo-differential phase detector is so called because the phase error information is contained in two signals, which must be combined in the loop filter, or an integrator. A very popular method is to have pulses that are normally at V_{cc} and pulsing low, but it is also possible to design the phase detector such that the signals are normally at ground and pulsing high. If ϕ_o/N is leading ϕ_{ref}, then the pseudo-differential phase detector output ϕ_v is pulsing low and output ϕ_r is predominately high with very narrow pulses. If ϕ_o/N is lagging ϕ_{ref}, then the pseudo-differential phase detector output ϕ_r is pulsing low and the output ϕ_v is predominately high with very narrow pulses. When the two input waveforms have identical phase, the pseudo-differential phase detector's outputs are both high with very narrow pulses. The pseudo-differential phase detector's waveforms are shown in Fig. 8.16. An advantage to using this type of phase-frequency detector is the reduced amount of filtering needed for attenuation of the reference sidebands. Another advantage is that with the introduction of the second integrator into the loop, the phase error between the VCO output and the reference signal approaches zero for a phase or frequency change, as was shown in the previous sections of this chapter. If the PLL being designed has a phase inversion in the feedback, as is the case when a high-side injection mixer, is used in the feedback path of the loop to reduce the loop's division ratio, the outputs of the pseudo-differential phase detector will have to be flipped in order for the loop to lock. The output spurious performance of the PLL is determined by the loop filter design.

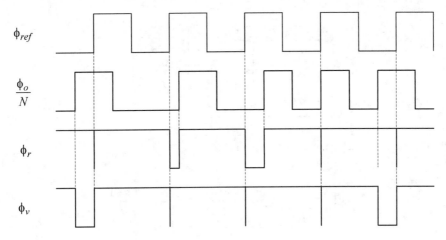

FIGURE 8.16 Pseudo-differential phase detector waveform.

8.8 Loop Filter Design

Once the designer has determined the type of PLL needed from the system requirements, the next step is to determine the configuration of the forward path of the PLL. In the design of a PLL, the level of the reference sideband spur on the VCO output is determined by a number of factors within the PLL. The amount of filtering needed for a desired reference sideband spur level can be calculated by utilizing the formulas presented in this section. Though these equations are empirical and by no means exact, they are a good starting point for determining the performance (i.e., filter attenuation) that is needed. In the basic form, the loop filter can take on three forms: (1) a lead-lag filter, (2) an active filter integrator, and (3) a charge pump passive filter. These three filters are shown in Fig. 8.17. It is common practice to add an elliptic filter to the output of these basic filters, in order to achieve the needed attenuation of the reference sidebands. The configuration of the loop filter is dependent upon the type of phase-frequency detector circuitry used.

8.8.1 Charge Pump Phase Detector

As mentioned before the charge pump phase detector has become very popular in commercial PLL ASIC applications and is covered in the following section. A typical filter used with a charge pump phase detector is shown in Fig. 8.18. In any application where wide tuning is needed, a higher tuning voltage must be supplied to the VCO. The higher tune voltage is supplied by adding an amplifier stage. A closer examination of the transfer function, $F_1(s)$, results in the determination of the component values. The transfer function of the filter, $F_1(s)$, is

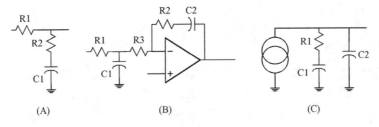

FIGURE 8.17 Loop filters: (A) lead-lag, (B) active filter integrator with zero, and (C) charge pump integrator.

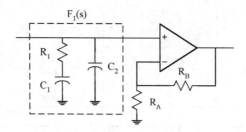

FIGURE 8.18   Basic forward path of a charge pump PLL.

$$F_1(s) = \frac{R_1 C_1 s + 1}{s(C_1 + C_2)\left(sR_1 \dfrac{C_1 C_2}{C_1 + C_2} + 1\right)} = K \frac{\tau_z s + 1}{s(\tau_p s + 1)} \tag{8.37}$$

where

$$K = \frac{1}{C_1 + C_2} \tag{8.38}$$

$$\tau_z = R_1 C_1 \tag{8.39}$$

$$\tau_p = R_1 \frac{C_1 C_2}{C_1 + C_2} \tag{8.40}$$

By plotting the transfer function, $F_1(s)$, as a function of frequency, the pole and zero break frequencies can easily be seen, as shown in Fig. 8.19.

In designing the loop filter for a specified loop bandwidth, the PLL's natural frequency will have to be adjusted by the filter gain. This will have to been done with the appropriate amount of phase margin.

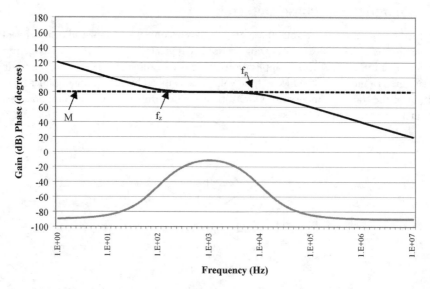

FIGURE 8.19   Transfer function of a charge pump filter.

The pole and zero location and the filter gain determine the component values for the filter.[3] The magnitude of $F_1(s)$ at the geometric mean of the pole and zero frequencies is defined as

$$M \equiv \left| F_1\left(j2\pi\sqrt{f_z f_p} \right) \right| \tag{8.41}$$

The component values for R_1, C_1, and C_2 may be calculated by the following formulas

$$R_1 = M\frac{f_p}{f_p - f_z} \tag{8.42}$$

$$C_1 = \frac{1}{2\pi f_p M} \tag{8.43}$$

$$C_2 = \frac{f_p - f_z}{2\pi f_p f_z M} \tag{8.44}$$

Once the component values of the filter, $F_1(s)$, are determined, the amount of additional filtering for attenuation of the reference sidebands will need to be calculated. It is important to note that the closed-loop transfer function will not predict the amount of filtering needed for attenuation of the reference side bands. The closed-loop response doesn't take into account all of the parameters that affect the level of the reference sideband signal on the output of the VCO. An approximate calculation of the needed filtering for a given reference sideband level may be calculated from the following formula

$$F(dB) = 20\log\left[\frac{\sin\left(n\pi \dfrac{V_{dcmax}}{R_{leak} I_{max}} \right)}{n\pi F_{ref}} m R_1 K_v I_{max} \right] + RSB(dB) \tag{8.45}$$

where n is the harmonic of the reference frequency to be filtered, V_{dcmax} is the maximum voltage of the charge pump, R_{leak} is the leakage resistance across the charge pump, I_{max} is the maximum charge pump current, F_{ref} is the reference frequency, m is either the gain or loss in the forward path (i.e., op amp/lead-lag network, etc.), and R_1 is the value of the resistor in the filter, $F_1(s)$.[4] An example of a charge pump phase detector utilizing the filter, $F_1(s)$, is given below.

Example 8.1
A charge pump phase detector utilizing the filter shown in Fig. 5.111 with the following parameters: maximum voltage of the charge pump of 5 Volts, maximum VCO gain of 40 MHz/Volt, a reference frequency of 1 MHz, a filter gain of 1, a leakage resistance of 100 kohm, a value of 200 ohms for R1, and the fundamental harmonic at a level of –70 dBc.

$$F(dB) = 20\log\left[\frac{\sin\left(\pi \dfrac{5}{1E5 \times 5E-3} \right)}{\pi \times 1E6} 200 \times 40E6 \times 5E-3 \right] + 70 = 62.04 \text{ dB} \tag{8.46}$$

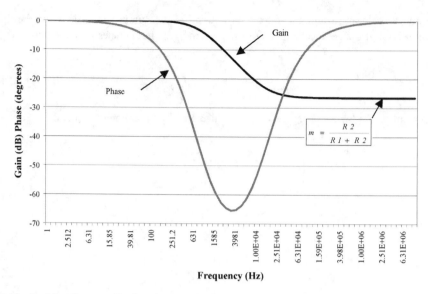

FIGURE 8.20   Lead-lag filter transfer function.

Approximately 62 dB of additional attenuation will be needed at the reference frequency of 1 MHz. If a simple single pole filter is used, the designer will have to take care that it doesn't introduce excessive phase shift at the loop bandwidth is not introduced. It is also worth noting that if the amplifier shown in Fig. 8.20 is replaced with any filter design, the loading affect of that filter will have an effect on the response of the integrator and lag network proceeding it. The designer may have to go to a higher order elliptic or Chebychev filter to minimize the amount of phase shift.

8.8.2   Proportional Phase Detector

The lead-lag filter shown in Fig. 8.17 (A) is commonly used with the proportional phase detector when a Type I loop is being designed. The transfer function for this filter is

$$F_1(s) = \frac{C_1 R_2 s + 1}{(R_1 + R_2) C_1 s + 1} \tag{8.47}$$

Using some typical values, the transfer function is plotted in Fig. 8.20. There is a zero located at

$$f_z = \frac{1}{2\pi C_1 R_2} \tag{8.48}$$

and a pole located at

$$f_p = \frac{1}{2\pi C_1 (R_2 + R_1)} \tag{8.49}$$

with the attenuation of the filter when $s \gg 1$ given as

$$m = \frac{R2}{R1+R2} \tag{8.50}$$

The use of a proportional phase detector often requires that a gain stage be added to the phase detector output in order to meet the system's frequency tuning bandwidth. This is dependent upon the gain of the VCO. The amplifier can be a discrete amplifier or an operational amplifier, dependent upon the phase noise performance needed. The level of filter attenuation needed at the reference frequency may be approximated by the following equation

$$F(dB) = 20\log\left[\frac{n^2\pi F_{ref}10^{\left(\frac{RSB}{20}\right)}}{mK_vV_{dd}}\right] \tag{8.51}$$

where V_{dd} is the maximum swing of the voltage into the LPF; V_{dd} is either directly out of the phase detector or after the added gain stage; K_v is the maximum gain of the VCO in Hz/Volt; n is the harmonic of interest; F_{ref} is the reference frequency in Hz. If a lead-lag network is used, m is the loss of the lead-lag network. RSB is the desired level of the reference side band at the VCO output. An example of a Type I loop utilizing an emitter follower amplifier with a lead-lag filter follows.

Example 8.2
A proportional phase detector in a Type I loop utilizing the lead-lag filter shown in Fig. 8.17 (A), and a discrete amplifier as shown in Fig. 8.15 (A) with a maximum voltage swing of 15 Volts, maximum VCO gain of 10 MHz/Volt, a reference frequency of 250 kHz, and a lead-lag loss of 0.047. The fundamental harmonic is desired to be at a level of –60 dBc, resulting in a filter needing a rejection of

$$F(dB) = 20\log\left[\frac{\pi \times 2.5E5 \times 10^{\left(\frac{-60}{20}\right)}}{0.047 \times 10E6 \times 15}\right] = -79.1 \tag{8.52}$$

In order to achieve –60 dBc sidebands, the filter following the discrete amplifier must have –79.1 dB of attenuation at 250 kHz. The next step is to determine the kind of low pass filter to be used following the lead-lag filter. The additional filter must meet the attenuation and minimize the amount of phase shift at the open-loop bandwidth. A chart, to aid in this decision is shown in Fig. 8.10, where the ratio of the filter's 45-degree point to frequency stop vs. the filter attenuation is plotted. As can be seen in Fig. 8.10, the minimum amount of phase shift for a –80 dB filter is a 7-pole elliptic.

8.8.3 Pseudo-Differential Phase Detector

A typical topology of the integrator used with the pseudo-differential phase detector (PDPD) is shown in Fig. 8.21. An advantage of the PDPD is the amount of attenuation needed following the integrator. A disadvantage is the effect the integrator has on the settling time. Because, the charge on the integrator capacitor, C_2, needs to change for every new output frequency of the PLL, dielectric absorption can increase the settling time. The use of this type of integrator is a common approach when the VCO has to have a wide frequency tuning bandwidth. The transfer function of this filter was

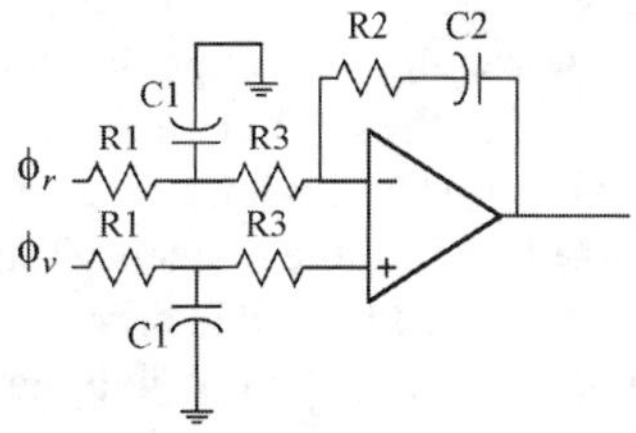

FIGURE 8.21 Pseudo-differential phase detector integrator.

given earlier in Eq. (8.27). The amount of filtering needed following the integrator can be calculated approximately by the following equation

$$F\left(dB\right) = 20\log\left[\frac{nF_{ref}10^{\left(\frac{RSB}{20}\right)}}{mK_vV_{off}}\right] \tag{8.53}$$

where V_{off} is the offset voltage of the op amp, K_v is the gain of the VCO in Hz/Volt, n is the harmonic of interest, F_{ref} is the reference frequency, m is the gain of the op amp, and RSB is the desired level of the reference side band at the VCO output.

Example 8.3
A PDPD in a Type II loop utilizing the circuit shown in Fig. 8.21 with a maximum op amp input offset voltage of 5 mVolts, maximum VCO gain of 10 MHz/Volt, a reference frequency of 250 kHz, an op amp gain of 1, and a desired fundamental harmonic at a level of –60 dBc. This results in a filter needing

$$F\left(dB\right) = 20\text{Log}\left[\frac{2.5E5 \times 10^{\left(\frac{-60}{20}\right)}}{20E6 \times 5E-3 \times 1}\right] = -46.0 \tag{8.54}$$

approximately 46 dB, of additional attenuation at the reference frequency of 250 kHz. In comparison to the example utilizing the proportional phase detector, the use of the integrator has reduced the amount of filtering needed by 33 dB.

8.9 Transient Response

So far the assumption has been made that the PLL is operating in a steady state. What is the PLL's response when a disturbance is introduced or what is the transient response? In most of the articles written on PLL design, a second order approximation is used to model the PLL's transient response to a change in phase or frequency. The most common is a change in the loop's division ratio to bring the PLL to a new frequency output. This change has two effects on the loop. One is, because the closed-loop response is dependent upon the value of the feedback divider, N, the loop bandwidth changes. The PLL has to acquire phase lock to the new VCO output frequency, which is known as the transient response. Of course, depending upon how large of change in frequency, the gain of the VCO will also affect the loop bandwidth. With the mathematical modeling software tools available to the designer today, the second order approximation is unnecessary, but is still useful in understanding the basic loop transient response.

In most systems, the PLL's transient phase error is the response of interest. Albeit many system specifications define the frequency error, Phillips[5] has shown the phase settling characteristic corresponds to the system performance better than the frequency settling characteristics. The transient phase response is the phase difference between the final value of the VCO and a steady-state signal, which has the same phase as the final value of the VCO. In the laboratory, the transient phase response is measured by mixing a signal generator and the VCO's output signal, both of which need to be phase-locked to the same frequency standard. The output of the mixer shows the phase difference and needs to be lowpass filtered in order to remove the summed output. As an example, Fig. 8.22 shows the result of a simulation of a PLL hopping between two frequencies. During the first hop, the loop has too little phase margin and there is excessive ringing. The second hop shows the phase response with a phase margin of 70°.

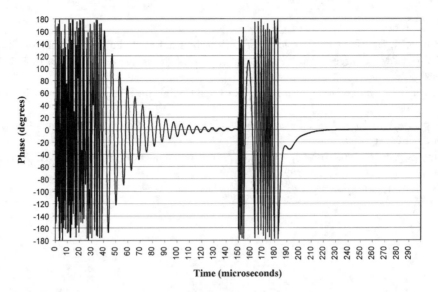

FIGURE 8.22   Phase settling response of a Type I second-order PLL.

There are many factors within the loop that can affect the transient response. The nonlinearity of the VCO gain, which typically is less at the high end of its tuning range, will affect the loop bandwidth and thereby the phase margin. The frequency range and modulation bandwidth of the VCO will impact the transient response. Prepositioning of the VCO's control voltage, either with a digital control word into a DAC or by summing another PLL's tune voltage, can be done to reduce the amount of overshoot and thereby reduce the settling time. To prevent additional noise from being introduced into the PLL, careful design of the prepositioning circuitry must be done. The discrete or sampling nature of the phase detector and divider need to be considered in wide loop bandwidth designs. The continuous time approximation is typically used in transient modeling, but a number of papers have been written that use z-transforms to model the loop's response.[6] Once again, if the loop bandwidth is kept low relative to the reference signal frequency, the discrete nature of the phase detector can be ignored. When a charge pump phase detector is used, the mismatch between current sources, I_ϕ and $-I_\phi$, the leakage current of the charge pump as well as the leakage current of the components used in the loop lowpass filter and board parasites, all have an effect on the transient response. If an operational amplifier is used in the loop filter design, the slew rate and voltage limits will impact the transient response.

8.10 Conclusion

In this chapter we have considered some of the fundamental design considerations that go into the design of a PLL. The design process is made up of a series of trade-offs (e.g., wide loop bandwidth for improved settling time, but narrow loop bandwidth for improved noise performance). There cannot be enough emphasis placed on the robustness of the PLL design. The designer needs to ensure that there is enough margin in the design parameters such that the loop works well over its operating temperature and other environmental conditions, as well as component tolerance. Modeling plays a key role in the development of the PLL and it is imperative that the designer have an understanding of those models and the limitations inherent in any mathematical model. It is a common practice to model higher order PLLs as second order systems to simplify the design process, but with the computer-aided design tools available to the designer today, this is an unnecessary simplification. When the designer is challenged by the system requirements, it is necessary to have the most comprehensive model possible.

References

1. Egan, W.F., *Frequency Synthesis by Phase Lock*, Robert E. Krieger Publishing, Malabar, FL, 1990.
2. Gardner, F.M., Charge-Pump Phase-Lock Loops, *IEEE Trans. Comm.*, COM-28, 1849–1858, Nov. 1980.
3. Opsahl, P.L., *Charge-Pump Filter Design*, Rockwell Collins, Inc., Frequency Control Team Internal Paper, 1999.
4. Mroch, A.B., *Charge-Pump Filter Attenuation*, Rockwell Collins, Inc., Frequency Control Team Internal Paper, 1999.
5. Phillips, D.E., Settling Time Specifications: Phase or Frequency?, *IEEE Military Communications Conference,* 3, 806–810, Oct. 1994.
6. Crawford, J.A., *Frequency Synthesizer Design Handbook*, Artech House, Inc., Boston, 1994.

9

Filters and Multiplexers

Richard V. Snyder
RS Microwave

9.1 Introduction

"Filter: a device or material for suppressing or minimizing waves or oscillations of certain frequencies"… per Webster. This definition, while accurate, is insufficient for microwave engineers. Microwave systems and components enhance and direct, as well as suppress waves and oscillations. Components such as circulators, mixers, amplifiers, oscillators, switches (in common with most complex systems) are in fact filters, in which inherent physical properties are represented as smaller, constituent networks embedded within larger "filtering" (response-determining) structures or systems. Inclusion of the concept of embedding is thus central to understanding microwave filters. To clarify "embedding": the terminals of a well-defined subnetwork are provided with a known interface to the rest of the system, thus providing selective suppression or enhancement of some oscillatory effect within the device or system. The subnetworks can be linear or nonlinear, passive or active, lumped or distributed, time dependent or not, reciprocal or nonreciprocal, chiral (handed) or nonchiral, or any combination of these or other properties of the basic constituent elements. The subnetworks are carefully defined (or characterized) so that the cascade response of a series of such subnetworks can be predicted (or analyzed) using software simulation tools employing a variety of methods, such as linear, harmonic balance, Volterra series, finite-element, method of moments, finite-difference time domain, etc. The careful definition normally involves the process called "synthesis," in which the desired response to a particular stimulus suggests a topological form for the subnetwork, followed by extraction of the specific elements of the subnetwork. The computed response of the synthesized network is compared to the desired response, with iteration as necessary using repeated synthesis or perhaps an optimization loop within the simulation tool. Hybrid combinations of these two iterative approaches are possible.

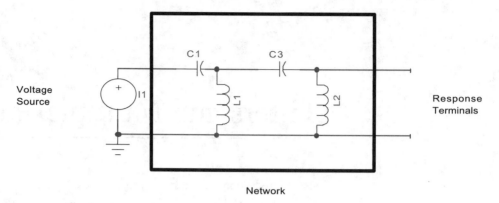

Network

FIGURE 9.1

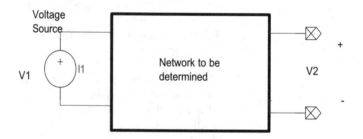

FIGURE 9.2

9.2 Analysis and Synthesis

The difference between prediction or "analysis" and definition or "synthesis" can be summarized as follows. The word analysis comes from the Greek *lysis*, a loosening, and *ana*, up; hence a loosening up of a complex. Synthesis, on the other hand, means the building up of a complex from parts or elements to meet prescribed excitation-response characteristics. Fig. 9.1 illustrates an example of the excitation network and response.

Another difference between analysis and synthesis must be considered. There is always a unique solution for an analysis, although it might be hard to find. Synthesis, on the other hand, might result in several networks with the specified response, or possibly no solution whatsoever. In general, solutions are not unique but some might be more realizable than others. Fig. 9.2 presents the general problem of synthesis. What combination of elements in Fig. 9.2 will give the prescribed response? It is important to realize that with a *finite* number of elements, in general the required response *cannot* be realized at all. Functions having a required variation over some band of frequencies and zero value for all other. frequencies cannot be represented by a rational function of the form of a quotient of polynomials. Thus, it is necessary to modify the response requirements to include some *tolerance*.

Figure 9.3 illustrates the imposition of a tolerance, or acceptable difference, between the desired response and the resultant response, for a particular synthesized characteristic. The approximation can take many possible forms. The approximations might require the magnitude squared of the voltage ratio to be the quotient of rational and even polynomials in ω. A typical quotient of polynomials might be as given in Fig. 9.3.

$$|G(\omega)|^2 = \frac{a_0\omega^6 + a_2\omega^4 + a_4\omega^2 + a_0}{b_0\omega^6 + b_2\omega^4 + b_4\omega^2 + b_0} \tag{9.1}$$

This is not a unique polynomial for realization. The are any number of other polynomial ratios of lower or higher degrees that may be used. The higher the degree of assumed polynomials, the better the approximation to the desired response, the smaller the tolerance region. The coefficients of Eq. (9.1) are determined by the solution of a set of simultaneous linear equations, as many simultaneous equations as there are unknown coefficients. In general, we cannot match the desired response characteristic at all points in the spectrum; rather, we must choose to match exactly at certain points and approximately over the remainder of the tolerance region. We can choose to match the points, the derivatives, or use other criteria that will be discussed herein. The response shown in Fig. 9.3 is amplitude vs. frequency. Typically, a network also has a desired time vs. frequency response. Generally these two requirements are interrelated and may not be specified independently. In very important classes of approximation, the amplitude and time responses are connected by the Hilbert transform, and thus to know either the amplitude or the time response is sufficient to enable determination of the other. This will be discussed in the section on Approximations. The polynomial in Eq. (9.1) has a simple dependence upon frequency ω (representing lumped elements in the network). Generally, microwave filters include distributed elements such as quarter-wave resonators, which display response characteristics dependent upon transcendental functions, such as $\tan(\omega)$. The resultant synthesis polynomials require more specialized techniques for element extraction. It is possible that real networks will include both lumped and distributed elements, with very complex transfer function polynomials. The synthesis process can thus be quite complex, and a comprehensive coverage is beyond the scope of this article.

9.3 Types of Transfer Function

Before approximating a particular transfer function, one must determine the type of transfer function desired. These functions can be defined in terms of amplitude or time, expressing either as functions of frequency. It is convenient to initially concentrate on amplitude vs. frequency transfer characteristics. Table 9.1 presents the four available types.

It should be understood that the above transfer functions in Table 9.1 are defined between any pair of input-output ports of a potentially multiport network. Transfer functions so defined are known as "two-port" transfer functions. We will initially restrict our efforts to such two-port circuits.

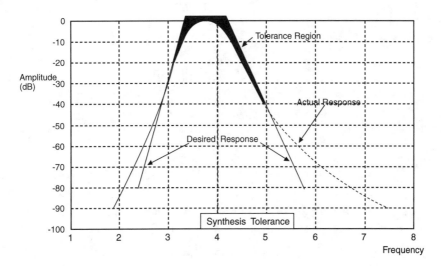

FIGURE 9.3

TABLE 9.1   Transfer Function Types

Transfer Function Type	Characteristics
Lowpass	Low loss region approximated over some bandwidth, prescribed rejection achieved at frequency some distance from the *highest* frequency in the low loss approximation region (Fig. 9.4a)
Highpass	Low loss region approximated over some bandwidth, prescribed rejection achieved at frequency some distance from the *lowest* frequency in the low loss approximation region (Fig. 9.4b)
Bandpass	Low loss region approximated over some bandwidth, prescribed rejection achieved at frequencies some distance from *both* the *highest and lowest* frequencies in the low loss approximation region (Fig. 9.4c)
Bandstop	Low loss approximated over two regions, extending downward toward DC and upward toward infinity. Prescribed rejection achieved at frequencies some distance *above* the lower low loss region and *below* the upper low loss region (Fig. 9.4d)

9.4   Approximations to Transfer Functions

It is not possible to achieve the flat passbands and abrupt transitions illustrated in Fig. 9.4 without using an infinite number of elements, each with zero resistance. We will discuss the properties of elements used to realize filters in a later section, but certainly the "Q" of available elements is less than infinity. Thus, some approximation to the idealized transfer functions must be made in order to implement a filter network falling within the allowable tolerance shown in Fig. 9.3. Essentially, the approximation procedure is directed toward writing mathematical expressions that approximate the ideal forms shown in Fig. 9.4. These expressions include polynomial functions that are substituted into the left side of Eq. (9.1) prior to element extraction. Some of the most common approximations will now be discussed. We will treat approximations to the amplitude response in some detail, and will briefly touch on approximations to phase or time delay.

9.4.1   Butterworth

The response function given by Eq. (9.2) is known as the nth order Butterworth or maximally flat form.

$$\left| G_{12}\left(j\omega \right) \right| = \frac{1}{\sqrt{1+\omega^{2n}}} \tag{9.2}$$

From binomial series expansion

$$\left(1 \pm x \right)^{-n} = nx + \frac{\left(n+1 \right)x^2}{2!} \mp \frac{n\left(n+1 \right)\left(n+2 \right)x^3}{3!} + \ldots, x^2 \le 1 \tag{9.3}$$

We see that near $\omega = 0$

$$\left(1+\omega^{2n} \right)^{-1/2} = 1 - 0.5\omega^{2n} + 0.375\omega^{4n-} 0.313\ \omega^{6n+\cdots} \tag{9.4}$$

and from this expression, the first 2n − 1 derivatives are zero at $\omega = 0$. Thus, the magnitude

$$\left| G_{12}\left(j\ 1 \right) \right| = 0.707 \text{ for all values of n.} \tag{9.5}$$

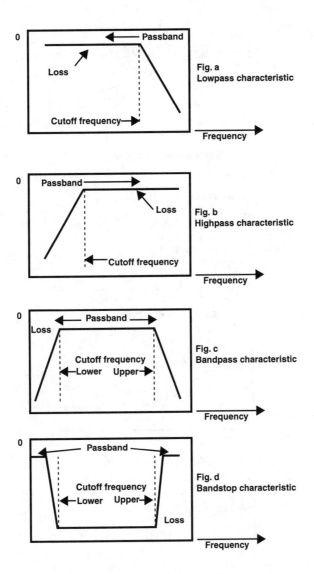

FIGURE 9.4

The pole locations corresponding to the Butterworth response may be determined using analytic continuation of the binomial series expansion above. The poles of this function are defined by the equation

$$1 + \left(-s^2\right)^n = 0. \tag{9.6}$$

The poles so defined are located on a unit circle in the s plane and have symmetry with respect to both the real and the imaginary axes. Only the left half plane poles are used to form what is known as the all-pole response function that will yield the response required in Eq. (9.1).

The form of the Butterworth response is shown in Fig. 9.5a for several values of n. The 2n − 1 zero derivatives ensure the "maximally flat" passband characteristic.

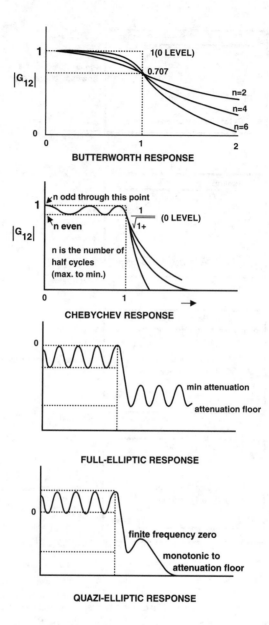

FIGURE 9.5

9.4.2   Chebychev

If a rippled approximation to the passband region of the ideal transfer function is acceptable, one can use the expression

$$|G_{12}|^2 = \frac{1}{1+\varepsilon^2 C_n^{2\omega} 2} \tag{9.7}$$

where $C_n(\omega)$ is the nth order Chebychev polynomial and $\varepsilon < 1$ is a real constant. These polynomials are defined in terms of a real variable ω as follows:

$$C_n\left(\omega\right) = \cos\left(n \cos^{-1} \omega\right) \tag{9.8}$$

The response of a Chebychev-approximated transfer function is shown is Fig. 9.5b. Analytic continuation can again be used to locate the poles, which will be found to be distributed on an ellipse, major and minor axes, respectively, the imaginary and real axes of the s-plane, s the normal Laplace transform variable $\sigma + j\omega$. (Remember that the Butterworth poles were distributed on a circle, same axes.) The reader is referred to many standard reference works for the details of element extraction, but the response of the Chebychev approximation will generally provide less passband performance but will achieve specified levels of stopband attenuation more quickly than the Butterworth approximation discussed previously. Butterworth and Chebychev transfer functions can be realized using single or resonant elements, with coupling only between adjacent elements. As such, the physical form for the network has the appearance of a ladder (see Fig. 9.1) and these circuits are known as "ladder" networks.

9.4.3   Elliptic Approximation

If one can utilize rippled approximations to both the amplitude of both passband and stopband regions, the resultant filter characteristics will display somewhat better passband performance coupled with steeper attenuation slopes, as compared to the Chebychev, but with attenuation slope characteristics with a level set on the minimum value of stopband. The response is shown in Fig. 9.5c. The stopband region contains finite-frequency transmission zeros. Filters designed with elliptic responses are derived from expressions containing elliptic functions. Such filters are sometimes termed "full-elliptic" or "Cauer parameter." They can be derived from many starting points, but it is important to note that for bandpass cases, narrow passbands are hard to achieve and for bandstop, narrow stopbands are difficult. Designs of this type require extra resonant elements and sometimes coupling between nonadjacent resonators. Typically, such designs are used to achieve specified minimum stopband levels in close proximity to the passband (5 to 10% away in frequency).

9.4.4   Quasi-Elliptic Approximation

This class of function is achieved with rippled approximation to the passband amplitude and a limited number of ripples (finite frequency transmission zeros) in the stopband. Typically, the filter stopband slope displays monotonicity beyond the few ripples. Filters in this category can achieve the improved passband response of elliptic designs, with stopband performance almost as steep as an elliptic, and also with maximum stopband levels considerably improved as compared to the full-elliptic approach. If the extra zeros are real axis, rather than real frequency, the filter will display improved passband flatness and more constant group delay. Filters in this category can be what is known as "non-minimum-phase," as the extra zeros can be located in the right half plane. The general response type is illustrated in Fig. 9.5d. These filters usually require coupling between nonadjacent resonators (sometimes this is called "cross-coupling"), but do not need the extra resonant elements nor do they display the realization difficulties of full-elliptic designs. Although this design approach has been known since the 1970s, recent advances in simulation tools and synthesis techniques have resulted in the emergence of this category as the "cutting edge" in filter design. Such filters are also known as "pseudo-elliptic."

9.4.5   Other Approximations

The aforementioned approaches have all started with approximating the amplitude portion of the transfer function. In some cases, it is desirable to approach the delay or phase as a function to be approximated. In the ladder-derived filters above, to know the amplitude is to have determined the phase/delay (and contrawise). In more complex structures (cross-coupled) it is possible to have some degree of control over both amplitude and phase/delay. It is also possible to adjust the amplitude or phase/delay properties with the cascade of an additional circuit, known as an "equalizer." When the adjustments are internal,

TABLE 9.2

Approximation Type	Salient Characteristics
Butterworth	Lowest loss at center frequency, bandwidth is defined as –3 dB relative to center, maximally flat (no ripples) in passband, stopband slope depends on order, good group delay performance, generally easy to realize as ladder structure. Poles distributed on a unit of the s-plane (Laplace variable space). This is an all-pole filter.
Chebychev	Rippled approximation to passband with reflections proportional to ripple level and higher than Butterworth, for the same order. Stopband slope depends on order, but steeper than Butterworth for the same order, poor group delay in passband (not constant), generally easy to realize as ladder structure. Poles distributed on an ellipse, major axis being the imaginary axis of the s-plane, with all poles within the unit circle of the s-plane. The ellipse intersects the Butterworth pole circle at two points on the imaginary axis. This is an all-pole filter.
Elliptic	Rippled approximation to both passband and stopband. Passband reflection proportional to ripple level, stopband slope proportional to stopband ripple level. Finite-frequency transmission zeros supplement the amplitude response attributable to the poles and affect the phase response. Ratio between stopband and passband widths can be smaller than for all-pole designs.
Quasi-Elliptic	Rippled approximation to passband, small number of ripples in stopband. Finite frequency or real-axis (imaginary frequency) zeros can be achieved, to emphasize either attenuation slopes, passband flatness and delay, or both to some extent. Easier to realize for narrow bandwidths than full-elliptic designs. Physical structure requires coupling between nonadjacent resonators and thus folding of the structure, for microwave implementations.
Max flat time delay (Bessel)	Analogous to Butterworth, but with the time-delay (group delay) function vs. frequency containing $2n-1$ zero derivatives. Poles lie on an ellipse-like path outside the s-plane unit circle (as contrasted to the Chebychev locations inside the unit circle). Provides flat, constant group delay within the passband but poor attenuation slopes.
Gaussian	A Taylor-series approximation to a Gaussian magnitude function $\left|G_{12}(j\omega)\right| = e^{-\omega 2/2}$ is used to extract filters that have optimum transient overshoot characteristics and thus display minimum ringing when excited by a pulsed input signal. The group delay is not as flat as that of the Bessel design nor are the stopband slopes as steep. In common with Bessel designs, the filter will display high reflections at frequencies away from center frequency.
Transitional	These are filters with transfer functions between those defined by the classical polynomials such as Chebychev, Butterworth, etc. An example is Gaussian-Chebychev.

as in the cross-coupled cases, the equalization is known as "self-equalized." In this case, the extra transmission zeros are located on the s-plane real axis (imaginary frequency). If the additional circuit is used for equalization of either amplitude or phase/delay, the descriptive term is "externally equalized." Some of the more common approximations are summarized in Table 9.2.

9.5 Element Types and Properties

There are many ways to classify the available elements. Perhaps the most basic is to characterize the element as "passive" (no D.C. required) or "active (D.C. required). Within the passive regime, classification includes "lumped," "distributed," "non-reciprocal" (and "reciprocal"), and combinations in which a particular element can display more than one of these characteristics.

"Lumped" elements are those that present capacitive, inductive, resistive, or gyrator responses. The element impedance is essentially a function of ω. Typically, the enumerated elements are predominantly capacitive, inductive, etc. but will also display bits of the other possibilities. For example, at low frequencies, the lead inductance of a capacitor is not important, but as frequency increases, the inductive reactance becomes a significant fraction of the element impedance, until at some frequency the capacitor behaves as a resonant circuit. It is possible to design networks with lumped element concepts all the way up to 100 Ghz or so, but the usual limitation is below 10 GHz. Lumped circuitry usually displays an intrinsically lowpass behavior *above* some frequency.

"Distributed" elements have impedance properties which are functions of tan (ω) or tanh (ω). These include quarter wavelength (or non-quarter wavelength) TEM mode resonators, waveguide resonators, cavities, dielectric resonators, and any structure built using essentially length-dependent techniques (as contrasted to length-independent but position-dependent lumped element circuits). The frequency range for "distributed" elements ranges from a few MHz to the terahertz range. Waveguide elements have the property that internal wavelength is not linearly related to actual free-space wavelength, and are thus termed "dispersive." Such elements display an intrinsically highpass response *below* a frequency known as the cutoff frequency (energy cannot freely propagate through the section below this frequency).

"Non-reciprocal" elements contain ferrimagnetic structures (circulators, isolators, various gyrators) and are used in conjunction with other elements. An additional hybrid element of interest is formed by a resonated short length of below-cutoff waveguide or other dispersive structure, and is termed an "evanescent" section. Such a below-cutoff section presents an essentially inductive equivalent circuit and can be resonated with a capacitor. The result is formation of a high-Q resonant circuit that can be embedded into a variety of filter circuits. These elements have impedance characteristics similar to lumped elements at frequencies well below cutoff, and similar to distributed, near cutoff.

Unloaded Q is an important property of any circuit element, and is a measure of the ability of the element to store energy without dissipation. High Q means low loss and is a desirable property. The above elements can be fabricated using superconductive material to obtain remarkably high unloaded Q values. Table 9.3 summarizes the properties of many common circuit elements. Application depends on various factors, including basic electrical specification, and ambient environment with concomitant difficulties associated with temperature, humidity, vibration, and shock. In general, lumped elements must be potted in place. Distributed and lumped elements have natural changes in impedance or resonant frequency as functions of temperature, which must be compensated using elements with opposite drift properties. Filters can be built that will be stable to no worse than 1 ppm per degree Centigrade, without the need for external stabilization. Vibration and shock must be damped or isolated from the circuitry. Humidity will affect resonant frequency as well as degrade performance over time. Typically, filter circuits are sealed to eliminate the presence of moisture and to prevent the intrusion of moisture as temperature changes, in a humid environment. Salt will degrade performance and must be eliminated through sealing

TABLE 9.3

Element Type	Frequency Range	Unloaded Q	Implementation
Inductor, lumped	Almost DC to 100 GHz	50–300 at room temperature, 1000s if superconductive	Coils (air and ferrite-loaded), helices, printed, shorted stubs, evanescent waveguide
Capacitor, lumped	To 100 GHz	50–1000 at room temperature, 1000 if superconductive	Multilayer, single layer, open stubs, coaxial
Resistor, lumped	DC–5 GHz	N/A (parasitic capacitance and inductance can be problems)	Metal, composition, chips
Stub or line, printed, TEM, suspended substrate, coplanar stripline, coplanar waveguide, finline, coaxial, other TEM or almost TEM lines	DC–100 GHz	100–500 at room, 1000 if superconductive	Microstrip, stripline, finline, CPS, CSS, SSS.
Evanescent	200 MHz–90 GHz	300–10,000 at room (no data on superconductive application	Below cutoff waveguide of various aspect ratios, machined sections resonated using various capacitive schemes
Dispersive (guided but non-TEM modes)	100 MHz to terahertz	1000–20,000 at room, 100,000 or more if superconductive	Waveguide, air or dielectric filled cavities with metal walls, dielectric resonators, multimode

and special plating systems. In general, filters can be built that combine the properties of the various lumped and distributed element types. This is a difficult, if not impossible synthesis problem but with the modern simulation and optimization tools, such globally-designed networks are practical and offer the optimum in electrical and environmental performance with associated production cost reduction. The cost reduction stems from the fact that performance over the full range of ambient environment can be predicted, with sensitivity to production tolerances easily taken into account prior to "cutting metal." Tolerances are thus fit to the problem at hand, with proper care taken and waste minimized.

9.6   Filter Implementations

Figure 9.6 presents a Filter Selection Guide applicable to current technology.

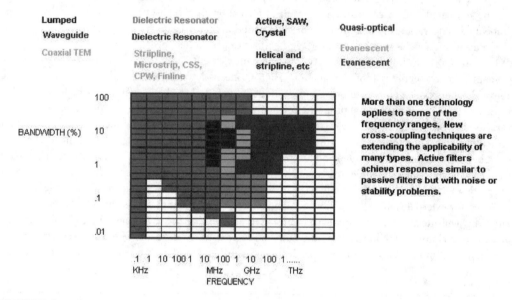

FIGURE 9.6

9.6.1   Multiplexers

The interconnection of more than one filter at a common junction results in a network termed a "multiplexer." With one common port and two individual ports, we have a "diplexer." With three individual ports, a "triplexer," and so forth through quadruplexer, quintaplexer, sextaplexer, etc. The individual networks can be lowpass, highpass, bandpass, or bandstop. The common connection presents significant difficulty, as without proper precaution, the interaction between the individual filters causes severe degradation of the desired path transfer function.

Many techniques have evolved for performing the interconnection. A multiplexer is normally used if a wide spectrum must be accessed equally and instantaneously. Conventionally, multiplexers have had the disadvantage of requiring at least 3 dB excess loss ("crossover" loss) at frequencies common to two channels. Thus, the passband characteristics for contiguous structures always showed an insertion loss variation over the passband of at least 3 dB. To construct any multiplexer, it is necessary to connect networks to the constituent filters such that each filter appears as an open circuit to each other filter (see Fig. 9.7). While this is simple for narrowband channels, it is difficult for broadband or contiguous filters. Normally, the filters and the multiplexing network are synthesized as a set, with computer optimization being used to simulate the results before construction begins. Some of the more common multiplexing techniques include line lengths, circulators, hybrids, and transformers.

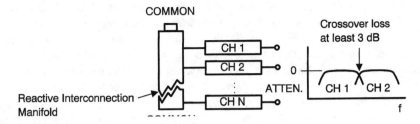

FIGURE 9.7

More recently, the multiplexer filter channels have been combined using power dividers (Fig. 9.8). This recent adaptation of always-available technology is due to newly available cheap and compact amplifier stages. Such gain blocks provide flat gain and low noise over wide bandwidths. In the case of two-way combining, conservation of energy means that the 3 dB insertion loss is still experienced, but on a flat-loss basis. Although each channel is subject to the additional 3 dB loss, it is essentially constant loss over each channel and thus the excess passband loss variation is less than 1 dB. Excess loss is defined as that loss not attributable to the individual channel filter roll off. This power divider based combining can be extended to triplexers (4.7 dB flat loss), quadruplexers (6 dB flat loss), etc. Because the loss variation is minimized, the overall insertion loss can frequently be made up using amplifiers, which display flat gain vs. frequency.

Filters can be multiplexed by parallel combination at both ends. For example, if two bandpass filters are multiplexed at both input and output, a network results that provides one input and one output, with two passbands essentially attenuating everything else. Such assemblies are useful in systems such as GPS that have two or more operating frequencies, with the requirement for isolation between the operating channels and adjacent, cluttered regions of the spectrum (Fig. 9.9). Another approach employs switched selection of

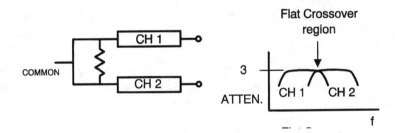

FIGURE 9.8

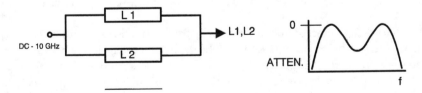

FIGURE 9.9

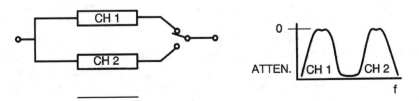

FIGURE 9.10

filters. Hybrid combinations using multiplexers with power dividers, switches, and amplifiers are now possible (see Fig. 9.10). The interactions of these essentially reactive components can cause undesirable degradation of stopbands or passbands, if precautions are not taken. Available computer simulation techniques are sufficiently sophisticated that accurate prediction of performance and dimensions minimizes the time required to develop and deliver such complex assemblies. Interconnection of subcomponents or submodules within multiplexers is sometimes difficult, with parasitic lengths causing degradation of performance. Although the computer can predict these problems, sometimes the parasitics reach levels for which compensation cannot be effected. It is possible to use blind-mate interconnection of submodules to minimize both parasitic interconnections and spurious crosstalk. Thus, the physical structure, including all interactions, can be predicted accurately and the unacceptable interactions and crosstalk eliminated using the mechanical elegance and electrical isolation of blind-mate internal connections.

Multiplexer development is impacted heavily by network synthesis and computer simulation techniques. As it becomes possible to synthesize combinations of lumped, distributed, and evanescent elements as well as predict and compensate their interactions, multiplexers will shrink in size, increase in order (number of channels), and display improved performance in insertion loss, isolation, and bandwidth.

9.7   Simulation and Synthesis Software

The process of simulation involves four separate, but related steps:

1. Synthesis and analysis of a theoretical network compliant to specification, under idealized terminating conditions and with idealized construction.
2. Representation of the synthesized network by an appropriate set of very accurate lumped elements. For any circuit, this involves modeling the physical structure and computing the lumped elements that best represent the actual, electromagnetic structure (i.e., solving Maxwell's equations inside the proposed filter structure).
3. Optimizing the filter response with the stipulated terminating impedances (i.e., the complex source and load impedance), using the representation of the circuit as computed in step 2.
4. Revising the physical structure, if required, by iterating the analysis portion of step 1.

The solutions to Maxwell's equations that allow for derivation of the lumped equivalents requires the comparison of a set of scattering parameters describing the physical structure (computed using E-M) to a set describing the characteristics of an assumed lumped element topology (computed using linear simulation). The difference between the two sets is reduced using optimization [7]. The data set is stored, and is used in an iterative manner as described in step 4. All physical structures can be described by a set of lumped elements of arbitrary complexity. Unfortunately, not every set of lumped parameters describes a physically realizable structure, so care must be taken to assume a "realizable" lumped circuit topology.

Traditional filter designs proceed from the basis of network synthesis. Over the last 90 years or so, the application of matrix, transform, complex variable theory, and advanced algebra has led to many clever network topologies. Numerical methods have also advanced the design process, not only simplifying the calculation process but enabling determination of the design suitability through the use of linear simulators that essentially compute the response of the synthesized structure so that the computed response may be compared to the desired response. If it is found that the synthesis is inadequate, the design can be iterated without the necessity for actual laboratory experimentation. Synthesis techniques have been developed to a very high degree for networks consisting of linear lumped elements or linear distributed elements, but to a much lesser extent for combinations of lumped and distributed elements. This is because the natural frequency variation for a lumped element is in terms of $j\omega$, while the variation of a distributed element is in terms of $\tan j\omega$. Thus, it is difficult to perform a synthesis that requires extraction of elements based upon the location of poles and zeros in a complex plane, when the coordinates of the complex plane are different for lumped and distributed structures.

9.8   Linear Simulators

The availability of linear simulators, combined with mathematical optimization, has reduced the need for advanced synthesis development (probably to the detriment of our profession and certainly to the dismay of many). The various elements can be readily combined and calculated in the simulator, as long as the elements can be described in transfer matrix (S-parameter) format. However, most physical elements have complex matrix descriptions because the elements are embedded into the surrounding structure in such a way as to respond to more than one mode of excitation. For example, a simple waveguide resonant cavity is analogous to an L-C tank circuit, but the waveguide cavity will resonate at more than one frequency based on field distribution. Thus, computation of the analogous (or equivalent) L-C values for the waveguide cavity requires knowledge of the excitation field. Combining microwave elements, such as cavities, probes, irises, etc. with each other (or for that matter with R-L-C elements), thus requires inclusion *of* the effects of the excitation field and the effects *upon* the field of each of the microwave elements encountered within the composite structure. Accomplishing this requires solutions to Maxwell's equations within the structure.

9.9   Electromagnetic (E-M) Simulators

Fortuitously, numerical methods have been applied to the solution of Maxwell's equations resulting in the development of what have come to be known as E-M simulators. These programs employ techniques such as finite elements in frequency or time domains, method of moments, spectral domain, etc., combined with advanced gridding methods and various structure generation software. Although quite advanced, most of these simulators are far too slow to use in conjunction with the mathematical optimization techniques that originally reduced the need for developing new and elegant synthesis techniques. It is well known [7–10] that frequency-dependent equivalent circuits can be derived that are adequate lumped representations of distributed structures to some degree of accuracy. When these structures are so represented, the equivalent circuits depend on the aforementioned mutual interaction of excitation and element. When the response modes are widely separated in the frequency or space domains, a single-mode computation provides a sufficiently accurate representation to enable the resultant lumped circuit to be used for computation of the approximate response of the distributed element or some combination of elements.

9.10   Synthesis Software

There have been a few software packages created that automate the design process to a large extent. However, most practitioners elect to create custom software to facilitate the transition from theory to practical filter networks. Some of the currently available most notable packages include Filter (Eagleware) and Filpro (Middle Eastern Technical University in Turkey) [11]. Packages that integrate linear and electromagnetic simulation are available from several sources, but inclusion of filter synthesis as an integrated package is rarely available (Eagleware has such an integrated package).

9.11   Active Filters

Since about 1970, it has been possible to simulate a high-Q inductance using a bipolar or FET transistor to convert the output capacitance into equivalent input inductance. The introduction of DC as an external power source acts to compensate for the loss properties of the inductor and make available the inductive element for inclusion into filter circuits. Over the years, other techniques have been developed for using active elements to realize high-Q filter circuits. These filters differ from the better known low frequency op-amp-based filters in that the synthesis generally is identical to that used for conventional passive RF filters, in which there is no requirement for constant voltage or constant current sources (typical impedances are 20 to 150 ohms). Such active filters and multiplexers have been built from 100 MHz to over

10 GHz. Stability, noise figure, thermal stability, and power consumption are problems yet to be fully overcome. Miniature passive filters suffer from poor performance due to the low Q of the available elements. In principle, miniature active filters can be constructed that will provide great selectivity, low loss, etc. with the price being the need for DC power. Applications to handheld cell phones are now to be found, with considerable progress reported. Combinations of active and passive devices are also possible, with the passive elements being used in such a way as to provide stability for the active, high-Q components.

References

1. H. Blinchikoff and A. Zverev, *Filtering in the Time and Frequency Domains*, John Wiley and Sons, New York, 1976.
2. C. Matthaei, L. Young and E.M.T. Jones, *Microwave Filters, Impedance-Matching Networks and Coupling Structures*, McGraw-Hill, New York, 1964. This is the so-called black-book of filters and should be purchased by any serious student.
3. S. Frankel, *Multiconductor Transmission Line Analysis*, Artech House, Boston, 1977.
4. Craven and Skedd, *Evanescent Mode Microwave Components*, Artech House, Boston, 1987.
5. M.E. Van Valkenburg, *Introduction to Modern Network Synthesis*, John Wiley and Sons, New York, 1960.
6. J. Malherbe, *Microwave Transmission Line Filters*, Artech House, Boston, 1979.
7. R. V. Snyder, Embedded-Resonator Filters, Proceedings of the ESA-ESTEC Conference on Filter CAD, ESA, The Netherlands, Nov. 6–8, 1995.
8. R. V. Snyder, Inverted Resonator Evanescent Mode Filters, IEEE-MTT-S Symposium Proceedings, San Francisco MTT IMS, 1996.
9. N. Marcuvitz, *Waveguide Handbook*, Vol. 10, MIT RadLab Series, 1948.
10. R. V. Snyder, Filter Design Using Multimode Lumped Equivalents Extracted from E-M Simulations, MTT/ED Workshop on Global Simulators, La Rochelle, France, May 27, 1998.
11. N. Yildirim, FILPRO Manual, November, 1996, METU, Ankara, Turkey.

10

RF Switches

Robert J. Trew
Virginia Tech University

10.1 Introduction

Microwave switches are control elements required in a variety of systems applications. They are used to control and direct, under stimulus from externally applied signals, the flow of RF energy from one part of a circuit to another. For example, all radars that use a common send and receive antenna require an RF switch to separate the send and receive signals, which often differ in amplitude by orders of magnitude. The large difference between the send and receive signals places severe demands upon the switching device, which must be able to sustain the high power of the transmitted signal, as well as have low loss to the returning signal. Isolation is very important in this application since the switch must be able to protect the sensitive receive circuit from the large RF transmitted power. The isolation requirement places severe restrictions upon the switch, and high power radars generally use gas discharge tubes to implement the switch function. Phased-array radars generally use semiconductor transmit/receive modules and use large numbers of switches. A phased-array radar, for example, may require thousands or tens of thousands of switches to permit precise electronic control of the radiated beam. The distributed nature of a phased-array permits the switches to operate at lower power, but the devices still need to operate at power levels on the order of 1 to 10 watts. In general, switches can be manually or electronically switched from one position to the next. However, most microwave integrated circuit applications require switching times that cannot be achieved manually, and electronic control is desirable. Integrated circuit implementation is ideal for switching applications since a large number of components can easily be accommodated in a relatively small area.

Electronically controlled switches can be fabricated using pin diodes [1,2] or transistors, generally GaAs MESFETs [3]. Both types of switches are commonly employed. Switches fabricated using pin diodes have often been used in radar applications [4], achieving insertion slightly over 1 db in L-bandwidth isolation greater than 35 db. Broadband operation can also be obtained [5] and 6 to 18 GHz bandwidth with insertion loss less than 2 db, isolation greater than 32 db, and CW power handling in excess of 6 watts has been reported using pin diodes connected in a shunt circuit configuration. Such switches have also demonstrated the ability to be optically controlled [6]. GaAs MESFETs are commonly used to fabricate RF switches suitable for use in integrated circuit applications [3,7]. High performance is achieved and a 1 watt SPDT switch with insertion loss of 0.6 db and isolation greater than 20 db has been reported [8]. These switches often use multi-gate GaAs MESFETs specifically designed for switching applications [9] that permit switching control of high RF power with low gate voltage. Such IC switches have demonstrated the ability to handle large power levels [10] and RF power on the order of 38 dbm can be effectively controlled. Switching at extremely high frequency is also possible by replacing

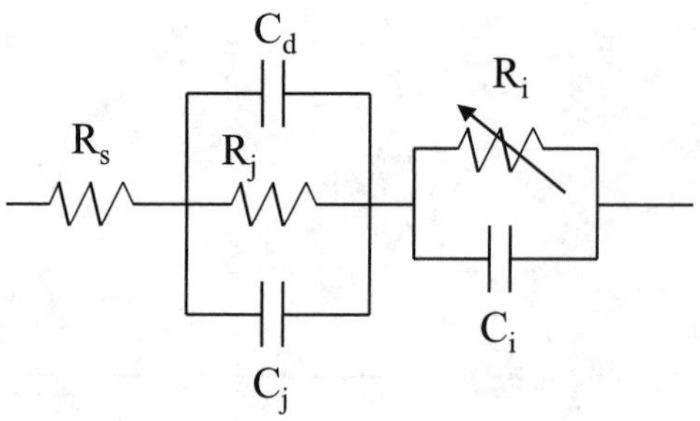

FIGURE 10.1   Schematic diagram for a PIN diode.

the GaAs MESFET with a HEMT, and high performance Q-band [11] and W-band [12] operation has been achieved. A comparison of the RF performance of MESFET and HEMT switches [13] indicates that HEMT devices generate more distortion than MESFET devices, but are useful at high millimeter-wave frequency. All semiconductor switches, whether fabricated using pin diodes or transistors, can be considered as two-state, one-port devices. Recently, a unified method for characterizing these networks has been presented [14].

10.2 PIN Diode Switches

A pin diode is a nonlinear device fabricated from a p^+nn^+ structure, as shown in Fig. 10.1. These devices are widely used in switch applications such as phase shifters [2] and have properties that result in low loss and high frequency performance. A pin diode can also be optically controlled [6], which is desirable for certain applications. The diode is a pn junction device with a lightly doped or undoped (intrinsic) region located between two highly doped contact regions. The presence of the intrinsic region yields operational characteristics very desirable for switching applications. That is, under reverse bias the intrinsic region produces very high values for breakdown voltage and resistance, thereby providing a good approximation to an "open" switching state. Both the breakdown voltage and off-state resistance are dependent upon the length of the intrinsic region, which is limited in design length only by transit-time considerations associated with the frequency of operation. Under forward bias, the conductivity of the intrinsic region is controlled or modulated by the injection of charge from the end regions and the diode conducts current, thereby providing the "on" switching state. The "on" resistance of the diode is controlled by the bias current and in forward bias, the diode has excellent linearity and low distortion.

An equivalent circuit for the PIN diode is shown in Fig. 10.2, and in operation the diode functions as a single-pole, double-throw (SPDT) switch, depending upon the bias state. Under reverse bias, the equivalent circuit reduces to that shown in Fig. 10.3, and under forward bias it reduces to the forward resistance R_f. The reverse bias resistance can be expressed as [3]

$$R_r = R_c + R_i + R_m \tag{10.1}$$

where R_c is the contact resistance of the metal semiconductor interfaces, R_i is the channel resistance of the intrinsic region, and R_m is the resistance of the contact metals. The resistance of the intrinsic region dominates and the reverse resistance becomes essentially that of the intrinsic region, which in terms of physical parameters can be expressed as

$$R_i \cong \frac{3(kT)L^2}{8qI_0L_a^2} \tag{10.2}$$

where L is the length of the intrinsic region, typically in the range of 1 to 100 μm. Depending upon design frequency, I_0 is the bias current, and La is the ambipolar diffusion length, which is a constant of

FIGURE 10.2 PIN diode equivalent circuit.

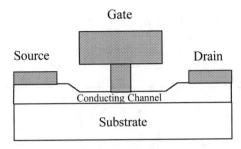

FIGURE 10.3 Reverse biased PIN diode equivalent circuit.

the material [3]. The other parameters have their usual meanings. Note that the reverse resistance, which can be in the kΩ range, is inversely proportional to bias current, and decreases with the magnitude of the applied bias current. The greatest off-state resistance, therefore, occurs under low reverse bias voltage. Under reverse bias the intrinsic region is essentially depleted of free charge, so the series capacitance is simply the capacitance of the intrinsic region, and can be expressed as

$$C_i = \frac{\varepsilon A}{L} \tag{10.3}$$

where A is the cross-sectional area of the diode. Note that the capacitance is constant under reverse bias.

Under forward bias the diode is dominated by the forward charge injection characteristics of the pn junction, and the diode can be represented as a resistance, with magnitude determined by the forward current. The on-state resistance can be expressed as

$$R_i = \frac{nkTA}{qI_0} \tag{10.4}$$

where n is the ideality factor for the diode (given in the diode specifications). The resistance of the diode in forward bias is inversely proportional to bias current, and the lowest resistance is obtained at high currents. The impedance of the diode can be tuned for RF circuit matching by adjustment of the bias current.

The rate at which the pin diode can be switched from a low-impedance, forward biased condition to a high-impedance, reverse biased condition, is determined by the speed at which the free charge can be extracted from the diode. Diodes with longer intrinsic regions and larger cross-sectional areas will store more charge, and require, therefore, longer times to switch. The actual switching time has two components: the time required to remove most of the charge (called the delay time) from the intrinsic region, and the time during which the diode is changing from a low- to a high-impedance state (called the transition time). The transition time depends upon diode geometry and details of the diode doping profile, but is not sensitive to the magnitude of the forward or reverse current. The delay

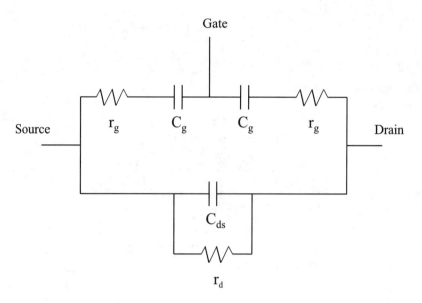

FIGURE 10.4   Schematic diagram for a MESFET switching element.

time is inversely proportional to the charge carrier lifetime. Diodes with short carrier lifetime have short delay times, but suffer from high values of forward bias resistance. High forward bias resistance increases the insertion loss for the diode, and this will produce attenuation of the signal through the device in the on-state.

10.3 MESFET Switches

A schematic diagram for a GaAs MESFET is shown in Fig. 10.4, and these devices are often used in switching applications. In general, a MESFET can be used in two different modes as passive or active elements. In the active mode the transistor is used as a three-terminal switch where the transistor is configured similar to an amplifier circuit. Either single-gate or dual-gate FETs can be used. The transistor is biased with a positive drain and a negative gate voltage, which are set so that the transistor is active. Switching action is accomplished by control of the transistor gain, which can be varied over several orders of magnitude. Dual-gate devices are particularly attractive for this application since the second gate can be used as a control port for efficient control of the gain.

In the passive mode of operation, the MESFET is configured to function as a passive two-terminal device, with the gate terminal acting as a port for only the control signal. That is, the RF signal is not applied to the gate and only travels between the drain and source terminals. The magnitude of the RF impedance between the drain and source terminals is controlled by a DC signal applied to the gate terminal. The drain-to-source impedance can be varied from a low value, obtained under open channel conditions when a zero potential is applied to the gate, to a high value, obtained when the gate is biased with a negative potential of sufficient amplitude to prevent current from flowing through the transistor. This occurs when the gate voltage achieves the transistor pinch-off voltage, which has a magnitude that is a function of the particular MESFET used.

In the passive mode the low-impedance state of the MESFET switch is dominated by the fully open conducting channel, and the open-channel resistance for the device is low. The equivalent circuit is essentially the "on" resistance for the transistor. In the high-impedance state the MESFET is dominated by the depleted channel or "off" resistance, which is large, and the switch has an equivalent circuit as shown in Fig. 10.5. The high-impedance state for the MESFET switch can be approximated with the simplified equivalent circuit shown in Fig. 10.6, where the "off" state resistance and capacitance are

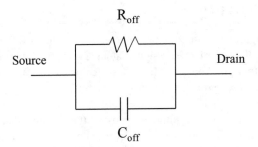

FIGURE 10.5   High-impedance, off-state equivalent circuit for a MESFET switch.

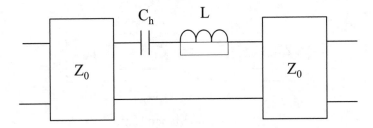

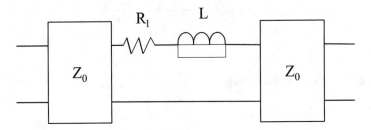

FIGURE 10.6   Simplified off-state equivalent circuit for a MESFET switch.

$$R_{off} = \frac{2r_d}{2 + r_d \omega^2 C_g^2 r_g} \qquad (10.5)$$

where ω is the radian frequency, and

$$C_{off} = C_{ds} + \frac{C_g}{2} \qquad (10.6)$$

Note that the "off" state resistance is an inverse function of frequency and, therefore, the magnitude of the blocking resistance decreases as frequency increases. The performance of the switch will degrade at high frequency and switch design becomes more difficult.

10.4 Switching Circuits

There are two basic configurations used for single-pole, double-throw (SPDT) switches that are commonly used to control the flow of microwave signals along a transmission line. The basic configurations can be fabricated using either pin diodes or MESFET switching elements, and are realized by utilization of the diode or transistor in a series or shunt connection to the transmission line. A simplified equivalent circuit for a series connected switch is shown in Fig. 10.7, and a shunt connected switch is shown in Fig. 10.8. The two configurations are complimentary in that the low-impedance state of the series switch permits signal flow, while the high-impedance state of the shunt switch permits signal flow. In the "off" state for both configurations, the microwave power incident upon the switching device is primarily reflected back toward the source. A small fraction of the incident power is dissipated in the switching

High Impedance State

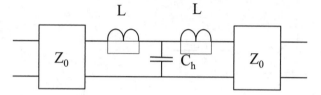

Low Impedance State

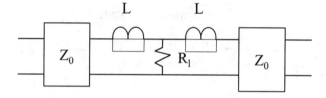

FIGURE 10.7   Simplified series connected switch circuit.

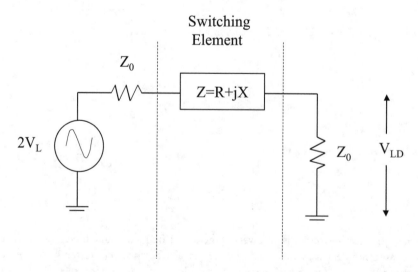

FIGURE 10.8   Simplified shunt connected switch circuit.

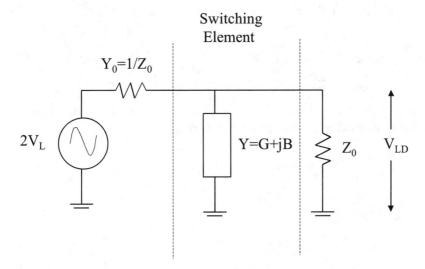

FIGURE 10.9 Equivalent circuit for a series connected switch.

FIGURE 10.10 Equivalent circuit for a shunt connected switch.

element and transmitted through the device toward the load. It is this fraction of the incident power that accounts for the insertion loss and the finite and nonideal isolation of the device. The fraction of microwave power that is transmitted through the device increases with frequency due to parasitic paths due to mounting, bonding, packaging, etc. elements, and switch isolation tends to degrade as operating frequency increases. It is possible, however, to minimize the parasitic signal flow by RF tuning and impedance compensation techniques

10.5 Insertion Loss and Isolation

Insertion Loss (IL) and isolation are important parameters that are used to characterize the performance of microwave switches. Insertion loss is defined as the ratio, generally in decibels, of the power delivered to the load in the "on" state of an ideal switch to the actual power delivered by the switch. The insertion loss can be calculated from consideration of the series and shunt equivalent circuits shown in Figs. 10.9 and 10.10. If V_L represents the voltage developed at the load for an ideal switch, the insertion loss can be written as,

$$IL = \left| \frac{V_L}{V_{LD}} \right|^2 \tag{10.7}$$

where, for the series configuration

$$V_{LD} = \frac{2V_L}{2 + Z/Z_0} \tag{10.8}$$

and

$$Z = R + jX \tag{10.9}$$

is the impedance of the switching device. The insertion loss is expressed as

$$IL = \left| \frac{2 + Z/Z_0^2}{2Z_0} \right| = \left| 1 + \frac{R + jX}{2Z_0} \right|^2 = 1 + \frac{R}{Z_0} + \frac{1}{4}\left(\frac{R}{Z_0}\right)^2 + \frac{1}{4}\left(\frac{X}{Z_0}\right)^2 \tag{10.10}$$

where R and X are the resistance and reactance of the switching device in the low-impedance state.
   For the shunt configuration, the voltage across the load is

$$V_{LD} = \frac{2V_L Y_0}{2Y_0 + Y} \tag{10.11}$$

and the insertion loss is written

$$IL = \left| \frac{2Y_0 + Y}{2Y_0} \right|^2 = \left| 1 + \frac{G + jB}{2Y_0} \right|^2 = 1 + \frac{G}{Y_0} = \frac{1}{4}\left(\frac{G}{Y_0}\right)^2 + \frac{1}{4}\left(\frac{B}{Y_0}\right)^2 \tag{10.12}$$

where $Y_0 = 1/Z_0$, and $Y = G + jB$.
   Isolation is a measure of the off-state performance of the switch. It is defined as the ratio of microwave power delivered to the load for an ideal switch in the "on" state, to the actual power delivered to the load when the switch is in the "off" state. In order to calculate isolation, the insertion loss expressions given above are used with the real and reactive terms for the device in the low-impedance state interchanged with the appropriate device parameters for the high-impedance state.

10.6 Switch Design

Switch design procedures are based upon the principle that the switching element in the "on" and "off" states can be considered as a reactance or susceptance that can be included in a filter configuration. Switch design, therefore, makes use of filter design procedures and all approaches to filter design can be used. The "on" and "off" state equivalent circuits are used to embed the switch element in the filter design. Generally, the "on" or low-insertion loss state is considered first, and the network is designed to yield the desired pass-band performance. The "off" state can be considered as a detuned network, and the impedances are adjusted to achieve the desired isolation. This approach to switch design may require several iterations until satisfactory performance in both the "on" and "off" states are achieved. Mounting and lead reactances are considered in the design and are absorbed and incorporated into the filter network. The actual filter element values may, therefore, differ in value from the design values. The performance of the insertion loss and isolation will vary with tuning and the lowest insertion loss and greatest isolation generally are obtained over narrow bandwidth. Increased bandwidth produces degradation in switch performance. Bias control circuits and thermal handling are accomplished in the same manner as for amplifier circuits.

References

   1. H.A. Watson, *Microwave Semiconductor Devices and Their Circuit Applications,* McGraw-Hill, New York, 1969.
   2. S.K. Koul and B. Bhat, Microwave and Millimeter Wave Phase Shifters, in *Semiconductor and Delay Line Phase Shifters,* Norwood, MA, Artech House, 1991.
   3. I. Bahl and P. Bhartia, *Microwave Solid State Circuit Design,* Wiley Interscience, New York, 1988.

4. M.E. Knox, P.J. Sbuttoni, J.J. Stangel, M. Kumar, and P. Valentino, Solid State 6x6 Transfer Switch for Cylindrical Array Radar, *1993 IEEE International Microwave Symposium Digest*, 1225–1228.

5. P. Omno, N. Jain, C. Souchuns, and J. Goodrich, High Power 6-18 Transfer Switch Using HMIC, *1994 International Microwave Symposium Digest*, 79–82.

6. C.K. Sun, C.T. Chang, R. Nguyen, and D.J. Albares, Photovoltaic PIN Diodes for RF Control — Switching Applications, *IEEE Trans. Microwave Theory Tech.*, 47, 2034–2036, Oct. 1999.

7. M. Shifrin, P. Katzin, and Y. Ayasli, High Power Control Components Using a New Monolithic FET Structure, *1989 IEEE Monolithic and Millimeter-Wave Integrated Circuits Symposium Digest*, 51–56.

8. T. Yamaguchi, T. Sawai, M. Nishida, and M. Sawada, Ultra-Compact 1 W GaAs SPDT Switch IC, *1999 IEEE International Microwave Symposium Digest*, 315–318.

9. H. Uda, T. Yamada, T. Sawai, K. Nogawa, and Yu. Harada, A High-Performance GaAs Switch IC Fabricated Using MESFET's with Two Kinds of Pinch-Off Voltages, *GaAs IC Symp. Digest*, 139–142, 1993.

10. M. Masuda, N. Ohbata, H. Ishiuchi, K. Onda, and R. Yamamoto, High Power Heterojunction GaAs Switch IC with P-1db of More Than 38 dbm for GSM Application, *GaAs IC Symp. Digest*, 229–232, 1998.

11. D.L. Ingram, K. Cha, K. Hubbard, and R. Lai, Q-Band High Isolation GaAs HEMT Switches, *IEEE GaAs IC Symposium Digest*, 289–292, 1996.

12. H. Takasu, F. Sasaki, H. Kawasaki, H. Tokuda, and S. Kamihashi, W-Band SPST Transistor Switches, *IEEE Microwave Guided Wave Letters*, 315–316, Sept. 1996.

13. R.H. Caverly, and K.J. Heissler, On-State Distortion in High Electron Mobility Transistor Microwave and RF Switch Control Circuits, *IEEE Trans. Microwave Theory Tech.*, 98–103, Jan. 2000.

14. I.B Vendik, O.G. Vendik, and E.L. Kollberg, Commutation Quality Factor of Two-State Switchable Devices, *IEEE Trans. Microwave Theory Tech.*, 802–808, May 2000.

11

RF Package Design and Development

Jeanne S. Pavio
Motorola SPS

11.1 Introduction

Successful RF and microwave package design involves adherence to a rigorous and systematic methodology in package development together with a multi-disciplined and comprehensive approach. This formal planning process and execution of the plan ultimately insures that the package and product will perform as expected, for the predicted lifetime duration in the customer's system, under the prescribed application conditions.

Probably the first concern is having a thorough and in-depth knowledge of the application and the system into which the microwave component or module will be placed. Once these are understood, then package design can begin. Elements that must be considered do not simply include proper electrical performance of the circuit within the proposed package. Mechanical aspects of the package design must be thoroughly analyzed to assure that the package will not come apart under the particular life conditions. Second, the substrates, components, or die within the package must not fracture or lose connection. Third, any solder, epoxy, or wire connections must be able to maintain their integrity throughout the thermal and mechanical excursions expected within the application. Once these elements are thoroughly investigated, the thermal aspects of the package must be simulated and analyzed to appropriately accommodate heat transfer to the system. Thermal management is probably one of the most critical aspects of the package design because it not only contributes to catastrophic circuit overload and failure in out-of-control conditions, but it could also contribute to reduced life of the product and fatigue failures over time. Thermal interactions with the various materials used for the package itself and within the package may augment mechanical stress of the entire package system, ultimately resulting in failure.

Once proper simulation and analysis have been completed from a mechanical and thermal point of view, the actual package design can be finalized. Material and electrical properties and parameters then become the primary concern. Circuit isolation and electromagnetic propagation paths within the package need to be thoroughly understood. In addition, impedance levels must be defined and designed for input and output to and from the package. New developments in package design systems have paved the way for rapid package prototyping through computer integrated manufacturing systems by tieing the design

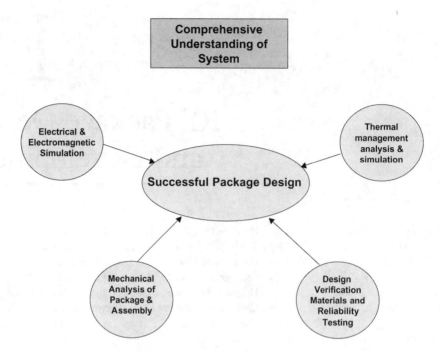

FIGURE 11.1　Elements of successful package design.

itself to the machining equipment that will form the package. These systems can prototype a part in plastic for further study or can actually build the prototypes in metal for delivery of prototype samples. Finally, design verification must take place. The verification process typically includes the various long-term reliability tests that gives the designer, as well as his or her customer confidence that the package and its contents will live through the predicted lifetime and application conditions. Other testing may be more specific, such as fracture testing, material properties tests, or precise design tolerance testing. Much of the final testing may also include system-level integration tests. Usually specific power levels are defined and the packages, fully integrated into the system, are tested to these levels at particular environmental conditions.

These are some of the key elements in RF and microwave package development. Although this is not an all-encompassing list, these elements are critical to success in design implementation. These will be explored in the following discussion, hopefully defining a clear path to follow for RF package design and development. Figure 11.1 depicts these key elements leading to successful package design.

11.2　Thermal Management

From an MTBF (mean time before failure) point of view, the thermal aspects of the circuit/package interaction are one of the most important aspects of the package design itself. This can be specifically due to actual heat up of the circuit, reducing lifetime. It may also be due to thermal effects that degrade performance of the materials over time. A third effect may be a materials/heat interaction that causes severe thermal cycling of the materials resulting in stress concentrations and degradation over time.

It is clear that the package designer must have a fully encompassing knowledge of the performance objectives, duty cycles, and environmental conditions that the part will experience in the system environment. The engineer must also understand the thermal material properties within the entire thermal path. This includes the die, the solder or epoxy attachment of that die, the package or carrier base, package system attachment, and material connection to the chassis of the system. There are relatively good

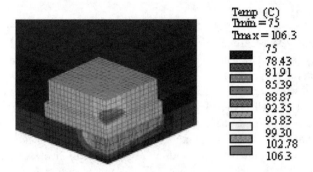

FIGURE 11.2　Ansys output showing thermal gradient across silicon die.

databases in the industry that provide the engineer with that information right at his or her fingertips. Among the many are the CINDAS [1] database and the materials' database developed at Georgia Institute of Technology. Other information may be gleaned from supplier datasheets or testing.

Thermal density within the package, and in particular, at the die level becomes an all-important consideration in the thermal management equation. To insure proper heat transfer and to eliminate any potential thermal failure modes (such as materials breakdown or diffusion and migration), analysis of heat transfer within the die must be completed at the die layout level. This thermal analysis will ultimately be parametrically incorporated into an analysis at the next level up, which may be at the circuit substrate or at the package level itself. The analysis is usually completed with standard finite element simulation techniques present in various software packages available in the industry. Ansys, MSC Nastran, Mechanica, Flowtherm, and Computational Fluid Dynamics (CFD) are some of those available. Material properties that are critical to input into the model would be thermal conductivity and the change in conductivity with temperature. Figure 11.2 shows a typical output of one of these software tools. The FEM simulation uses $1/4$ model symmetry. In this particular figure, the analysis demonstrates the thermal gradient across a silicon die, which is an 8 W power amplifier transistor, solder attached to a via structure, with 75°C applied to the bottom of the heat sink. The die junction is at 106.3°C.

It is through such simulated analysis that the entire heat transfer methodology of component to system can be developed. Assuming that there is good correlation between simulated and verified results, the engineer can then gain confidence that the product will have a reasonable lifetime within the specific application. The correlation is typically achieved through the use of infrared microscopy techniques. A number of these infrared microscopes are available in the industry. Usually, the component or module is fixtured on a test station under the infrared camera. The camera is focused on the top surface of the die, which is the heat-generating element. As power is applied to the component or module, the die begins to heat up to a steady-state level. The heat can be measured under RF or DC power conditions. A measurement is done of the die surface temperature. At the same time, a thermocouple impinges on the bottom of the case or package and makes a temperature measurement there. With the maximum die junction temperature (Tjmax in °C), the case temperature (Tc in °C) and the dissipated power in watts, the packaged device junction to case thermal resistance (in °C/W) can be calculated from the following equation:

$$\theta jc = \left(Tjmax - Tc\right)\big/Pdis$$

In a fully correlated system, the agreement between simulated and measured results is usually within a few percentage points.

11.3 Mechanical Design

The mechanical design usually occurs concurrently with thermal analysis and heat transfer management. In order to adequately assess the robustness of the package system and its elements, an in-depth understanding of all the material properties must be achieved. In addition to the mechanical properties such as Young's modulus and stress and strain curves for materials, behavior of those materials under thermal loading conditions must be well understood. Once again, these material properties can be found in standard databases in the industry as mentioned above. Typically, the engineer will first insure that the packaging materials under consideration will not cause fracture of the semiconductor devices or of the substrates, solder joints, or other interconnects within the package. This involves knowing the Coefficient of Thermal Expansion (CTE) for each of these materials, and understanding the processing temperatures and the subsequent temperature ramp up that will be experienced under loading conditions in the application. The interaction of the CTEs of various materials may create a mismatched situation and create residual stresses that could result in fracture of any of the elements within the package. The engineer also assesses the structural requirements of the application and weight requirements in order to form appropriate decisions on what materials to use. For instance, a large microwave module may be housed within an iron-nickel (FeNi) package that sufficiently addresses all of the CTE concerns of the internal packaged elements. However, this large, heavy material might be inappropriate for an airborne application where a lightweight material such as AlSiC (aluminum silicon carbide) would be more suitable.

It is not sufficient to treat the packaged component or module as a closed structure without understanding and accounting for how this component or module will be mounted, attached, or enclosed within the actual system application. A number of different scenarios come to mind. For instance, in one situation a packaged component may be soldered onto a printed circuit board (PCB) of a wireless phone. Power levels would not be of concern in this situation, but the mechanical designer must develop confidence through simulation that the packaged RF component can be reliably attached to the PCB. He or she also must insure that the solder joints will not fracture over time due to the expansion coefficient of the PCB compared to the expansion of the leads of the package. Finally, the engineer must comprehend the expected lifetime in years of the product. Cost is obviously a major issue in this commercial application. A solution may be found that is perfectly acceptable from a thermo-mechanical perspective, but it may be cost prohibitive for a phone expected to live for three to five years and then be replaced.

Another scenario on the flip side of the same application is the power device or module that must be mounted into a base station. Here, obviously, the thermal aspects of the packaged device become all important. And great pain must be taken to insure that an acceptable heat transfer path is clearly delineated. With the additional heat from the power device and within the base station itself, heat degradation mechanisms are thoroughly investigated both with simulation techniques and with rigorous testing. It is common for the RF power chains within base station circuits to dissipate 100 to 200 watts each. Since the expected lifetime of base stations may be over fifteen years, it would be a great temptation for a mechanical engineer to utilize optimum heat transfer materials for the package base, such as diamond for instance, with a thermal conductivity of 40.6 W/in°C. A high power device attached to diamond would operate much cooler than a device attached to FeNi or attached to ceramic. Since, over time, it is the heat degradation mechanisms that eventually cause failure of semiconductor devices, a high power die mounted over a diamond heat sink would be expected to have a much longer lifetime than one mounted over iron nickel or over ceramic. However, once again, the cost implications must enter into the equation. Within the multifunctioned team developing the package and the product, a cost trade-off analysis must be done to examine cost comparisons of materials vs. expected lifetimes. The mechanical package designer may develop several simulations with various materials to input into the cost-reliability matrix. It is necessary that such material substitutions can be done easily and effectively in the parametric model that was initially developed.

The mechanical analysis must encompass attachment of the RF or microwave component or module to the customer system. As we have discussed, in a base station, the thermal path is all important. In order to provide the best heat transfer path, engineers may inadvertently shortcut mechanical stress concerns, which then compromise package integrity. An example was a system mounting condition initially created for the eight-watt power device shown in Fig. 11.2. This semiconductor die was packaged on a copper lead frame to which plastic encapsulation was applied. The lead frame was exposed on the bottom side of the device to insure that there would be a good thermal path to the customer chassis. The copper leads were solder attached (using the typical lead-tin, PbSn, solder) to the printed circuit board. At the same time, the bottom of the device was solder attached to a brass heat sink, as shown in Fig. 11.3, which was then screw mounted to the aluminum chassis to provide thermal transfer to the chassis.

The CTE mismatch of materials resulting in residual stresses during thermal excursions, caused the plastic to rip away from the copper leads. It was the expansion of the aluminum chassis impacting the brass heat sink that created both tensile and shear forces on the leads of the device. The brass heat sink, in effect, became a piston pushing up at the center of the component. The stress levels in the plastic mold compound, which resulted in the failure of the mold/copper interface, can be seen in Fig. 11.4.

Through subsequent simulation, a solution was found that provided the proper heat transfer for the eight-watt device as well as mechanical stability over time and temperature. This was verified through thousands of hours of temperature cycle testing and device power conditioning over temperature excursions.

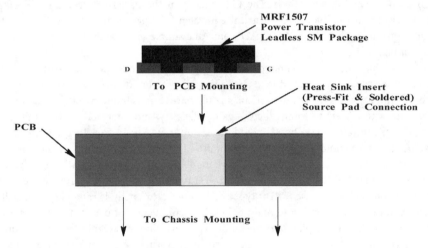

FIGURE 11.3   Eight watt power device attached to brass heat sink.

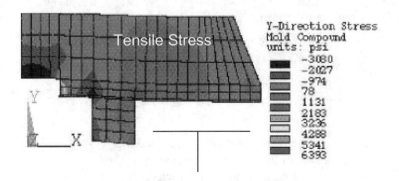

FIGURE 11.4   Modeled stresses in plastic mold compound resulting in failure.

11.4 Package Electrical and Electromagnetic Modeling

Quite obviously, the electrical design cannot stop at the circuit model for the silicon or GaAs die itself. Particularly at higher frequencies, such as those in the RF or microwave arena, the electromagnetic propagation due to all circuit elements create interactions, interference, and possibly circuit oscillations if these electrical effects are not accounted for and managed. Of course the customer's initial requirement will be that a packaged device, component, or module have a specific impedance into and out of their system. Typically, this has been 50 ohms for many microwave systems. It can be achieved through properly dimensioned microstrip input and output leads, through coaxial feeds, or through stripline to microstrip connections that feed into the customer system. These are modeled using standard industry software such as that provided by Hewlett Packard or Ansoft. The next consideration for the package designer is that all of the circuit functions that require isolation are provided that isolation. This can be accomplished through the use of actual metal wall structures within the package. It can also be done by burying those circuit elements in cavities surrounded by ground planes or through the use of solid vias all around the functional elements. These are only some of the predictive means of providing isolation. The need for isolating circuit elements and functions is ascertained by using full wave electromagnetic solvers such as HFSS, Sonnet, or other full wave tools. The EM analysis of the packaged structure will output an S parameter block. From this block, an electrical equivalent circuit can then be extracted with circuit optimization software such as Libra, MDS, ADS, etc. An example of an equivalent circuit representation can be seen in Fig. 11.5.

After proper circuit isolation is achieved within the package, the designer must insure that there will not be inductive or capacitive effects due to such things as wire bonds, leads, or cavities. Wire bonds, if not controlled with respect to length in particular, could have serious inductive effects that result in poor RF performance with respect to things such as gain, efficiency, and intermodulation distortion, etc. In the worst case, uncontrolled wire bonds could result in circuit oscillation. In the same way, RF and microwave performance could be severely compromised if the capacitive effects of the leads and other capacitive elements are not accounted for. These are modeled with standard RF and microwave software tools, and then the materials or processes are controlled to maintain product performance within specifications. Most software tools have some type of "Monte Carlo" analysis capability in which one can alter the material or process conditions and predict the resulting circuit performance. This is especially useful if the processes have been fully characterized and process windows are fully defined and understood. The Monte Carlo analysis then can develop expected RF performance parameters for the characterized process within the defined process windows.

11.5 Design Verification, Materials, and Reliability Testing

After all of the required simulation and package design has been completed, the time has come to begin to build the first prototypes to verify the integrity of the design. During the simulation phase, various

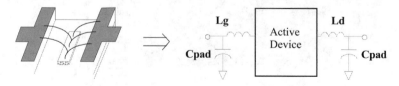

Parasitic effects are represented by:
- Leadframe capacitance, Cpad
- Total gate side wire inductance, Lg
- Total drain side wire inductance, Ld

FIGURE 11.5   Equivalent circuit representation of simple package.

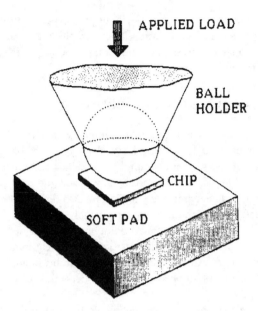

FIGURE 11.6 Technique used to measure fracture strength of semiconductor die.

material property studies may have been undertaken in order to insure that the correct properties are input into the various models. These may be studies of dielectric constant or loss on a new material, fracture studies to determine when fracture will occur on a uniquely manufactured die or on a substrate, or thermal studies, such as laser flash, to determine the precise thermal conductivity of a material. Figure 11.6 shows one technique used to measure the fracture strength of a GaAs or silicon die. A load is applied to a fixtured sphere, which then impacts the die at a precise force level. From the test, the critical value of the force to break the die is recorded. Then this force is converted to the maximum die stress via the following well-known [5] equation:

$$\sigma_t = \frac{3W}{2\pi m t^2}\left[(m+1)\ln\left(\frac{a}{r_o}\right)+(m-1)\left(1-\left(\frac{r_o}{a}\right)^2\right)\right]$$

After various material properties tests and all simulations and models have been completed, initial prototypes are built and tested. This next phase of tests typically assess long-term reliability of the product through thermal cycling, mechanical shock, variable random vibration, long-term storage, and high temperature and high humidity under biasing conditions. These, as well as other such tests, are the mainstay of common qualification programs. The levels of testing and cycles or hours experienced by the packaged device are often defined by the particular final application or system. For instance, a space-qualified product will require considerably more qualification assessment than a component or module going into a wireless handset that is expected to live 3 to 5 years. The temperature range of assessment for the space-qualified product may span from cryogenic temperatures to +150°C. The RF component for the wireless phone, on the other hand, may simply be tested from 0°C to 90°C.

In high power applications, often part of the reliability assessment involves powering up the device or module after it is mounted to a simulated customer board. The device is powered up and down at a specific duty cycle, through a number of cycles often as the ambient progresses through a series of thermal excursions. This represents what the RF packaged component would experience in the customer system, though usually at an accelerated power and/or temperature condition. Lifetime behavior can then be predicted, using standard prediction algorithms such as Black's equation, depending on the test results.

The RF product and packaging team submit a series of prototype lots through final, standard quali-
fication/verification testing. If all results are positive, then samples are usually given to the customer at
this time. These will undergo system accelerated life testing. The behavior of the system through this
series of tests will be used to predict expected life cycle.

11.6 Computer-Integrated Manufacturing

As mentioned above, there are tools available in the industry that can be used to rapidly develop
prototypes directly from the package design files. These prototypes may be constructed of plastic or of
various metals for examination and further assessment. Parametric Technologies offers such design and
assembly software modules, although they are not by any means the only company with this type of
software. The package design is done parameterically in Pro-E so that elements of the design can easily
be changed and/or uploaded to form the next higher assembly. The package design elements then go
through a series of algorithms to which processing conditions can be attached. These algorithms translate
the information into CNC machine code which is used to operate equipment such as a wire EDM for
the cutting of metals. Thus, a lead frame is fashioned automatically, in a construction that is a perfect
match to the requirements of the die to be assembled. A process flow chart for this rapid prototyping
scenario is shown in Fig. 11.7.

Computer-integrated manufacturing is also a highly effective tool utilized on the production floor,
once the designed package has been accepted by the customer and is ready for production implementation. Here it is utilized for automated equipment operation, for statistical process control (SPC), for
equipment shut down in out-of-control situations, etc. Coupled with neural networks, computer-integrated manufacturing can also be used for advanced automated process optimization techniques.

11.7 Conclusions

The development and design of packages for RF and microwave applications must involve a rigorous
and systematic application of the proper tools and methodology to create a design that "works the first
time" and every time for the predicted lifetime of the product. This encompasses an in-depth knowledge
of the system requirements, the environmental conditions, and the mounting method and materials to
be used for package assembly into the customer's system. Then modeling and simulation can take place.
Often, in order to understand material properties and to use these more effectively in the models, material
studies are done on specific parameters. These are then inserted into electrical, mechanical, and thermal

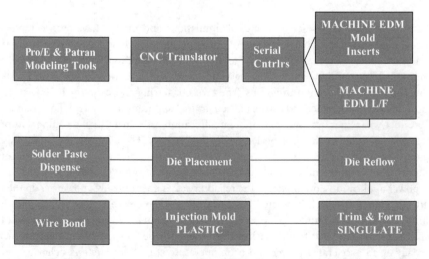

FIGURE 11.7   Rapid prototyping system.

models, which must be completed for effective package design. After a full set of models is completed, verification testing of the design can be done on the first prototypes. Rapid prototyping is made simple through techniques that automatically convert design parameters into machine code for operation of machining equipment. Computer-integrated manufacturing is a highly effective technique that can be utilized at various levels of the product introduction. In package design, it is often used for rapid prototyping and as a tool for better understanding the design. At the production level, it is often used for automated equipment operation and for statistical process control.

Verification testing of the prototypes may include IR scanning to assess thermal transfer. It may include instron testing to test the integrity of a solder interface or of a package construction. It may include power cycling under DC or RF conditions to insure that the packaged design will work in the customer application.

The final phase of assessment is the full qualification of the RF packaged device or module. This certifies to the engineer, and ultimately to the customer that the packaged product can live through a series of thermal cycles, through high temperature and high humidity conditions. It certifies that there will be no degradation under high temperature storage conditions. And it certifies that the product will still perform after appropriate mechanical shock or vibration have been applied. Typically, predictive lifetime assessment can be made using performance to accelerated test conditions during qualification and applying these results to standard reliability equations.

These package design elements, when integrated in a multidisciplined approach, provide the basis for successful package development at RF and microwave frequencies.

References

1. CINDAS = Center for Information and Data Analysis; Operated by Purdue University; Package Materials Database created under SRC (Semiconductor Research Corporation) funding.
2. G. Hawkins, H. Berg, M. Mahalingam, G. Lewis, and L. Lofgran "Measurement of silicon strength as affected by wafer back processing," International Reliability Physics Symposium, 1987.
3. T. Liang, J. Pla, and M. Mahalingam, Electrical Package Modeling for High Power RF Semiconductor Devices, Radio and Wireless Conference, IEEE, Aug. 9-12, 1998.
4. R.J. Roark, *Formulas for Stress and Strain*, 4th Edition, McGraw-Hill, New York, 219.

12

Guided Wave Propagation and Transmission Lines

W.R. Deal
Malibu Networks

V. Radisic
HRL Laboratory, LLC

Y. Qian
University of California

T. Itoh
University of California

12.1 Introduction

At higher frequencies where wavelength becomes small with respect to feature size, it is often necessary to consider an electronic signal as an electromagnetic wave and the structure where this signal exists as a waveguide. A variety of different concepts can be used to examine this wave behavior. The most simplistic view is transmission line theory, where propagation is considered in a simplistic 1-D manner and the cross-sectional variation of the guided wave is entirely represented in terms of distributed transmission parameters in an equivalent circuit. This is the starting point for transmission line theory that is commonly used to design microwave circuits. In other guided wave structures, such as enclosed waveguides, it is more appropriate to examine the concepts of wave propagation from the perspective of Maxwell's equations, the solutions of which will explicitly demonstrate the cross-sectional dependence of the guided wave structure.

Most practical wave guiding structures rely on single-mode propagation, which is restricted to a single direction. This allows the propagating wave to be categorized according to its polarization properties. A convenient method is classifying the modes as TEM, TE, or TM. TEM modes have both the electric and magnetic field transverse in the direction of propagation. Only the magnetic field transverses in the direction of propagation in TM modes, and only the electric field transverses in the direction of propagation in TE modes.

In this chapter, we first briefly examine the telegrapher's equation, which is the starting point for transmission line theory. The simple transmission line model accurately describes a number of guided wave structures and is the starting point for transmission line theory. In the next section, enclosed waveguides including rectangular and circular waveguides will be discussed. Relevant concepts such as cutoff frequency and modes will be given. In the final section, four common planar guided wave structures will be discussed. These inexpensive and compact structures are the foundation for the modern commercial RF front end.

12.2 TEM Transmission Lines, Telegrapher's Equations, and Transmission Line Theory

In this section, the concept of guided waves in simple TEM-guiding structures will be explored in terms of the simple model provided by Telegrapher's Equations, also referred to as transmission line equations. Telegrapher's equations demonstrate guided wave properties in terms of lumped equivalent circuit parameters available for many types of simple two-conductor transmission lines, and are valid for all types of TEM waveguides if their corresponding equivalent circuit parameters are known. These parameters must be found from Maxwell's equations in their fundamental form. Finally, properties and parameters for several types of two-wire TEM transmission line structures are introduced.

A transmission line or waveguide is used to transmit power and information from one point to another in an efficient manner. Three common types of transmission lines that support TEM guided waves are shown in Figure 12.1(a–c), including the parallel-plate transmission line, two-wire line, and coaxial transmission line. The parallel-plate transmission line consists of a dielectric slab sandwiched between two parallel conducting plates of width *w*. More practical, commonly used variations of this structure at microwave and millimeter-wave frequencies include microstrip and stripline, which will be briefly discussed in the final part of Section 12.4. A two-wire transmission line, consisting of two parallel conducting lines separated by a distance *d* is shown in Fig. 12.1b. This is commonly used for power distribution at low frequencies. Finally, the coaxial transmission line consists of two concentric conductors separated by a dielectric layer. This structure is well shielded and commonly used at high frequencies well into the microwave range.

The telegrapher's equations form a simple and intuitive starting point for the physics of guided wave propagation in these structures. An equivalent circuit model is shown in Fig. 12.2 for a two-conductor transmission line of differential length Δz in terms of the following four parameters:

R, resistance per unit length of both conductors (Ω/m).
L, inductance per unit length of both conductors (H/m).
G, conductance per unit length (S/m).
C, capacitance per unit length of both conductors (F/m).

These parameters represent physical quantities for each of the relevant transmission lines. For each of the structures shown in Fig. 12.1(a–c), R represents conductor losses, L represents inductance, G represents dielectric losses, and C represents the capacitance between the two lines.

Returning to Fig. 12.2, the quantities $v(z,t)$ and $v(z + \Delta z,t)$ represent change in voltage along the differential length of transmission line, while $i(z,t)$ and $i(z + \Delta z,t)$ represent the change in current. Writing Kirchoff's voltage law and current laws for the structure, dividing by Δz, and applying the fundamental theorem of calculus as $\Delta z \rightarrow 0$, two coupled differential equations known as the telegrapher's equations are obtained:

$$-\frac{\partial v(z,t)}{\partial z} = Ri(z,t) + L\frac{\partial i(z,t)}{\partial t} \tag{12.1}$$

$$-\frac{\partial i(z,t)}{\partial z} = Gi(z,t) + D\frac{\partial v(z,t)}{\partial t} \tag{12.2}$$

However, typically we are interested in signals with harmonic time dependence ($e^{j\omega t}$). In this case, the time harmonic forms of the telegrapher's equations are given by

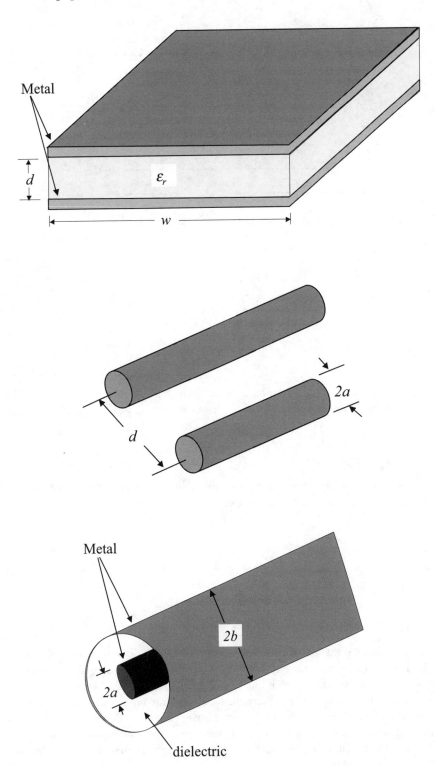

FIGURE 12.1 Three simple TEM-type transmission line geometries including (a) parallel-plate transmission line, (b) two-wire line, and (c) coaxial line.

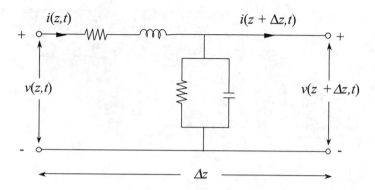

FIGURE 12.2   Distributed equivalent circuit model for a transmission line.

$$-\frac{dV(z)}{dz} = (R + j\omega L)I(z) \tag{12.3}$$

$$-\frac{dI(z)}{dz} = (G + j\omega C)V(z) \tag{12.4}$$

The constant γ is defined as the propagation constant with real and imaginary parts, α and β, corresponding to the attenuation constant (Np/m) and phase constant (rad/m) in the following manner

$$\gamma = \alpha + j\beta = \sqrt{(R + j\omega L)(G + j\omega C)} \tag{12.5}$$

This may then be substituted into the telegrapher's equations, which may then be solved for $V(z)$ and $I(z)$ to yield the following one-dimensional wave equations:

$$\frac{d^2 V(z)}{dz^2} - \gamma^2 V(z) = 0 \tag{12.6}$$

$$\frac{d^2 I(z)}{dz^2} - \gamma^2 I(z) = 0 \tag{12.7}$$

The form of this equation is the well-known wave equation. This indicates that the transmission line will support a guided electromagnetic wave traveling in the z-direction. The telegrapher's equations use a physical equivalent circuit and basic circuit theory to demonstrate the wave behavior of an electromagnetic signal on a transmission line. Alternatively, the same result can be obtained by starting directly with Maxwell's equations in their fundamental form, which may be used to derive the wave equation for a propagating electromagnetic wave. In this case, the solution of the wave equation will be governed by the boundary conditions. Similarly, the parameters R, L, G, and C are determined by the geometry of the transmission line structures.

Returning to the telegrapher's equations, several important facts may be noted. First, the characteristic impedance of the transmission line may be found by taking the ratio of the forward traveling voltage and current wave amplitudes, and is given in terms of the equivalent circuit parameters as

$$Z_0 = \sqrt{\frac{R + j\omega L}{G + j\omega C}} \qquad (12.8)$$

In the case of a lossless transmission line, this reduces to $Z_o = \sqrt{L/C}$. The phase velocity, also known as the propagation velocity, is the velocity of the wave as it moves along the waveguide. It is defined as

$$v_p = \frac{\omega}{\beta} \qquad (12.9)$$

In the lossless case, this reduces to:

$$v_p = \frac{1}{\sqrt{LC}} = \frac{1}{\sqrt{\mu\varepsilon}} \qquad (12.10)$$

This shows that the velocity of the signal is directly related to the medium. In the case of an air-filled, purely TEM mode, the wave will propagate at the familiar value $c = 3 \times 10^8$ m/s. Additionally, it provides the relationship between L, C and the medium in which the wave is guided. Therefore, if the properties of the medium are known, it is only necessary to determine either L or C. Once C is known, G may be determined by the following relationship:

$$\frac{G}{C} = \frac{\sigma}{\varepsilon} \qquad (12.11)$$

Note that σ is the conductivity of the medium, not of the metal conductors. The final parameter, the series resistance R, is determined by the power loss in the conductors. Simple approximations for the transmission line parameters R, L, G, and C for the three types of transmission lines shown in Figs. 12.1(a–c) are well known and are shown in Table 12.1. Note that μ, ε, and σ relate to the medium separating the conductors, and σ_c refers to the conductor. Once the equivalent circuit parameters are determined, the characteristic impedance and propagation constant of the transmission line may be determined. Note that R_s represents the surface resistance of the conductors, given as

$$R_s = \sqrt{\frac{\pi f \mu_c}{\sigma_c}} \qquad (12.12)$$

12.3 Guided Wave Solution from Maxwell's Equations, Rectangular Waveguide, and Circular Waveguide

A waveguide is any structure that guides an electromagnetic wave. In the preceding section, several simple TEM transmission structures were discussed. While these structures do support a guided wave, therm waveguide more commonly refers to a closed metallic structure with a fixed cross-section within which a guided wave propagates, as shown for the arbitrary cross-section in Fig. 12.3. The guide is filled with a material of permittivity ε and permeability μ, and is defined by its metallic wall parallel to the z-axis. These structures demonstrate lower losses than the simple transmission line structures of the first section, and are used to transport power in the microwave and millimeter-wave frequency range. Ohmic losses are low and the waveguide is capable of carrying large power levels. Disadvantages are bulk, weight, and limited bandwidth, which cause planar transmission lines to be used wherever possible in modern

TABLE 12.1 Transmission Line Parameters for Parallel-Plate, Two-Wire Line and Coaxial Transmission Lines

	Parallel-Plate Waveguide	Two-Wire Line	Coaxial Line
R (Ω/m)	$\dfrac{2}{w}R_s$	$\dfrac{R_s}{\pi a}$	$\dfrac{R_s}{2\pi}\left(\dfrac{1}{a}+\dfrac{1}{b}\right)$
L (H/m)	$\mu\dfrac{d}{w}$	$\dfrac{\mu}{\pi}\cosh^{-1}\left(\dfrac{D}{2a}\right)$	$\dfrac{\mu}{2\pi}\ln\left(\dfrac{b}{a}\right)$
G (S/m)	$\sigma\dfrac{w}{d}$	$\dfrac{\pi\sigma}{\cosh^{-1}\left(D/2a\right)}$	$\dfrac{2\pi\sigma}{\ln\left(b/a\right)}$
C (F/m)	$\varepsilon\dfrac{w}{d}$	$\dfrac{\pi\varepsilon}{\cosh^{-1}\left(D/2a\right)}$	$\dfrac{2\pi\varepsilon}{\ln\left(b/a\right)}$

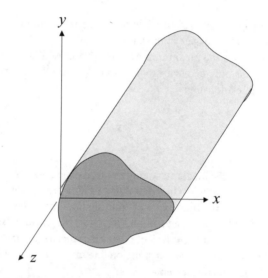

FIGURE 12.3 Geometry of enclosed waveguide with arbitrary cross-section. Propagation is in the z direction.

communications circuits. However, a wide variety of components are available in this technology, including high performance filters, couplers, isolators, attenuators, and detectors.

Inside this type of enclosed waveguide, an infinite number of distinct solutions exist, each of which is referred to as a *waveguide mode*. At a given operating frequency, the cross-section of the waveguide and the type of material in the waveguide determine the characteristics of these modes. These modes are usually classified by the longitudinal components of the electric and magnetic fields, E_z and H_z, where propagation is in the z direction. The most common classifications are TE (Transverse Electric), TM (Transverse Magnetic), EH, and HE modes. The basic characteristics are described in the next two paragraphs. The TEM modes that were discussed in the previous section do not propagate in this type of metallic enclosed waveguide. This is because a TEM mode requires two conductors to propagate, where a conventional enclosed waveguide has only a single enclosing conductor.

The two most common waveguide modes are the TE and TM modes. TE modes have no component of E in the z direction, which means that E is completely transverse to the direction of propagation. Similarly, TM modes have no component of H in the z direction.

EH and HE modes are hybrid modes that may be present under certain conditions, such as a waveguide partially filled with dielectric. In this case, pure TE and TM are unable to satisfy all of the necessary boundary conditions and a more complex type of modal solution is required. With both EH and HE, neither E nor H are zero in the direction of propagation. In EH modes, the characteristics of the transverse fields are controlled more by H_z than by E_z. HE modes are controlled more by E_z than by H_z. These types of hybrid modes may also be referred to as LSE (Longitudinal Section Electric) and LSM (Longitudinal Section Magnetic). It should be noted that most commonly used waveguides are homogenous, being entirely filled with material of a single permittivity (which may of course be air) and these types of modes will not be present.

Inside a homogenous waveguide, E_z and H_z satisfy the scalar wave equation inside the waveguide:

$$\left(\frac{\partial^2}{\partial x^2} + \frac{\partial^2}{\partial y^2}\right)E_z + h^2 E_z = 0 \tag{12.13}$$

$$\left(\frac{\partial^2}{\partial x^2} + \frac{\partial^2}{\partial y^2}\right)H_z + h^2 H_z = 0 \tag{12.14}$$

Note that h is given as:

$$h^2 = \omega^2 \mu \varepsilon + \gamma^2 = k^2 + \gamma^2 \tag{12.15}$$

The wavenumber, k, is for the material filling the waveguide. For several simple homogenous waveguides with commonly used waveguide geometries, applying boundary equations on the walls of the waveguide may be used to solve these equations to obtain closed form solutions. The resulting modal solution will possess distinct eigenvalues determined by the cross-section of the waveguide. One important result obtained from this procedure is that waveguide modes, unlike the fundamental TEM mode that propagates in two-wire structures at any frequency, will have a distinct cutoff frequency. It may be shown that the propagation constant varies with frequency as

$$\gamma = \alpha + j\beta = h\sqrt{1 - \left(\frac{f}{f_c}\right)^2} \tag{12.16}$$

where the cutoff frequency, f_c is given by:

$$f_c = \frac{h}{2\pi\sqrt{\mu\varepsilon}} \tag{12.17}$$

By inspection of Eq. (12.26), and recalling the $\exp(j\omega t - \gamma z)$ dependence of the wave propagating in the $+z$ direction (for propagation in the $-z$ direction, replace z with $-z$), the physical significance of the cutoff frequency is clear. For a given mode, when $f > f_c$, the propagation constant γ is imaginary and the wave is propagating. Alternatively, when $f < f_c$, the propagation constant γ is real and the wave decays exponentially. In this case, modes operated below the cutoff frequency attenuate rapidly and are therefore referred to as evanescent modes. In practice, a given waveguide geometry is seldom operated at a frequency where more than one mode will propagate. This fixes the bandwidth of the waveguide to operate at some point above the cutoff frequency of the fundamental mode and below the cutoff frequency of the second order mode, although in some rare instances higher order modes may be used for specialized applications.

The guided wavelength is also a function of the cross-section geometry of the waveguide structure. The guided wavelength is given as:

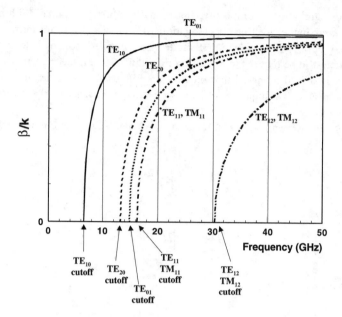

FIGURE 12.4 β/k diagram for WR-90 waveguide illustrating the concept of higher mode propagation and cutoff frequency.

$$\lambda_g = \frac{\lambda_0}{\sqrt{1-\left(\dfrac{f_c}{f}\right)^2}} \qquad (12.18)$$

Note that λ_0 is the wavelength of a plane wave propagating in an infinite medium of the same material as the waveguide. Two important facts may be noted about this expression. First, at frequencies well above the cutoff frequency, $\lambda_g \approx \lambda$. Secondly, as $f \to f_c$, $\lambda \to \infty$, further illustrating that the mode does not propagate. This is another reason that the operating frequency is always chosen above the cutoff frequency. This concept is graphically depicted in Fig. 12.4, a β/k diagram for a standard WR-90 waveguide. At the cutoff frequency, the phase constant goes to zero, indicating that the wave does not propagate. At high frequencies, β approaches the phase constant in an infinite region of the same medium. Therefore, β/k approaches one.

The wave impedance of the waveguide is given by the ratio of the magnitudes of the transverse electric and magnetic field components, which will be constant across the cross-section of the waveguide. For a given mode, the wave impedance for the TE and TM modes are given as:

$$Z_{TE} = \frac{E_T}{H_T} = \frac{j\omega\mu}{\mu} \qquad (12.19)$$

$$Z_{TM} = \frac{E_T}{H_T} = \frac{\gamma}{j\omega\varepsilon} \qquad (12.20)$$

E_T and H_T represent the transverse electric and magnetic fields. Note that at frequencies well above cutoff, the wave impedance for both the TE and TM modes approaches $\sqrt{\mu/\varepsilon}$, the characteristic impedance of a

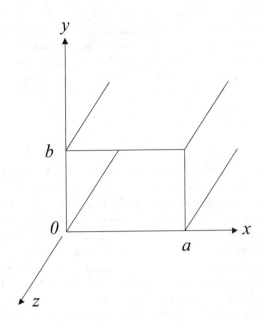

FIGURE 12.5   Geometry of a rectangular waveguide.

plane wave propagating in an infinite medium of the same material as the waveguide. Further, as $f \to f_c$, then $Z_{TE} \to \infty$ and $Z_{TM} \to 0$, again demonstrating the necessity of choosing an operating point well above cutoff.

A variety of geometries are used for waveguides, the most common being the rectangular waveguide, which is used in the microwave and well into the millimeter-wave frequency regime. Shown in Fig. 12.5, it is a rectangular metallic guide of width a and height b. Rectangular waveguide propagate both TE and TM modes. For conciseness, the field components of the TE_{mn} and TM_{mn} modes are presented in Table 12.2. From the basic form of the equations, we see that the effect of the rectangular cross-section

TABLE 12.2   Field Components for Rectangular Waveguide

	TE	TM
E_z	0	$E_0 \sin\left(\dfrac{m\pi x}{a}\right)\sin\left(\dfrac{n\pi y}{b}\right)e^{-\gamma_{mn} z}$
H_z	$H_0 \cos\left(\dfrac{m\pi x}{a}\right)\cos\left(\dfrac{n\pi y}{b}\right)e^{-\gamma_{mn} z}$	0
E_x	$H_0 \dfrac{j\omega\mu n\pi}{h_{mn}^2 b}\cos\left(\dfrac{m\pi x}{a}\right)\sin\left(\dfrac{n\pi y}{b}\right)e^{-\gamma_{mn} z}$	$-E_0 \dfrac{\gamma_{mn} m\pi}{h_{mn}^2 a}\cos\left(\dfrac{m\pi x}{a}\right)\sin\left(\dfrac{n\pi y}{b}\right)e^{-\gamma_{mn} z}$
H_x	$H_0 \dfrac{\gamma_{mn} m\pi}{h_{mn}^2 a}\sin\left(\dfrac{m\pi x}{a}\right)\cos\left(\dfrac{n\pi y}{b}\right)e^{-\gamma_{mn} z}$	$H_0 \dfrac{j\omega\varepsilon n\pi}{h_{mn}^2 b}\sin\left(\dfrac{m\pi x}{a}\right)\cos\left(\dfrac{n\pi y}{b}\right)e^{-\gamma_{mn} z}$
E_y	$-H_0 \dfrac{j\omega\mu m\pi}{h_{mn}^2 a}\sin\left(\dfrac{m\pi x}{a}\right)\cos\left(\dfrac{n\pi y}{b}\right)e^{-\gamma_{mn} z}$	$-E_0 \dfrac{\gamma_{mn} n\pi}{h_{mn}^2 b}\sin\left(\dfrac{m\pi x}{a}\right)\cos\left(\dfrac{n\pi y}{b}\right)e^{-\gamma_{mn} z}$
H_y	$H_0 \dfrac{\gamma_{mn} n\pi}{h_{mn}^2 b}\cos\left(\dfrac{m\pi x}{a}\right)\sin\left(\dfrac{n\pi y}{b}\right)e^{-\gamma_{mn} z}$	$-E_0 \dfrac{j\omega\varepsilon m\pi}{h_{mn}^2 a}\cos\left(\dfrac{m\pi x}{a}\right)\sin\left(\dfrac{n\pi y}{b}\right)e^{-\gamma_{mn} z}$
h_{mn}	$\sqrt{\left(\dfrac{m\pi x}{a}\right)^2 + \left(\dfrac{n\pi y}{b}\right)^2} = 2\pi f_c \sqrt{\mu\varepsilon}$	$\sqrt{\left(\dfrac{m\pi x}{a}\right)^2 + \left(\dfrac{n\pi y}{b}\right)^2} = 2\pi f_c \sqrt{\mu\varepsilon}$

is a standing wave dependence determined by the dimensions of the cross-section, *a* and *b*. Further, *h* (and therefore the propagation constant, γ) are determined by *a* and *b*. The dimensions of the waveguide are chosen so that only a single mode propagates at the desired frequency, with all other modes cut off. By convention, *a* > *b* and a ratio of *a/b* = 2.1 is typical for commercial waveguide types.

The dominant mode in rectangular waveguide is the TE$_{10}$ mode, which has a cutoff frequency of:

$$f_{c_{10}} = \frac{1}{2a\sqrt{\mu\varepsilon}} = \frac{c}{2a} \tag{12.21}$$

The concept of cutoff frequency is further illustrated in Fig. 12.4, a β/*k* diagram for a lossless WR-90 waveguide (note that in the lossless case, the propagation constant will be equal to *j*β). It is apparent that higher order modes may propagate as the operating frequency increases. At the cutoff frequency, β is zero because the guided wavelength is infinity. At high frequencies, the ratio β/*k* approaches one.

A number of variations of the rectangular waveguide are available, including single and double-ridged waveguides, which are desirable because of increased bandwidth. However, closed solutions for the fields in these structures do not exist and numerical techniques must be used to solve for the field distributions, as well as essential design information such as guided wavelength and characteristic impedance. Additionally, losses are typically higher than standard waveguides.

The circular waveguide is also used in some applications, although not nearly as often as rectangular geometry guides. Closed form solutions for the fields in a circular geometry, perfectly conducting waveguide with an inside diameter of 2*a* are given in Table 12.3. Note that these equations use a standard cylindrical coordinate system with ρ the radial distance from the *z*-axis, and ϕ is the angular distance measured from the *y*-axis. The axis of the waveguide is aligned along the *z*-axis. For both the TE$_{mn}$ and TM$_{mn}$ modes, any integer value of *n* ≥ 0 is allowed, and $J_n(x)$ and $J_n'(x)$ are Bessel functions of order *n* and its first derivative. As with the rectangular waveguide, only certain values of *h* are allowed. For the TE$_{mn}$ modes, the allowed values of the modal eigenvalues must satisfy the roots of $J_n'(h_{mn}a) = 0$, where *m* signifies the root number and may range from one to infinity with *m* = 1 the smallest root. Similarly, for the TM$_{mn}$ modes, the values of the modal eigenvalues are the solutions of $J_n(h_{mn}a) = 0$. The dominant mode in the circular waveguide is the TE$_{11}$ mode, with a cutoff frequency given by:

TABLE 12.3   Field Components for Circular Waveguide

	TE	TM
E_z	0	$E_0 J_n(h_{nm}\rho)\cos(n\phi)e^{-\gamma_{nm}z}$
H_z	$H_0 J_n(h_{nm}\rho)\cos(n\phi)e^{-\gamma_{nm}z}$	0
E_ρ	$H_0 \dfrac{j\omega\mu n}{h_{nm}^2\rho} J_n(h_{nm}\rho)\sin(n\phi)e^{-\gamma_{nm}z}$	$-E_0 \dfrac{\gamma_{nm}}{h_{nm}} J_n'(h_{nm}\rho)\cos(n\phi)e^{-\gamma_{nm}z}$
H_ρ	$-H_0 \dfrac{\gamma_{nm}}{h_{nm}} J_n'(h_{nm}\rho)\cos(n\phi)e^{-\gamma_{nm}z}$	$-E_0 \dfrac{j\omega\varepsilon n}{h_{nm}^2\rho} J_n(h_{nm}\rho)\sin(n\phi)e^{-\gamma_{nm}z}$
E_φ	$-H_0 \dfrac{j\omega\mu}{h_{nm}} J_n'(h_{nm}\rho)\cos(n\phi)e^{-\gamma_{nm}z}$	$E_0 \dfrac{\gamma_{nm}}{h_{nm}^2\rho} J_n(h_{nm}\rho)\sin(n\phi)e^{-\gamma_{nm}z}$
H_φ	$H_0 \dfrac{\gamma_{nm}}{h_{nm}^2} J_n(h_{nm}\rho)\sin(n\phi)e^{-\gamma_{nm}z}$	$-E_0 \dfrac{j\omega\varepsilon}{h_{nm}} J_n'(h_{nm}\rho)\cos(n\phi)e^{-\gamma_{nm}z}$

TABLE 12.4   Cutoff Frequencies
for Several Lower Order Waveguide
Modes for Circular Waveguide

$f_c/f_{c_{10}}$	Modes
1.0	TE_{11}
1.307	TM_{01}
1.66	TE_{21}
2.083	TE_{01}, TM_{11}
2.283	TE_{31}
2.791	TE_{21}
2.89	TE_{41}
3.0	TE_{12}

Note: Frequencies have been normalized to the cutoff frequency of the
TE_{10} mode.

$$f_{c_{11}} = \frac{0.293}{a\sqrt{\mu\varepsilon}} \qquad\qquad (12.22)$$

The cutoff frequencies for several of the lowest order modes are given in Table 12.4, referenced to the cutoff frequency of the dominant mode.

12.4 Planar Guiding Structures

Planar guiding structures are composed of a comparatively thin dielectric substrate with metallization on one or both planes. By controlling the dimensions of the metallization, a variety of passive components, transmission lines, and matching circuits can be constructed using photolithography and photoetching. Further, active devices are readily integrated into planar guiding structures. This provides a low-cost and compact way of realizing complicated microwave and millimeter-wave circuits. Microwave integrated circuits (MICs) and monolithic microwave integrated circuits (MMICs) based on this concept are commonly available.

A variety of planar transmission lines have been demonstrated, including microstrip, coplanar waveguide (CPW), slotline, and coplanar stripline. The cross-section of each of these planar transmission lines is shown in Figs. 12.6(a–d). Once the dielectric substrate is chosen, characteristics of these transmission lines are controlled by the width of the conductors and/or gaps on the top planes of the geometry. Of these, the microstrip is by far the most commonly used planar transmission line. CPW is also often used, with slotlines and coplanar striplines being the least common at microwave frequencies, for a variety of reasons that will briefly be discussed later. In this section, we will describe the basic properties of planar transmission lines. Because of its prevalence, the microstrip will be described in detail and closed form expressions for the design of the microstrip will be given.

12.4.1   Microstrip

As seen in Fig. 12.6(a), the simplest form of microstrip consists of a single conductor on a grounded dielectric slab. Microstrip is the most common type of planar transmission line used in microwave and millimeter-wave circuits, with a great deal of design data freely available. A broad range of passive components may be designed with the microstrip, including filters, resonators, diplexers, distribution networks, and matching components. Additionally, three terminal active components can be integrated by using vias to ground. However, this may introduce considerable inductances at high frequencies.

The fundamental mode of propagation for this type of planar waveguide is often referred to as quasi-TEM, because of its close resemblance to pure TEM modes. In fact, noting that the majority of the power

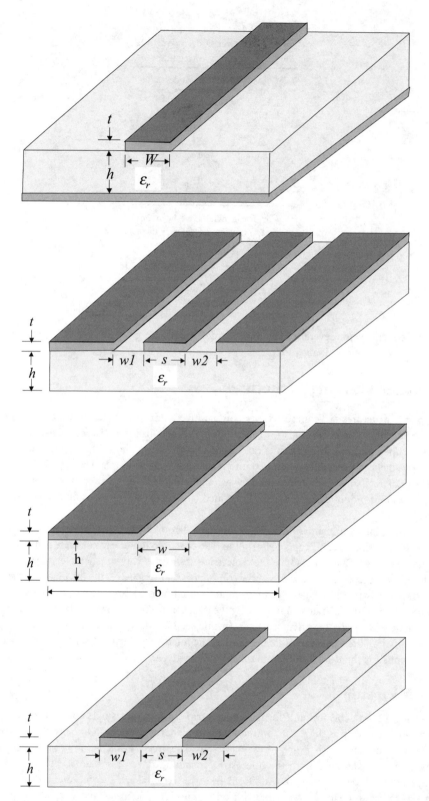

FIGURE 12.6 Cross-section of four of the most popular types of planar guiding structures, including (a) microstrip, (b) coplanar waveguide, (c) slotline, and (d) coplanar stripline.

is confined in the region bounded by the width of the microstrip, the basic characteristics of microstrip are quite similar to the parallel-strip transmission line of Fig. 12.1(a). Because of the presence of the air-dielectric interface, it is not a true TEM mode. The use of the dielectric between the ground and top conductor confines the majority of the fields in this region, but some energy may radiate from the structures. Using a high permittivity substrate and shielding the structure helps to minimize this factor. Microstrip is capable of carrying moderate power levels (a 50 Ω microstrip line on 25 mil alumina can handle several kW of power), is broadband, and enables realization of a variety of circuit topologies, both active and passive.

To design the basic microstrip line, it is necessary to be able to determine characteristic impedance and effective permittivity, preferably as a function of frequency. A wide variety of approximations have been presented in the literature, with most techniques using a quasi-static approximation for the characteristic impedance, Z_0, at low frequencies, and a dispersion model for the characteristic impedance as a function of frequency, $Z_0(f)$ in terms of Z_0. One fairly accurate and simple model commonly used to obtain Z_0 and the effective permittivity, ε_{re}, neglecting the effect of conductor thickness is given as[1]:

$$Z_0 = \frac{\eta}{2\pi\sqrt{\varepsilon_{re}}} \ln\left(\frac{8h}{W} + 0.25\frac{W}{h}\right) \quad \text{for} \left(\frac{W}{h} \le 1\right) \tag{12.23}$$

$$Z_0 = \frac{\eta}{\sqrt{\varepsilon_{re}}} \left\{\frac{W}{h} + 1.393 + 0.667 \ln\left(\frac{W}{h} + 1.444\right)\right\}^{-1} \quad \text{for} \left(\frac{W}{h} \ge 1\right) \tag{12.24}$$

Note that η is 120π-Ω, by definition. The effective permittivity is given as:

$$\varepsilon_{re} = \frac{\varepsilon_r + 1}{2} + \frac{\varepsilon_r - 1}{2} F\left(W/h\right) \tag{12.25}$$

$$F\left(W/h\right) = \left(1 + 12h/W\right)^{-1/2} + 0.04\left(1 - W/h\right)^2 \quad \text{for} \left(\frac{W}{h} \le 1\right)$$

$$F\left(W/h\right) = \left(1 + 12h/W\right)^{-1/2} \quad \text{for} \left(\frac{W}{h} \ge 1\right)$$

With these equations, one can determine the characteristic impedance in terms of the geometry. For a desired characteristic impedance, the line width can be determined from:

$$W/h = \frac{8\exp\left(A\right)}{\exp\left(2A\right) - 2} \quad \text{for } A > 1.52 \tag{12.26}$$

$$W/h = \frac{2}{\pi}\left\{B - 1 - \ln\left(2B - 1\right) + \frac{\varepsilon_r - 1}{2\varepsilon_r}\left[\ln\left(B - 1\right) + 0.39 - \frac{0.61}{\varepsilon_r}\right]\right\} \quad \text{for } A > 1.52 \tag{12.27}$$

where

$$A = \frac{Z_0}{60}\left\{\frac{\varepsilon_r+1}{2}\right\}^{1/2} + \frac{\varepsilon_r-1}{\varepsilon_r+1}\left\{0.23+\frac{0.11}{\varepsilon_r}\right\}$$

$$B = \frac{60\pi^2}{Z_0\sqrt{\varepsilon_r}}$$

Once Z_0 and ε_{re} have been determined, effects of dispersion may also be determined using expressions from Hammerstad[2] and Jensen for $Z_0(f)$ and Kobayashi[3] for $\varepsilon_{re}(f)$. To illustrate the effects of dispersion, the characteristic impedance and effective permittivity of several microstrip lines on various substrates are plotted in Figs. 12.7(a–b) using the formulas from the previously mentioned papers. The substrates

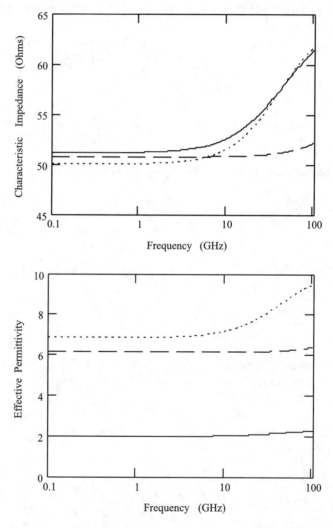

FIGURE 12.7 Dispersion characteristics of 50 Ω line on three substrates (solid line is $\varepsilon_r = 2.33$, $h = 31$ mils, $W = 90$ mils, dotted line is $\varepsilon_r = 10.2$, $h = 25$ mils, $W = 23$ mils and the dashed line is $\varepsilon_r = 9$, h = 2.464 mils, W = 2.5 mils). Shown in (a), the impedance changes significantly at high frequencies for the thicker substrates as does the effective permittivity shown in (b).

indicated by the solid ($\varepsilon_r = 2.33$, h = 31 mils, W = 90 mils) and dashed ($\varepsilon_r = 10.2$, h = 25 mils, W = 23 mils) lines in these figures are typical for those that might be used in a hybrid circuit at microwave frequencies. We can see in Fig. 12.7(a) that the characteristic impedance is fairly flat until X-band, above which it may be necessary to consider the effects of dispersion for accurate design. The third line in the figure is an alumina substrate ($\varepsilon_r = 9$, h = 2.464 mils, W = 2.5 mils) on a thin substrate. The characteristic impedance is flat until about 70 GHz, indicating that this thin substrate is useful at higher frequency operation. The effective permittivity as a function of frequency is shown in Fig. 12.7a. Frequency variation for this parameter is more dramatic. However, it must be remembered that guided wavelength is inversely proportional to the square root of the effective permittivity. Therefore, variation in electrical length will be less pronounced than the plot suggests.

In addition to dispersion, higher frequency operation is complicated by a number of issues, including decreased Q-factor, radiation losses, surface wave losses, and higher order mode propagation. The designer must be aware of the limitations of both the substrate on which he is designing and the characteristic impedance of the lines he is working with. In terms of the substrate, a considerable amount of energy can couple between the desired quasi-TEM mode of the microstrip and the lowest order surface wave mode of the substrate. In terms of the substrate thickness and permittivity, an approximation for determining the frequency where this coupling becomes significant is given by the following expression.[4]

$$f_T = \frac{150}{\pi h}\sqrt{\frac{2}{\varepsilon_r - 1}\arctan(\varepsilon_r)} \qquad (12.28)$$

Note that f_T is in gigahertz and h is in millimeters. In addition to the quasi-TEM mode, microstrip will propagate undesired higher order TE and TM-type modes with cutoff frequency roughly determined by the cross-section of the microstrip. The excitation of the first mode is approximately given by the following expression.[4]

$$f_c = \frac{300}{\sqrt{\varepsilon_r}\left(2W + 0.8h\right)} \qquad (12.29)$$

Again, note that f_c is in gigahertz, and h and W are both in millimeters. This expression is useful in determining the lowest impedance that may be reliably used for a given substrate and operating frequency. As a rule of thumb, the maximum operating frequency should be chosen somewhat lower. A good choice for maximum frequency may be 90% of this value or lower.

A variety of techniques have also been developed to minimize or characterize the effects of discontinuities in microstrip circuits, a variety of which are shown in Figs. 12.8(a,b) including a microstrip bend and a T-junction. Another common effect is the fringing capacitance found at impedance steps or open-circuited microstrip stubs.

The microstrip bend allows flexibility in microstrip circuit layouts and may be at an arbitrary angle with different line widths at either end. However, by far the most common is the 90° bend with equal widths at either end, shown on the left of Fig. 12.8a. Due to the geometry of the bend, excess capacitance is formed causing a discontinuity. A variety of techniques have been used to reduce the discontinuity by eliminating a sufficient amount of capacitance, including the mitered bend shown on the right. Note that another way of reducing this effect is to use a curved microstrip line with sufficiently large radius to minimize the effect. A second type of discontinuity commonly encountered by necessity in layouts is the T-junction, shown in Fig. 12.8b, which is formed at a junction of two lines. As with the bend, excess capacitance is formed, degrading performance. The mitered T-junction below is used to reduce this problem. Again, a variety of other simple techniques have also been developed.

Fringing capacitance will be present with microstrip open-circuited stubs and at impedance steps. With the open-circuited stub, this causes the electrical length of the structure to be somewhat longer.

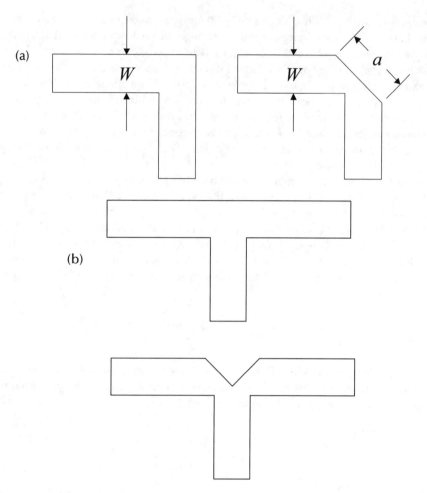

FIGURE 12.8 Two common microstrip discontinuities encountered in layout, including (a) the microstrip bend and (b) the T-junction.

For an impedance step, the lower impedance line will also appear to be electrically longer. The simplest way of compensating for this problem is by modeling the capacitance and effective length of the fringing fields. Again, a variety of simple models have been developed to perform this task, most based on quasi-static approximations. A commonly used expression for the length extension of an open end based on empirical data is given by the following expression.[5]

$$\frac{\Delta l_{oc}}{h} = 0.412 \frac{\varepsilon_{re} + 0.3}{\varepsilon_{re} - 0.258} \left[\frac{W/h + 0.264}{W/h + 0.8} \right] \qquad (12.30)$$

This expression is reported to yield relatively accurate results for substrates with permittivity in the range of 2 to 50, but is not as accurate for wide microstrip lines. For the impedance step, a first order approximation for determining the excess length of the impedance step is to multiply the open-end extension, $\Delta l_{oc}/h$ by an appropriate factor to obtain a useful value, i.e., $\Delta l_{step}/h \approx \Delta l_{oc} (w_1/w_2 - 1)/h$.

Because of the prevalence of microstrip, modern microwave CAD tools typically have extensive libraries for microstrip components, including discontinuities effects.

12.4.2   Coplanar Waveguide (CPW)

Coplanar Waveguide (CPW), shown in Fig. 12.6b, consists of a signal line and two ground planes on a dielectric slab with metallization on one side. For a given substrate, characteristic impedance is determined by the signal line width, s, and the two gaps, w_1 and w_2. This structure often demonstrates better dispersion characteristics than microstrip. Additionally, three terminal devices are easily integrated into this uniplanar transmission line that requires no vias for grounding. For this reason, parasitics are lower than microstrip making CPW a good choice for high frequency operation where this is a primary design concern.

The three-conductor line shown in Fig. 12.6b supports two fundamental modes, including the desired CPW-mode and an undesired coupled slotline mode if the two ground planes separating the signal line are not kept at the same potential. For this reason, wires or metal strips referred to as *air bridges* are placed at discontinuities where mode conversion may occur.

Packaging may be a problem for this type of structure, because the bottom plane of the dielectric may come in close proximity with other materials, causing perturbations of the transmission line characteristics. In practice, this is remedied by using *grounded* or *conductor-backed* CPW (CB-CPW) where a ground plane is placed on the backside for electrical isolation. At high frequencies, this may present a problem with additional losses through coupling to the parallel-plate waveguide mode. These losses can be minimized using vias in the region around the transmission line to suppress this problem.

Although CPW was first proposed by Wen[6] in 1969, acceptance of CPW has been much slower than microstrip. For this reason, simple and reliable models for CPW are not as readily available as for microstrip. A compilation of some of the more useful data can be found in Reference 6.

12.4.3   Slotline and Coplanar Stripline

Two other types of planar transmission lines are slotline and coplanar stripline (CPS). These structures are used less often than either microstrip or CPW, but do find some applications. Both of these structures consist of a dielectric slab with metallization on one side. Slotline has a slot of width w etched into the ground plane. CPS consists of two metal strips of width w_1 and w_2 separated by a distance s on the dielectric slab. Due to their geometry, both of these structures are balanced transmission line structures, and are useful in balanced circuits such as mixers and modulators. Only limited design information is available for these types of transmission lines.

The slotline mode is non-TEM and is almost entirely TE. However, no cutoff frequency exists as with the waveguide TE modes discussed previously in this section. Microwave circuits designed solely in slotline are seldom used. However, slotline is sometimes used in conjunction with other transmission line types such as microstrip or CPW for increased versatility. Examples of these include filters, hybrids, and resonators. Additionally, slotline is sometimes used in planar antennas, such as the slot antenna or some kinds of multilayer patch antennas.

The CPS transmission line has two conductors on the top plane of the circuit, allowing series or shunt elements to be readily integrated into CPS circuits. CPS is often used in electro-optic circuits such as optic traveling wave modulators, as well as in high-speed digital circuits. Due to its balanced nature, CPS also makes an ideal feed for printed dipoles. Difficulties (or benefits, depending on the application) with CPS include high characteristic impedances.

References

1. E. Hammerstad, Equations for microstrip circuit design, *Proc. European Microwave Conf.*, 1975, 268–272.
2. E. Hammerstad and O. Jensen, Accurate models for microstrip computer-aided design, *IEEE MTT-S Int. Microwave Symp. Dig.*, 1980, 407–409.
3. M. Kobayashi, A dispersion formula satisfying recent requirements in microstrip CAD, *IEEE Trans.*, MTT-36, August 1988, 1246–1250.

4. G.D. Vendelin, Limitations on stripline Q, *Microwave J.,* 13, May 1970, 63–69.
5. R. Garg and I.J. Bahl, Microstrip discontinuities, *Int. J. Electron.,* 45, July 1978, 81–87.
6. C.P. Wen, Coplanar waveguide: A surface strip transmission line suitable for non-reciprocal gyro-magnetic device applications, *IEEE Trans.,* MTT-23, 1975, 541–548.
7. K.C. Gupta, R. Garg, I. Bahl, and R. Bhartia, *Microstrip Lines and Slotlines*, Artech House, Inc., Norwood MA, 1996.

13

Linear Measurements

Ron E. Ham
Consulting Engineer

13.1 Introduction

Microwave and RF measurements can be classified in two distinct but often overlapping categories: signal measurements and network measurements. Signal measurements include observation and determination of the characteristics of waves and waveforms. These parameters can be obtained in the time, frequency, or modulation domain. Network measurement determines the terminal and signal transfer characteristics of devices and systems with any number of ports.

13.2 Signal Measurements

Signal measurements are taken in any one or more of three measurement planes as illustrated in Fig. 13.1. The most common measurement at low frequencies is in the time domain where the amplitude of a signal waveform is observed with respect to time. The instrument used for this is an oscilloscope. By continuing to observe the amplitude of the signal over a small frequency range, the spectral components of the signal are obtained. This measurement is normally made with a spectrum analyzer. Determining the instantaneous frequency of a signal versus time is a modulation domain measurement.

13.2.1 Time Domain

Observation of RF and microwave signals with an analog oscilloscope is limited by the speed of response of the instrument circuits and of the display. Building such an instrument for operation beyond a few hundred megahertz is very difficult and expensive. For observing very high-speed waveforms signal sampling techniques are incorporated.

A sampling oscilloscope measures the value of a waveform at a particular time and digitally stores the sample data for display. If the sampling can be performed fast enough, the entire waveform shape can be recreated from the sample data. This is done at relatively low frequencies; however, as the frequency increases it is not possible to capture enough points during one waveform occurrence. By delaying subsequent samples to be taken a very short time later during the occurrence of another cycle of the waveform, the recurring waveform shape can ultimately be reconstructed from the stored digital data.

Note the qualification that the waveform shape that can be measured by the high-speed sampling technique must be recurring. This makes capturing a onetime occurrence very difficult and, even if points in a gigahertz waveform can be captured, this does not mean that one cycle of the waveform can be captured.

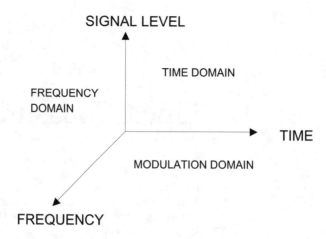

FIGURE 13.1   Signals are characterized by three different types of measurements.

13.2.2   Frequency Domain

The number of measurements that must be made on a signal over a specified period of time is a function of the stability and modulation placed on the signal. The exact measurement of the frequency of a stable and spectrally pure signal is performed with a frequency counter and measurements are normally made a few times per second. Direct counting circuits are available well into the lower microwave frequency range. At high microwave frequencies counters use conversion oscillators and mixers to heterodyne the signal down in frequency to where it can be directly counted. Microprocessor controllers and knowledge of the exact frequency of the conversion oscillators enables an exact signal frequency to be calculated.

A spectrum analyzer [1] is used to make frequency domain measurements of complex signals and signals with characteristics that vary with time. This is basically a swept frequency filter with a detector to determine the signal amplitude within the bandwidth of the filter and some means of displaying or storing the measured information. To increase the selectivity and dynamic range of such a basic instrument, heterodyne conversions are used. Figure 13.2 is the block diagram of a typical microwave spectrum analyzer.

The first intermediate frequency is chosen to permit a front-end filter to eliminate the image from the first mixer. In this case, 300 MHz is chosen because the tunable filter, usually a YIG device, will have considerable attenuation at the image frequency 600 MHz away from the desired signal. The second intermediate frequency is chosen because reasonably selective filters can be constructed to enable resolving signal components that are close to each other. Additionally, detector and signal processing components, such as digital signal processors, can be readily constructed at the lower frequency.

Because the normal frequency range required from a microwave spectrum analyzer is many octaves wide, multiple first conversion oscillators are required; however, this is an extremely expensive approach. Spectrum analyzers use a harmonic mixer for the first conversion and the first filter is tuned to eliminate the products that would be received due to the undesired harmonics of the conversion oscillator. Note the list of harmonic numbers (n) and the resulting tuned frequency of the example analyzer. As the harmonic number increases the sensitivity of the analyzer decreases because the harmonic mixer efficiency decreases with increasing n.

The most important spectrum analyzer specifications are:

1. Frequency tuning range — to include all of the frequency components of the signal to be measured.
2. Frequency accuracy and stability — to be more stable and accurate than the signal to be measured.
3. Sweep width — the band of frequencies over which the unit can sweep without readjustment.
4. Resolution bandwidth — narrow enough to resolve different spectral components of the signal.
5. Sensitivity and/or noise figure — to observe very small signals or small parts of large signals.

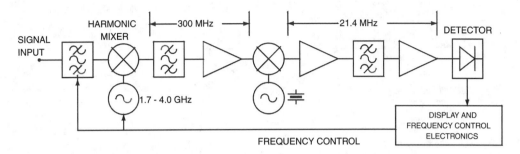

FIGURE 13.2   Simplified block diagram of a microwave spectrum analyzer.

OSCILLATOR HARMONIC NUMBER (n)	EFFECTIVE OSCILLATOR FREQUENCY (GHz)	TUNED RF FREQUENCY (GHz)
1	1.7 - 4.0	2.0 - 4.3
2	3.4 - 8.0	3.7 - 8.3
3	5.1 - 12.0	5.4 - 12.3
4	6.8 - 14.0	7.1 - 14.3
5	8.5 - 20.0	8.8 - 20.3

6. Sweep rate — maximum sweep rate is established by the settling time of the filter that sets the resolution bandwidth.
7. Dynamic range — the difference between the largest and smallest signal the analyzer can measure without readjustment.
8. Phase noise — a signal with spectral purity greater than that of the analyzer conversion oscillators cannot be characterized.

Spectrum analyzers using other than swept frequency techniques can be made. For example, high speed sampling methods used with digital signal processors (DSP) calculating the Fast Fourier Transform (FFT) are readily implemented; however, the speed of operation of the logic circuits limits the upper frequency of operation. This is a common method of intermediate frequency demodulation and the usable frequency will move upward with semiconductor development.

13.2.3   Modulation Domain

Modulation domain measurements [2] yield the instantaneous frequency of a signal as a function of time. Two examples of useful modulation domain data are the instantaneous frequency of a phase-locked oscillator as the loop settles and the pulse repetition rate of a fire control radar as it goes from search mode (low pulse repetition frequency or PRF) to lock and fire mode (high PRF).

A modulation domain analyzer establishes the exact time at which a desired event occurs and catalogs the time. The event captured in a phase-locked oscillator is the zero crossing of the oscillator output voltage. For a radar it is the leading edge of each pulse. From this information the event frequency is calculated. Various other modulation domain analyzers can be made with instantaneous frequency correlators and frequency discriminators.

13.3 Network Measurements

Low frequency circuit design and performance evaluation is based upon the measurement of voltages and currents. Knowing the impedance level at a point in a circuit to be the ratio of voltage to current, a voltage or current measurement can be used to calculate power. By measuring voltage and current as a complex quantity, yielding complex impedances, this method of circuit characterization can be used at relatively high frequencies even with the limitations of nontrivial values of circuit capacitive and inductive

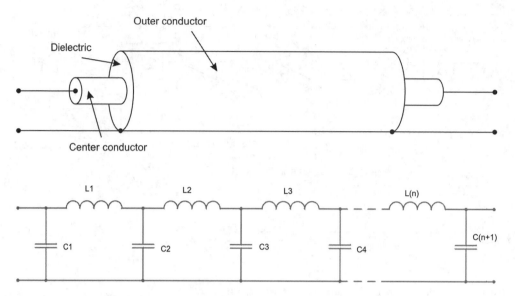

FIGURE 13.3   Examples of transmission lines: (a) coaxial; (b) lumped element.

parasitics. When the parasitics can no longer be treated as lumped elements, distributed circuit concepts must be used.

A simple transmission line such as the coaxial line in Fig. 13.3a can, if physically very small in all dimensions with respect to a wavelength, be modeled as a lumped element circuit as shown in Fig. 13.3b; however, as the size of the line increases relative to the wavelength, it becomes necessary to use an extremely complex lumped element model or to use the transmission line equations for the distributed line. The concept of a transmission line accounts for the transformation of impedances between circuit points and for the time delay between points that must be considered when the circuit size approaches a significant fraction of a wavelength of the frequency being measured; hence, RF and microwave measurements are primarily based upon transmission line concepts and measurements.

The basic quantities measured in high frequency circuits are power, impedance, port-to-port transfer functions of n-port devices, frequency, and noise [3, 4].

13.3.1   Power

Microwave power cannot be readily detected with equipment used at lower frequencies such as voltmeters and oscilloscopes [5]. The RF and microwave utility of these instruments are limited by circuit parasitics and the resultant limited frequency response. Central to all microwave measurements is the determination of the microwave power available at ports in the measurement circuit. To facilitate measurements, a characteristic impedance or reference resistance is assumed. The instruments used to measure microwave and RF power typically have a 50-ohm input and output impedance at the frequency being measured.

Diode detectors sense the amplitude of a signal. By establishing the input impedance of a diode detector, the power of a signal at a test port can be measured. The diode detector shown in Fig. 13.4 allows current to pass through the diode when the diode is forward biased and prevents current from flowing when the diode is reverse biased. The average of the current flow when forward biased results in a DC output from the lowpass RC filter that is proportional to the amplitude of the input voltage. Note that as the diode junction area must be small to minimize the parasitic junction capacitance that would short the signal across the diode, the load resistor must be a relatively large value to minimize the diode current; therefore, the impedance seen looking into the diode detector is established primarily by the resistor placed across the detector input. If the input voltage is less than that where the diode current becomes linearly proportional to the input voltage, the diode is in a predominantly square law region and the voltage out

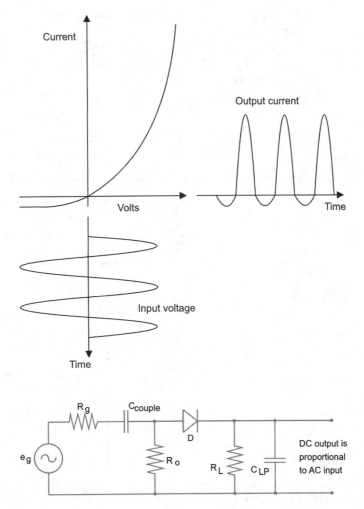

FIGURE 13.4 Diode detector: (a) diode detector waveforms; (b) diode detector circuit.

of the detector is proportional to the input power in decibels. This square law range typically extends over a 50-dB range from –60 dBm to –10 dBm in a 50-ohm system. Diodes are used in the linear range up to about 10 dBm. The one significant disadvantage of the diode detector is the temperature sensitivity of the diode. The diode detector response can be very fast, but it cannot easily be used for accurate power measurement.

The most accurate RF and microwave power measuring devices are thermally dependent detectors. These detectors absorb the power and by either measuring the change in the detector temperature or the change in the resistance of the detecting device with a change in temperature, the power absorbed by the detector can be accurately determined.

The primary thermally dependent detectors are the bolometer and the thermistor. They are placed across the transmission media as a matched impedance termination. A bridge as shown in Fig. 13.5(a) can be used to detect a change in the resistance of the bolometer. To increase the detector sensitivity, two units can be placed in parallel for the RF/microwave signal and in series for the change in DC resistance as shown in Fig. 13.5(b). Unfortunately, this circuit can also be used as a thermometer; therefore, an identical pair of bolometer detectors are normally placed in close thermal proximity but only one of the detectors is used to detect signal power. The other detector is used to detect environmental temperature changes so that the difference in temperature change is due to the signal power absorbed in the upper detector.

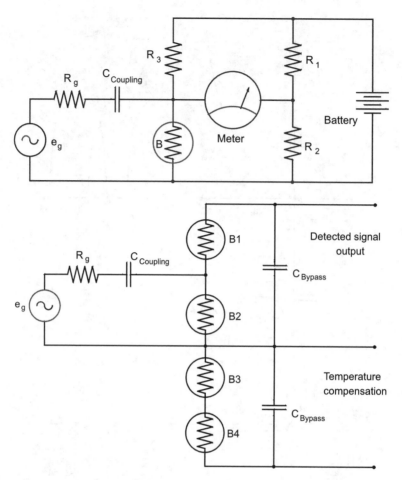

FIGURE 13.5 Thermally dependent detector circuits: (a) bolometer in a bridge circuit; (b) temperature compensated bolometer head; (c) self-balancing bridge circuit.

To maintain a constant impedance looking into the bolometer elements, a bias current is passed through the elements to increase their temperature above operational ambient. The resistance of the detectors is compared to a fixed resistance in a bridge. The bridge error is used to adjust the bias current in the bolometers. The bias energy that must be removed from the detector to maintain a constant resistance is equal to the amount of signal energy absorbed by the detector; therefore, the meter can be calibrated in power by knowing the amount of bias power applied to the detector. Figure 13.5(c) is a simplified example of a self-balancing bridge circuit.

13.3.2 Impedance

Consider a very simple transmission line, two parallel pieces of wire spaced a uniform distance and in free space, as shown in Fig. 13.6. A DC voltage with a source resistance R_g and series switch is connected to terminal 1 and a resistor R_L is placed across terminal two.

First, let the length of the wires be zero. Close the switch. If the load resistor R_L is equal to the source resistance R_g, the condition necessary for maximum power transfer from a source to a load, then the voltage across the load R_L is $e_g/2$. This is the voltage that will be measured from a signal generator when the output is terminated in its characteristic impedance, commonly called R_o. The signal power from the signal generator, and also the maximum available power from the generator, is then e_g^2/R_g. If R_L is a short circuit the output voltage is zero. If R_L is an open circuit the output voltage is 2 times $e_g/2$ or e_g.

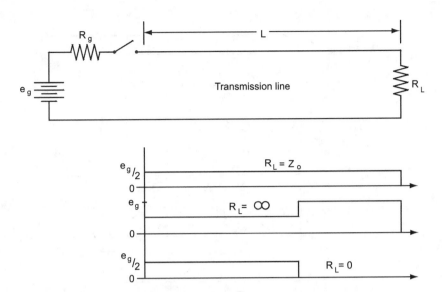

FIGURE 13.6 Switched DC line voltage at time > length/velocity for various impedances at the end of the line.

Now let the line have a length, L. When the switch is closed, a traveling wave of voltage moves toward the load resistor at the speed of light, c. At time t, the wave has moved down the line a distance ct. A wave of current travels with the wave of voltage. If the characteristic impedance of this parallel transmission line is Z_o and the load resistance is equal to Z_o, then the current traveling with the voltage wave has a value at any point along the line of the value of the voltage at that point divided by Z_o. For this special case, when the wave reaches the load resistor, all of the energy in the wave is dissipated in the resistor; however, if the resistor is not equal to Z_o there is energy in the wave that must go someplace as it is not dissipated in the load resistor.

This mismatch between the characteristic impedance of the line and the terminating load resistor results in a reflected wave that travels back toward the voltage source. If the load resistor is a short circuit, the voltage at the end of the line must equal zero at all times. The only way for this to occur is for the reflected voltage at the end of the wire to be equal to -1 times the incident voltage at that same point. If the load is an open circuit the reflected voltage will be exactly equal to the incident voltage; hence the sum of the incident and reflected voltages will be twice the value of the incident voltage at the end of the line. Note the similarity of these three cases to those of the zero length line.

Now replace the DC voltage source and switch with a sinusoidal voltage source as in Fig. 13.7. The voltages shown are the RMS values of the vector sum of the incident and reflected waves. As the source voltage varies, the instantaneous value of the sinusoidal voltage between the wires travels down the wires. The ratio of the traveling voltage wave to the traveling current wave is the characteristic impedance of the transmission line. If the terminating impedance is equal to the line characteristic impedance, there is no wave reflected back toward the generator; however, if the termination resistance is any value other than Z_o there is a reflected wave. If R_L is a real impedance and greater than Z_o, the reflected wave is 180° out of phase with the incident wave. If R_L is a real impedance and is less than Z_o, the reflected wave is in phase with the incident wave. The amount of variation of R_L from Z_o determines the magnitude of the reflected wave. If the termination is complex, the phase of the reflected wave is neither zero nor 180°.

Assuming the generator impedance R_s equal to the line characteristic impedance, so that a reflected wave incident on the generator does not cause another reflected wave, sampling the voltage at any point along the transmission line will yield the vector sum of the incident and reflected waves. With a matched impedance ($R_L = Z_o$) termination the magnitude of the AC voltage along the line is a constant. With a short circuit termination, the voltage magnitude at the load will be zero and, moving back toward the generator, the

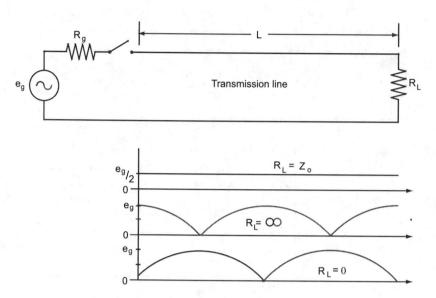

FIGURE 13.7 Waveforms on the line for a sinusoidal source and various impedances at the end of the line.

voltage one-half wavelength from the end of the line will also be zero. With an open circuit there is a voltage maxima at the end of the line and a minima on the line one-quarter wavelength back toward the generator.

The complex reflection coefficient Γ is the ratio of the reflected wave to the incident wave; hence it has a magnitude ρ between 0 and 1 and an angle θ between $+180°$ and $-180°$. The reflection coefficient as a function of the measured impedance Z_L with respect to the measurement system characteristic impedance Z_o is

$$\Gamma = \frac{Z_L - Z_o}{Z_L + Z_o} = \rho\left(\sin\theta + j\cos\theta\right)$$

13.3.2.1 Slotted Line

Determination of the relative locations of the minima and maxima along the line, or similarly the determination of the magnitude of waves traveling toward and away from the load resistor, is the basis for the measurement of RF and microwave impedance and the most basic instrument used for making this measurement is the slotted line. The slotted line is a transmission line with a slit in the side that enables a probe to be inserted into the transmission mode electromagnetic field as shown in Fig. 13.8. A diode detector placed within the sliding probe provides a DC voltage that is proportional to the magnitude of the field in the slotted line. As the probe is moved along the line, the minimum and

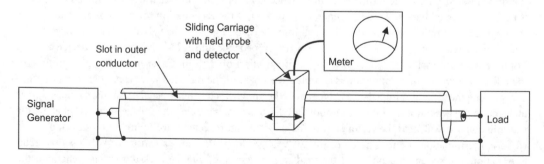

FIGURE 13.8 A slotted line is used to measure the impedance of an unknown load.

maximum field positions and magnitudes can be determined. The ratio of the maximum field magnitude to the minimum field magnitude is the standing wave ratio (SWR). SWR is normally stated as a scalar quantity and is

$$SWR = \frac{1+\rho}{1-\rho}$$

Before placing an unknown impedance at the measurement terminal of the slotted line, the line is calibrated with a short circuit. This establishes a measurement plane at the short circuit. Any measurement made after calibrating with this reference short is made at the plane of the short circuit. A phase reference is located at the position on the slotted line of a minimum voltage measurement. The distance between two minimum voltage measurement locations is one-half wavelength at the measurement frequency.

If the short circuit is replaced with an open circuit, the minimum voltage locations along the line are shifted by one-quarter wavelength. The difference between the phase of a reflected wave of an open and a short circuit is 180°; hence, the distance between two minimum measurements represents 360° of phase shift in the reflected wave. Note that it is very difficult to use an open circuit for a reference at high frequencies because fringing and radiated fields at the end of the transmission line result in phase and amplitude errors in the reflected wave.

The impedance to be measured now replaces the calibrating short circuit. The new minimum voltage location is found by moving the detector carriage along the slotted line. The distance the minimum voltage measurement moves from the short circuit reference location is ratioed to 180° at a quarter of a wavelength shift (For example, a minimum shift of one-eighth wavelength results from a reflection coefficient phase shift of 90°). This is the phase difference between the forward and reflected waves on the transmission line. Either way the minimum moves from the short circuit calibrated reference point is a shift from 180° back toward 0°. If the shift is toward the load, then the actual phase of the reflection coefficient is −180° plus the shift. If the shift is toward the generator from the reference point, the actual phase of the reflection coefficient is 180° minus the shift.

The best method of visualizing complex impedances as a function of the complex reflection coefficient is the Smith Chart [6, 7, 8]. A simplified Smith Chart is shown in Fig. 13.9. The distance from the center of the chart to the outside of the circle is the reflection coefficient ρ. The minimum value of ρ is 0 and the maximum value is 1. If there is no reflection, the impedance is resistive and equal to the characteristic impedance of the transmission line or slotted line. If the reflected wave is equal to the incident wave, the reflection coefficient is one and the impedance lies on the circumference of the circle. If the angle of the reflection coefficient is zero or 180°, the impedance is real and lies along the central axis. Reflection coefficients with negative angles have capacitive components in the impedance and those with positive angles have inductive components.

13.3.2.2 Directional Coupler

Slotted lines must be on the order of a wavelength long. Additionally, they do not lend themselves to computer-controlled or automatic measurements. Another device for measuring the forward and reflected waves on a transmission line is the directional coupler [9]. Physically this is a pair of open transmission lines that are placed close enough for the fields generated by a propagating wave in one line to couple to the other line, hence inducing a proportional wave in the second line. The coupler is a four-port device. Referencing Fig. 13.10, a wave propagating to the right in line one couples to line 2 and propagates to the left. A wave propagating in line 1 to the left couples to line 2 and propagates to the right; therefore, the outputs from ports 3 and 4 are proportional to the forward and reverse wave propagating in line 1.

The primary specifications for a coupler are its useful frequency range, the attenuation of the coupled wave to the coupled ports (coupling), and the attenuation of a signal traveling in the opposite direction to the desired signal at the desired signal's coupled port (directivity). For example, a 10-dB coupler with

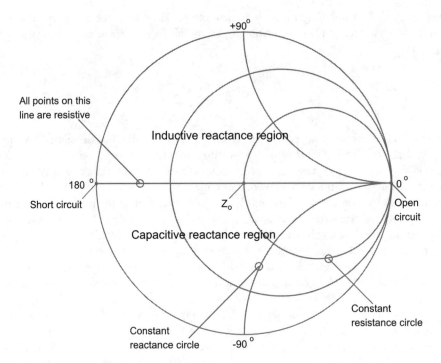

FIGURE 13.9   The Smith Chart is a plot of all nonnegative real impedances.

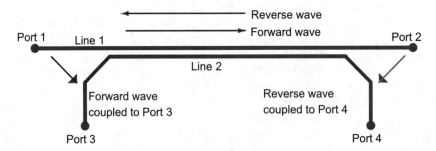

FIGURE 13.10   A directional coupler separates forward and reverse waves on a transmission line.

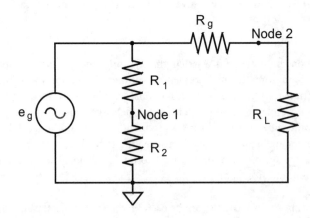

FIGURE 13.11   A resistive bridge can be used to measure the reverse wave on a transmission line.

a 10-dBm signal propagating in the forward direction in line 1 will output a 10-dBm signal at port 3. If the directivity of the coupler is 30 dB there will also be a –40-dBm signal resulting from the forward wave at the reverse wave port, port 4. If the forward wave is properly terminated with the system impedance, there will be no reverse wave on line 1; hence, there will not be an output at port 4 due to a reverse wave.

Note that power must be conserved through the coupler. Therefore, if in the example above, 1.54 dBm is coupled from the forward signal in line 1 to port 3, there will be only a 90-dBm output from port 2. This power must be taken into account in the measurement. The greater the attenuation to the coupled ports, the less the correction will be. Normally 20- or 30-dB couplers are used so the correction is minimal and, in many cases, small enough to be ignored.

By measuring the power from the forward and reverse coupled ports, the magnitude of the reflection coefficient and the SWR can be calculated. Typically the most common indication of the quality of the power match of a device being measured is the attenuation of the reflected wave. This is

$$RL = 10 * \log_{10}\left(\frac{P_{Forward}}{P_{Reverse}}\right)$$

As power is proportional to voltage squared, when the termination resistance is equal on all ports, the return loss can also be expressed as a voltage ratio

$$RL(dB) = 10 * \log_{10}\left(\frac{V_{Forward}^2 / R_o}{V_{Reverse}^2 / R_o}\right) = -20 * \log_{10}\left(\frac{V_{Reverse}}{V_{Forward}}\right) = -20 * \log_{10}(\rho)$$

Hence the return loss is the magnitude of the reflection coefficient ρ in decibels.

13.3.2.3 Resistive Bridge

The directional coupler is functionally equivalent to a bridge circuit, the primary difference being that the only losses in the transmission line coupler are from parasitics and can be designed to be very small. Referencing Fig. 13.11, the voltage drop across R_g when R_L equals R_g is $e_g/2$. For this case, the equivalent reflected wave amplitude is zero. By summing circuit voltages it is found that

$$\Gamma = \frac{e_g - e_g/2}{e_g/2} = \frac{V_{Reverse}}{V_{Forward}}$$

By placing a series circuit of two equal resistors across e_g, node 1 has a voltage of $e_g/2$. The voltage between node 1 and node 2 is equal to the reflected wave. Note that this is the standard resistive bridge circuit.

13.3.3 Network Analyzers

General RF and microwave network analyzers (NWA) measure scattering parameters (s-parameters). These measurements use a source with a well-defined impedance equal to the system impedance and all ports of the device under test (DUT) are terminated with the same impedance. The output port being measured is terminated in the test channel of the network analyzer that has an input impedance equal to the system characteristic impedance. Measurement of system parameters with all ports terminated minimizes the problems caused by short-circuit, open-circuit, and test-circuit parasitics that cause considerable difficulty in the measurement of Y- and h-parameters at very high frequencies. S-parameters can be converted to Y- and h-parameters.

Figure 13.12 illustrates a two-port device under test. If the generator is connected to port 1 and a matched load to port 2, the incident wave to the DUT is V_1^+. A wave reflected from the device back to

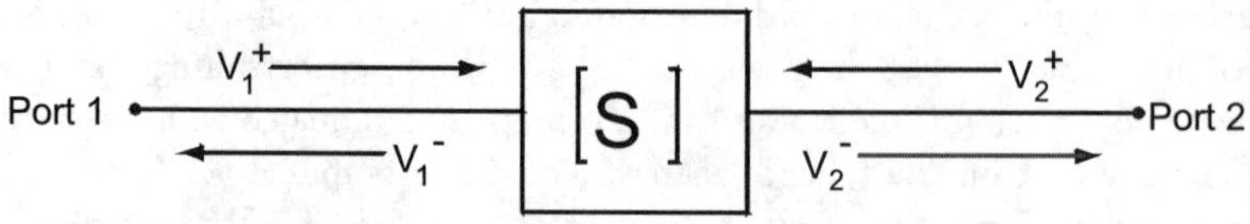

FIGURE 13.12   S-parameters are defined by forward and reverse voltage waves.

port one is V_1^-. A signal traveling through the DUT and toward port 2 is V_2^-. Any reflection from the load (zero if it is truly a matched load) is V_2^+. The s-parameters are defined in terms of these voltage waves:

$s_{11} = V_1^-/V_1^+ =$ Input terminal reflection coefficient, Γ_1

$s_{21} = V_2^-/V_1^+ =$ Forward gain or loss

By moving the signal generator to port 2 and terminating port 1, the other two port s-parameters are measured:

$s_{12} = V_1^-/V_2^+ =$ Reverse gain or loss

$s_{22} = V_2^-/V_2^+ =$ Output terminal reflection coefficient, Γ_2

The s-matrix is then

$$[S] = \begin{bmatrix} s_{11} & s_{12} \\ s_{21} & s_{22} \end{bmatrix}$$

where

$$\begin{bmatrix} V_1^- \\ V_2^- \end{bmatrix} = [S] \begin{bmatrix} V_1^+ \\ V_2^+ \end{bmatrix}$$

13.3.3.1 Scalar Analyzer

A scalar network analyzer, Fig. 13.13, with resistor-loaded diode probes or power meters is used to measure scalar return loss and gain. Diode detectors are either used in the square law range as power detectors or logarithmic amplifiers are used in the analyzer to produce nominally a 50 dB dynamic range of measurement. A spectrum analyzer with a tracking test generator can be used as a scalar analyzer with up to 90 dB of dynamic range.

Gains and losses are calculated in scalar analyzers by adding and subtracting relative power levels in decibels. Note that this can only establish the magnitude of the reflection coefficient so that an absolute impedance cannot be measured. To establish the impedance of a device, the phase angle of the reflected wave relative to the incident wave must be known. To measure the phase difference between the forward and reflected wave, a phase meter or vector network analyzer is used.

13.3.3.2 Vector Heterodyne Analyzer

Accurate direct measurement of the phase angle between two signals at RF and microwave frequencies is difficult; therefore, most vector impedance analyzers downconvert the signals using a common local oscillator. By using a common oscillator the relative phase of the two signals is maintained. The signal is ultimately converted to a frequency where rapid and accurate comparison of the two signals yields their phase difference. In these analyzers the relative amplitude information is maintained so that the amplitude measurements are also made at the low intermediate frequency (IF).

The vector network analyzer (VNA) is a multichannel phase-coherent receiver with a tracking signal source. When interfaced with various power splitters and couplers, the channels can measure forward, reverse, and transmitted waves. As the phase and amplitude information is available on each channel,

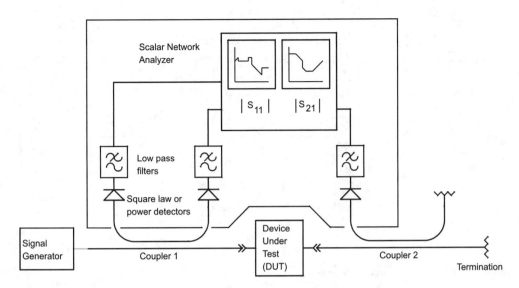

FIGURE 13.13   A scalar network analyzer can measure the magnitude of gain and return loss.

parameters of the device being measured can be computed. The most common VNA configuration measures the forward and reflected waves to and from a two-port device. From these measurements, the two-port scattering matrix can be computed.

The automatic vector network analyzer performs these operations under the supervision of a computer requiring the operator to input instructions relating to the desired data. The computer performs the routine "housekeeping."

The use of computers also facilitates extensive improvement in measurement accuracy by measuring known high-quality components, calculating nonideal characteristics of the measurement system, and applying corrections derived from these measurements to data from other devices. In other words, the accuracy of a known component can be transferred to the measurement accuracy of an unknown component

With the measurement frequency accurately known and the phase and amplitude response measured and corrected, the Fourier transform of the frequency domain yields the time domain response. A very useful measurement of this type transforms the s_{11} frequency domain data to a time domain response with the same information as time domain reflectometry; that is, deviations from the characteristic impedance can be seen over the length of the measured transmission media.

The simplified block diagram of a typical multichannel VNA is shown in Fig. 13.14. There are two channels fed from the test set. The inputs are converted first to a low intermediate frequency such as 20 MHz and then to 100 kHz before being routed to phase detectors. The first conversion oscillator is followed by a comb generator and the oscillator is phase locked to the mixer output so the unit will frequency track the test source.

Multiple methods of generating the conversion oscillator voltages are used. Low RF frequency analyzer signal generators commonly generate a test signal plus another output that is frequency offset by the desired IF frequency. This can be done with offset synthesizers or by mixing a common oscillator with a stable oscillator at the IF frequency and selecting the desired mixing product using phasing or filtering techniques.

For microwave analyzers, because of the high cost of oscillators and the wide frequency coverage required, a more common method of generating conversion oscillators is to use a low frequency oscillator and a very broadband frequency multiplier. A harmonic of the low frequency conversion oscillator is offset by an oscillator equal to the IF frequency and the conversion oscillator is then phase locked to the

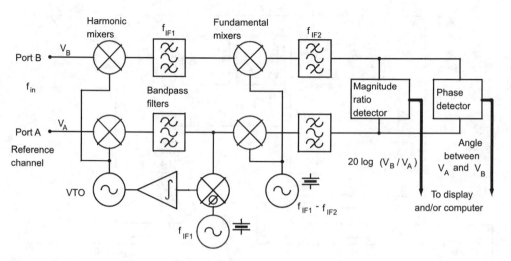

FIGURE 13.14   A vector network analyzer measures complex ratios.

reference channel of the NWA. The reference channel signal is normally the forward wave voltage derived from a directional coupler in an impedance measurement.

    The outputs of the synchronous detectors supply the raw data to be converted to a format compatible with the computer. Corrections and manipulation of the data to the required output form is then done by the processors.

    The test set supplies the first mixer inputs with the sampled signals necessary to make the desired measurement and there are many possible configurations. The most versatile is the two-port scattering matrix test set. This unit enables full two-port measurements to be made without the necessity of changing cable connections to the device. The simplified block diagram of a two-port s-parameter test set is shown in Fig. 13.15. The RF/microwave input is switched between port 1 and port 2 measurements. In each case the RF is split into a reference and test channel. The reference channel is fed directly to a reference channel converter. The test channel feeds the device under test by way of a directional coupler. The

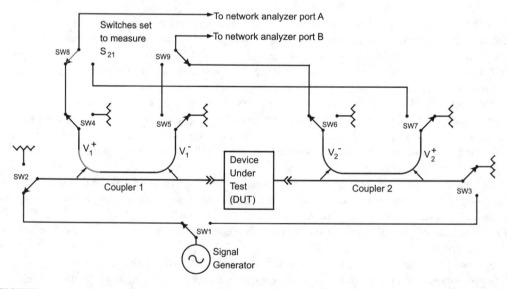

FIGURE 13.15   A two-port s-parameter test set can measure all four s-parameters without moving the DUT.

coupler output sampling the reflected power is routed to the test channel converter. Sampled components of incident and reflected power to both the input and output of the test device are available for processing.

In a full two-port measurement, multiple error terms can be identified, measured, and then used to translate the accuracy of calibration references to the measured data from the device under test. For example, if the load used is not ideal, there will be some reflection back into the DUT. If the source generator impedance is not ideal, any reflections from the input of the DUT back to the generator will result in a further contribution to the incident DUT voltage. The couplers are also nonideal and have phase and amplitude errors.

By measuring the full two-port s-parameters of a set of known references such as opens, shorts, matched loads, known lengths of transmission lines, and through and open circuited paths, a system of equations can be derived that includes the error terms. If 8 error terms are identified, then 8 equations with 8 unknowns can be derived. The error terms can then be solved for and applied to the results of the measurement of an unknown two-port device to correct for measurement system deviations from the ideal.

13.3.3.3 Vector Six-Port Analyzer

A combination of couplers and power dividers, having 0°, 90°, and 180° differences in their output signals can be used to construct a circuit with multiple outputs where the power from the outputs can be used in a system of *n* equations with *n* unknowns. An example of this circuit is shown in Fig. 13.16. In a

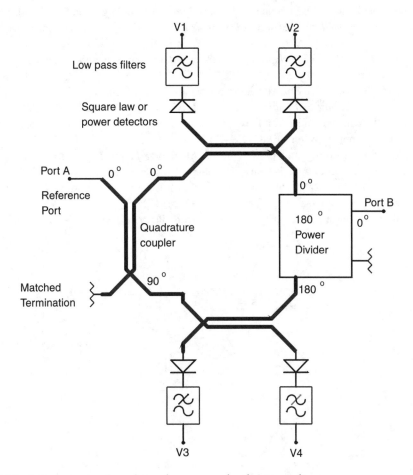

FIGURE 13.16 A six-port network can be used as a narrow band vector analyzer.

properly designed circuit, among the solutions to the system of equations will be the magnitudes and relative phase of the forward and reflected wave. The optimum number of ports for such a device is six; hence, a passive six-port device with diode or power detectors on four of the ports can be used as a vector impedance analyzer [10].

The six-port analyzer has limited bandwidth, usually no more than an octave, because the couplers and power dividers [11] have the same limitation in frequency range to maintain the required amplitude and phase characteristics; however, the low cost of the six-port analyzer makes it attractive for narrowband and built-in test applications.

Typically, measurement test set deviations from the ideal are even more prevalent with the six-port analyzer than for the frequency converting VNA; therefore, use of known calibration elements and the application of the resultant error correction terms is very important for the six-port VNA. The derivation of the error terms and their application to measurement correction is virtually the same for the two analyzers.

References

1. M. Engelson and F. Telewski, *Spectrum Analyzer Theory and Applications*, Artech House, Dedham, MA, 1974.
2. Agilent Technologies, Inc., *Operating Reference Manual for HP 53310A Modulation Domain Analyzer*, Agilent, Santa Clara, CA.
3. S. F. Adam, *Microwave Theory and Applications*, Prentice-Hall, Englewood Cliffs, NJ, 1969.
4. T. S. Laverghetta, *Modern Microwave Measurements and Techniques*, Artech House, Dedham, MA, 1989.
5. J. G. Webster (ed.), *Wiley Encyclopedia of Electrical and Electronics Engineering*, Vol. 13, John Wiley & Sons, New York, 1999, 84–90.
6. P. H. Smith, Transmission line calculator, *Electronics*, 12, 29, January 1939.
7. F. E. Terman, *Electronic and Radio Engineering*, McGraw-Hill, New York, 1955, 100.
8. S. Ramo, J. R. Winnery, and T. Van Duzer, *Fields and Waves in Communications Electronics*, 2nd ed., John Wiley & Sons, New York, 1988, 229–238.
9. G. L. Matthaei, L. Young, and E. M. T. Jones, *Microwave Filters, Impedance-Matching Networks, and Coupling Structures*, Artech House, Dedham, MA, 1980, 775–842.
10. G. F. Engen, A (Historical) Review of the Six-Port Measurement Technique, *IEEE Transactions on Microwave Theory and Technique*, December 1997, 2414–2417.
11. P. A. Rizzi, *Microwave Engineering: Passive Circuits*, Prentice Hall, Englewood Cliffs, NJ, 1988, 367–404.

14

Network Analyzer Calibration

14.1 Introduction

Vector network analyzers (VNA) find very wide application as a primary tool in measuring and characterizing circuits, devices, and components. They are typically applied to measure small signal or linear characteristics of multi-port networks at frequencies ranging from RF to beyond 100 GHz (submillimeter in wavelength). Although current commercial VNA systems can support such measurements at much lower frequencies (a few Hz), higher frequency measurements pose significantly more difficulties in calibrating the instrumentation to yield accurate results with respect to a known or desired electrical reference plane. For example, characterization of many microwave components is difficult since the devices cannot easily be connected directly to VNA-supporting coaxial or waveguide media. Often, the device under test (DUT) is fabricated in a noncoaxial or waveguide medium and thus requires fixturing and additional cabling to enable an electrical connection to the VNA (Fig. 14.1). The point at which the DUT connects with the measurement system is defined as the DUT reference plane. It is generally the point where it is desired that measurements be referenced. However, any measurement includes not only that of the DUT, but contributions from the fixture and cables as well. Note that with increasing frequency, the electrical contribution of the fixture and cables becomes increasingly significant. In addition, practical limitations of the VNA in the form of limited dynamic range, isolation, imperfect source/load match, and other imperfections contribute systematic error to the measurement. To lessen the contribution of systematic error, remove contributions of cabling and fixturing, and therefore enhance measurement accuracy, the VNA must first be calibrated though a process of applying and measuring standards in lieu of the DUT.

Basic measurements consist of applying a stimulus and then determining incident, reflected, and transmitted waves. Ratios of these vector quantities are then computed via post processing yielding network scattering parameters (S-parameters). Most VNAs support measurements on one- and two-port networks, although equipment is commercially available that supports measurements on circuits with more than two ports as well as on differential networks.

14.2 VNA Functionality

A highly simplified block diagram illustrating the functionality of a vector network is provided in Fig. 14.2. Generally, a VNA includes an RF switch such that the RF stimulus can be applied to either port 1 or 2, thereby allowing full two-port measurements without necessitating manual disconnection of the DUT and reversing connections. RF couplers attached at the input and output ports allow measuring reflected voltages. With the RF signal applied in the forward direction (i.e., to port 1), samples of the incident (a_1) and reflected signals at port 1 (b_1) are routed to the receiver. The transmitted signal b_2 reaching port 2 is also directed to the receiver. The receiver functions to downconvert these signals to a lower frequency, which enables digitization and post-processing. Assuming ideal source and load terminations such that a_2 is equal to zero, two scattering parameters can be defined:

$$S_{11} = \frac{b_1}{a_1} \text{ and } S_{21} = \frac{b_2}{a_1}.$$

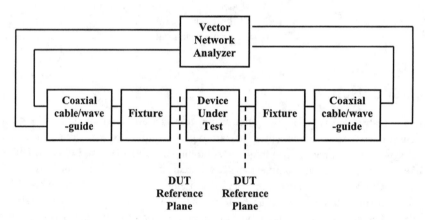

FIGURE 14.1   Typical measurement setup consisting of a device under test embedded in a fixture connected to the vector network analyzer with appropriate cables.

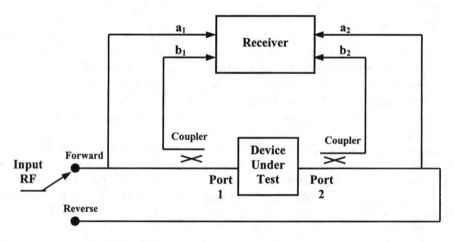

FIGURE 14.2   High simplified VNA block diagram.

In reverse operation, the RF signal is directed to port 2 and samples of signals a_2, b_2, and b_1 are directed to the receiver. Assuming ideal source and load terminations such that a_1 is equal to zero, the remaining two scattering parameters are defined:

$$S_{22} = \frac{b_2}{a_2} \text{ and } S_{12} = \frac{b_1}{a_2}.$$

14.3 Sources of Measurement Uncertainties

Sources of uncertainty or error in VNA measurements are primarily the result of systematic, random, and drift errors. The latter two effects tend to be unpredictable and therefore cannot be removed from the measurement. They are the results of factors such as system noise, connector repeatability, temperature variations, and physical changes within the VNA. Systematic errors, however, arise from imperfections within the VNA, are repeatable, and can be largely removed through a process of calibration. Of the three, systematic errors are generally the most significant, particularly at RF and microwave frequencies. In calibration, such errors are quantified by measuring characteristics of known devices (standards). Hence, once quantified, systematic errors can be removed from the resulting measurement. The choice of calibration standards is not necessarily unique. Selection of a suitable set of standards is often based on such factors as ease of fabrication in a particular medium, repeatability, and the accuracy to which the characteristics of the standard can be determined.

14.4 Modeling VNA Systematic Errors

A mathematical description of systematic errors is accomplished using the concept of error models. The error models are intended to represent the most significant systematic errors of the VNA system up to the reference plane — the electrical plane where standards are connected (Fig. 14.1). Hence, contributions from cables and fixturing in the measurement, up to the reference plane, are accounted for as well.

A flow graph illustrating a typical error model for one-port reflection measurements is depicted in Fig. 14.3. The model consists of three terms, E_{DF}, E_{RF}, and E_{SF}. The term S_{11M} represents the reflection coefficient measured by the receiver within the VNA. The term S_{11} represents the reflection coefficient of the DUT with respect to the reference plane (i.e., the desired quantity).

The three error terms represent various imperfections. Term E_{DF} accounts for directivity in that the measured reflected signal does not consist entirely of reflections caused by the DUT. Limited directivity of the coupler and other signal leakage paths result in other signal components vectorally combining with the DUT reflected signal. Term E_{SF} accounts for source match in that the impedance at the reference plane is not exactly the characteristic impedance (generally 50 ohms). Term E_{RF} describes frequency tracking imperfections between reference and test channels.

A flow graph illustrating a typical error model for two-port measurement, accounting for both reflection and transmission coefficients is depicted in Fig. 14.4. The flow graph consists of both forward (RF signal applied to port 1) and reverse (RF signal applied to port 2) error models. The model consists of twelve terms, six each for forward and reverse paths. Three more error terms are included in addition to those shown in the one-port model, (E_{LF}, E_{TF}, and E_{XF} for the forward path, and similarly E_{LR}, E_{TR}, and E_{XR} for the reverse path). As before, reflection as well as transmission coefficients measured by the receiver within the VNA are denoted with an M subscript (e.g., S_{21M}). The desired two-port S-parameters referenced with respect to port 1/2 reference planes are denoted as S_{11}, S_{21}, S_{12}, and S_{22}. The transmission coefficients are ratios of transmitted and incident signals. Error term E_{LF} accounts for measurement errors resulting from an imperfect load termination. Term E_{TF} describes transmission frequency tracking errors. The term E_{XF} accounts for isolation in that a small component of the transmitted signal reaching the receiver is due to finite isolation where it reaches the receiver without passing through the DUT. The error coefficients for the reverse path are similarly defined.

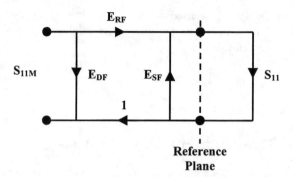

FIGURE 14.3   Typical one-port VNA error model for reflection coefficient measurements.

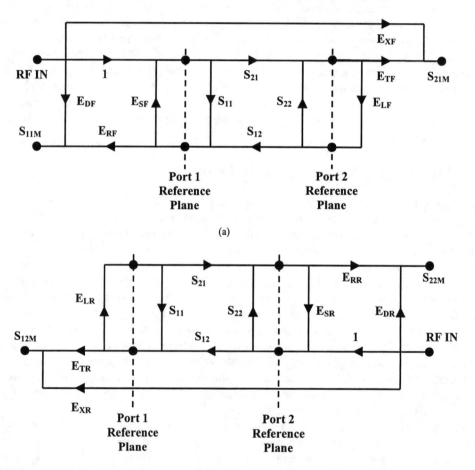

FIGURE 14.4   Typical two-port VNA error model; (a) forward model, and (b) reverse model.

14.5 Calibration

From the above discussion, it is possible to mathematically relate uncorrected scattering parameters measured by the VNA (S_M) to the above-mentioned error terms and the S-Parameters exhibited by the DUT (S). For example, with the VNA modeled for one-port measurements as illustrated in Fig. 14.3, the reflection coefficient of the DUT (S_{11}) is given by:

$$S_{11} = \frac{S_{11M} - E_{DF}}{E_{SF}\left(S_{11M} - E_{DF}\right) + E_{RF}}$$

Similarly, for two-port networks, DUT S-parameters can be mathematically related to the error terms and uncorrected measured S-parameters. DUT parameters S_{11} and S_{21} can be described as functions of S_{11M}, S_{21M}, S_{12M}, S_{22M} and the six forward error terms. Likewise, S_{12} and S_{22} are functions of the four measured S-parameters and the six reverse error terms. Hence, when each error coefficient is known, the DUT S-parameters can be determined from uncorrected measurement.

Therefore, calibration is essentially the process of determining these error coefficients. This is accomplished by replacing the DUT with a number of standards whose electrical properties are known with respect to the desired reference plane (the reader is referred to [1-5] for additional information). Additionally, since the system is frequency dependent, the process is repeated at each frequency of interest.

14.6 Calibration Standards

Determination of the error coefficients requires the use of several standards, although the choice of which standards to use is not necessarily unique. Traditionally, short, open, load, and through (SOLT) standards have been applied, especially in a coaxial medium that facilitates their accurate and repeatable fabrication. Electrical definitions for ideal and lossless SOLT standards (with respect to port 1 and 2 reference planes) are depicted in Fig. 14.5. Obviously, and especially with increasing frequency, it is impossible to fabricate standards such that they are (1) lossless and (2) exhibit the defined reflection and transmission coefficients

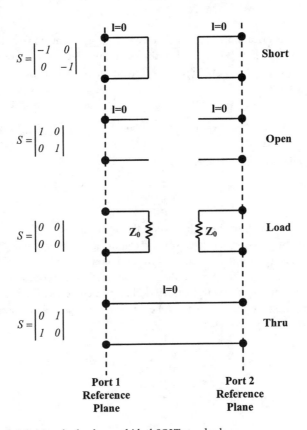

FIGURE 14.5 Electrical definition for lossless and ideal SOLT standards.

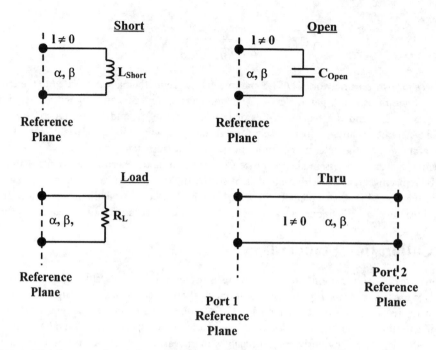

FIGURE 14.6 High frequency descriptions of SOLT standards generally consider nonzero length transmission lines, loss mechanisms, and fringing field effects associated with the open standard.

at these reference planes. Fabrication and physical constants dictate some nonzero length of transmission line must be associated with each (Fig.14.6). Hence, for completeness, the characteristics of the transmission line must be (1) known, and (2) included in defining the parameters of each standard. Wave propagation is described as

$$V(z) = Ae^{-\gamma z} + Be^{\gamma z}$$

where γ is the propagation constant defined as

$$\gamma = \alpha + j\beta$$

Assuming the electrical length of the transmission line associated with the standards is short, losses become small and perhaps α can be neglected without significant degradation in accuracy. Alternatively, commercial VNA manufacturers often describe the transmission line in terms of a delay coefficient with a small resistive loss component. The open standard exhibits further imperfections since the electric field pattern at the open end tends to vary with frequency. The open-end effect is often described in terms of a frequency-dependent fringing capacitance (C_{Open}) expressed in terms of a polynomial expansion taking the form:

$$C_{Open} = C_0 + C_1 F + C_2 F^2 + + C_3 F^3 + \ldots$$

where C_0, C_1, ... are coefficients and F is frequency.

The load termination largely determines forward and reverse directivity error terms (E_{DF} and E_{DR}). Considering the error models in Figs. 14.3 and 14.4, with the load standard applied on port 1, forward directivity error takes the following form:

$$E_{DF} = S_{11M} - \frac{S_{11Load} E_{RF}}{1 - E_{SF} S_{11Load}}$$

where $S_{11\,Load}$ is the actual reflection coefficient of the load standard. Ideally, the load standard should exhibit an impedance of Z_0 (characteristic impedance) and thus a reflection coefficient of zero (i.e., $S_{11\,Load} = 0$) in which case E_{DF} becomes the measured value of S_{11} with the load standard connected to port 1. High quality coaxial-based fixed load standards exhibiting high return loss over broad bandwidths are generally commercially available, especially at RF and microwave frequencies. At higher frequencies and/or where the electrical performance of the fixed load terminations is inadequate, sliding terminations are employed. Sliding terminations use mechanical methods to adjust the electrical length of a transmission line associated with the load standard. Neglecting losses in the transmission line, the above expression forms a circle in the S_{11} measurement plane as the length of the transmission line is varied. The center of the circle defines error term E_{DF} (Fig. 14.7).

Often it is desirable to characterize devices in noncoaxial media. For example, measuring the characteristics of devices and circuits at the wafer level by connecting microwave probes directly to the wafer. Other situations arise where components cannot be directly probed but must be placed in packages with coaxial connectors and it is desirable to calibrate the fixture/VNA at the package/fixture interface. Although fabrication techniques favor SOLT standards in coax, it is difficult to realize them precisely in other media such as microstrip and hence non-SOLT standards are more appropriate. Presently, standards based on one or more transmission lines and reflection elements have become popular for RFICs and MMICs. Fundamentally, they are more suitable for MMICs and RFICs since they rely on fabricating transmission lines (in microstrip, for example), where the impedance of the lines can be precisely determined based on physical dimensions, metalization, and substrate properties. The TRL (thru, reflect, line) series of standards have become popular as well as variations of it such as LRM (line, reflect, match), and LRL (line, reflect, line) to name but a few. In general, TRL utilizes a short length thru (sometime assumed zero length), a highly reflective element, and a nonzero length transmission line. One advantage of this technique is that a complete electrical description of each standard is not necessary. However, each standard is assumed to exhibit certain electrical criteria. For example, the length of the thru generally

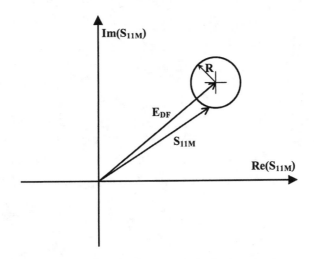

FIGURE 14.7 Characterizing directivity error terms using a sliding load termination.

must be known, or alternatively, the thru may in many cases be fabricated such that its physical dimensions approach zero length at the frequencies of interest and are therefore insignificant. The characteristic impedance of the line standard is particularly important in that it is the major contributor in defining the reference impedance of the measurement. Its length is also important. Lengths approaching either $0°$ or multiples of $180°$ (relative to the length of the thru) are problematic and lead to poor calibrations. The phase of the reflection standard is not critical, although its phase generally must be known to within one-quarter of a wavelength.

In the interest of reducing hardware cost, a series of VNAs are commercially available based on a receiver architecture containing three rather than four sampling elements. In four-sampling receiver architecture, independent measurements are made of a_1, b_1, a_2, and b_2. Impedance contributions of the internal switch that routes the RF stimulus to port 1 for forward measurements and to port 2 for reverse measurements can be accounted for during the calibration process. In a three-sampling receiver architecture, independent measurements are made on b_1, b_2 and on a combined a_1 and a_2. This architecture is inherently less accurate than the former in that systematic errors introduced by the internal RF switch are not fully removed via TRL calibration, although mitigating this effect to some extent is possible [5]. However, it should be noted that this architecture provides measurement accuracy that is quite adequate for many applications.

References

1. Staudinger, J., A two-tier method of de-embedding device scattering parameters using novel techniques, Master Thesis, Arizona State University, May 1987.
2. Lane, R., De-Embedding Device Scattering Parameters, *Microwave J.*, Aug. 1984.
3. Fitzpatrick, J., Error Models For Systems Measurement, *Microwave J.*, May 1978.
4. Operating and Programming Manual For the HP8510 Network Analyzer, Hewlett Packard, Inc., Santa Rosa, CA.
5. Metzger, D., Improving TRL* Calibrations of Vector Network Analyzers, *Microwave J.*, May 1995.

15

Noise Measurements

Alfy Riddle
Macallan Consulting

15.1 Fundamentals of Noise

15.1.1 Statistics

Noise is a random process. There may be nonrandom system disturbances we call noise, but this section will consider noise as a random process. Noise can have many different sources such as thermally generated resistive noise, charge crossing a potential barrier, and generation-recombination (G-R) noise [1]. The different noise sources are described by different statistics, the thermal noise in a resistor is a Gaussian process while the shot noise in a diode is a Poisson process. In the cases considered here, the number of noise "events" will be so large that all noise processes will have essentially Gaussian statistics and so be represented by the probability distribution in Eq. (15.1).

$$p(x) = 1/(2\pi\sigma^2) e^{-x^2/2\sigma^2} \tag{15.1}$$

The statistics of noise are essential for determining the results of passing noise through nonlinearities because the nonlinearity will change the noise distribution [2]. Noise statistics are useful even in linear networks because multiple noise sources will require correlation between the noise sources to find the total noise power. Linear networks will not change the statistics of a noise signal even if the noise spectrum is changed.

15.1.2 Bandwidth

The noise energy available from a hot resistor is given in Eq. (15.2), where $h = 6.62 \times 10^{-34}$ J s, T is in degrees Kelvin, and $k = 1.38 \times 10^{-23}$ J/degree K [1]. N is in joules, or watt-seconds, or W/Hz, which is noise power spectral density. For most of the microwave spectrum $hf \ll kT$ so Eq. (15.2) reduces to Eq. (15.3).

$$N = hf/(e^{hf/kT} - 1) \tag{15.2}$$

FIGURE 15.1   Equivalent thermal noise sources: (a) voltage and (b) current.

$$N \cong kT \qquad\qquad f \ll kT/h \qquad\qquad (15.3)$$

The noise power available from the hot resistor will be the integration of this energy, or spectral density, over the measuring bandwidth as given in Eq. (15.4).

$$P = {}_{f1}\!\int^{f2} N \, df \qquad\qquad (15.4)$$

As the frequency increases, N reduces so the integration in Eq. (15.4) will be finite even if the frequency range is infinite. Note that for microwave networks using cooled circuits, quantum effects can become important at relatively low frequencies because of the temperature-dependent condition in Eq. (15.3). For a resistor at microwave frequencies and room temperature, N is independent of frequency so the total power available is simply $P = kT(f_2 - f_1)$, or $P = kTB$, as shown in Eq. (15.5), where B is the bandwidth. Figure 15.1 shows a resistor with an available thermal power of kTB, which can be represented either as a series voltage source with $e_n^2 = 4\,kTRB$ or a shunt current source with $i_n^2 = 4\,kTB/R$, where the squared value is taken to be the mean-square value. At times it is tempting to represent e_n as $\sqrt{(4\,kTB)}$, but this is a mistake because e_n is a random variable, not a sinusoid. The process of computing the mean-square value of a noise source is important for establishing any possible correlation with any other noise source in the system [1,16]. Representing a noise source as an equivalent sinusoidal voltage can result in an error due to incorrect accounting of correlation.

$$P = kT\,B \qquad\qquad f \ll kT/h \qquad\qquad (15.5)$$

When noise passes through a filter we must repeat the integration of Eq. (15.4). Two useful concepts in noise measurement are noise power per hertz and equivalent noise bandwidth. Noise power per hertz is simply the spectral density of the noise, or N in the above equations because it has units of watt-seconds or joules. Spectral densities are also given in V^2/Hz and A^2/Hz. The equivalent noise bandwidth of a noise source can be found by dividing the total power detected by the maximum power detected per hertz, as shown in Eq. (15.6).

$$B_e = P/\mathrm{Max}\{N\} \qquad\qquad (15.6)$$

The noise equivalent bandwidth of a filter is especially useful when measuring noise sources with a spectrum analyzer. The noise equivalent bandwidth of a filter is defined by integrating its power transfer function, $|H(f)|^2$, over all frequency and dividing by the peak of the power transfer function, as shown in Eq. (15.7).

$$B_e = {}_0\!\int^\infty \left|H(f)\right|^2 df \Big/ \mathrm{Max}\left\{\left|H(f)\right|^2\right\} \qquad\qquad (15.7)$$

Power meters are often used with bandpass filters in noise measurements so that the noise power has a well-defined range. The noise equivalent bandwidth of the filter can be used to convert the noise power back to a power/Hz spectral density that is easier to use in computations and comparisons. As an example, a first-order bandpass filter has a $B_e = \pi/2 \, B_{-3}$, where B_{-3} is the –3 dB bandwidth. The noise equivalent bandwidth is greater than the 3 dB bandwidth because of the finite power in the filter skirts.

15.2 Detection

The most accurate and traceable measurement of noise power is by comparison with thermal standards [3]. In the everyday lab the second best method for noise measurement is a calibrated power meter preceded by a filter of known noise bandwidth. Because of its convenience, the most common method of noise power measurement is a spectrum analyzer. This most common method is also the most inaccurate because of the inherent inaccuracy of a spectrum analyzer and because of the nonlinear processes used in a spectrum analyzer for power estimation. As mentioned in the section on statistics, nonlinearities change the statistics of a noise source. For example, Gaussian noise run through a linear envelope detector acquires a Rayleigh distribution as shown in Fig. 15.2.

The average of the Rayleigh distribution is not the standard deviation of the Gaussian, so a detector calibrated for sine waves will read about 1 dB high for noise. Spectrum analyzers also use a logarithmic amplifier that further distorts the noise statistics and accounts for another 1.5 dB of error. Many modern spectrum analyzers automatically correct for these nonlinear errors as well as equivalent bandwidth when put in a "Marker Noise" mode [4].

15.3 Noise Figure and Y-Factor Method

At high frequencies it is far easier to measure power flow than it is to measure individual voltage and current noise sources. All of a linear device's noise power can be considered as concentrated at its input as shown in Fig. 15.3 [1].

We can lump all of the amplifier noise generators into an equivalent noise temperature with an equivalent input noise power per hertz of $N_a = kT_e$. As shown in Fig. 15.3, the noise power per hertz available from the noise source is $N_S = kT_S$. In system applications the degradation of signal-to-noise ratio (SNR) is a primary concern. We can define a figure of merit for an amplifier, called the noise factor (F), which describes the reduction in SNR of a signal passed through the amplifier shown in Fig. 15.3 [5]. The noise factor for an amplifier is derived in Eq. (15.8).

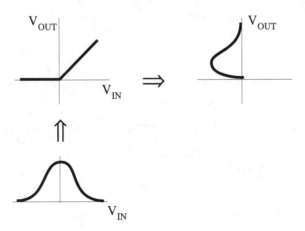

FIGURE 15.2   Nonlinear transformation of Gaussian noise.

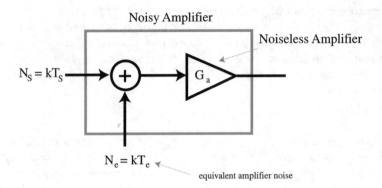

FIGURE 15.3   System view of amplifier noise.

$$F = SNR_{IN}/SNR_{OUT} = S_{IN}/kT_S \Big/ \Big(G_a S_{IN} \Big/ \big(G_a k(T_S + T_e)\big)\Big) = (T_S + T_e)/T_S = 1 + T_e/T \quad (15.8)$$

Eq. (15.8) is very simple and only contains the amplifier equivalent temperature and the source temperature. F does vary with frequency and so is measured in a narrow bandwidth, or spot. Note that F is not a function of measurement bandwidth. Eq. (15.8) also implies that the network is tuned for maximum available gain, which happens by default if all the components are perfectly matched to 50 ohms and used in a 50-ohm system.

Device noise factor can be measured with the setup shown in Fig. 15.4 [1]. The Y-Factor method takes advantage of the fact that as the source temperature is varied, the device noise output, N_O, varies yet the device noise contribution remains a constant. Figure 15.5 shows that as T_S changes the noise power measured as the power meter changes according to Eq. (15.9).

$$N_O(T_S) = \big(k\,T_S\,G_a + k\,T_e\,G_a\big)B \qquad (15.9)$$

The value of $N_O(T_S = 0)$ gives the noise power of the device alone. By using two known values of T_S, a cold measurement at $T_S = T_{COLD}$, and a hot measurement at $T_S = T_{HOT}$, the slope of the line in Fig. 15.5 can be derived. Once the slope is known, the intercept at $T_S = 0$ can be found by measuring $N_O(T_{COLD})$ and $N_O(T_{HOT})$. The room temperature, T_O, is also needed to serve as a reference temperature for the device noise factor, F. For a room temperature $F = 1 + T_e/T_O$, we can define $T_e = (F - 1)\,T_O$. The following equations derive the Y-Factor method. Equation (15.10) is the basis for the Y-Factor method.

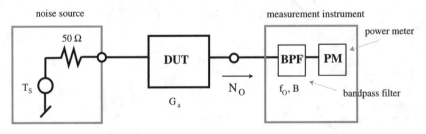

FIGURE 15.4   Test setup for Y-factor method.

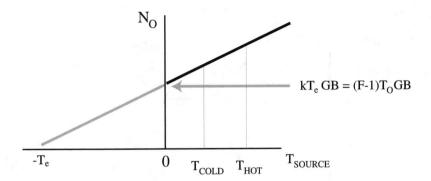

FIGURE 15.5 Output noise power vs. source temperature.

$$Y = \frac{N_O(T_{HOT})}{N_O(T_{COLD})} = \frac{\left(k\,T_{HOT}\,G_a + (F-1)k\,T_O\,G_a\right)B}{\left(k\,T_{COLD}\,G_a + (F-1)k\,T_O\,G_a\right)B} = \frac{T_{HOT} + (F-1)\,T_O}{T_{COLD} + (F-1)\,T_O} \qquad (15.10)$$

Solving for F we get Eq. (15.11) which can be solved for T_e as shown in Eq. (15.12).

$$F = 1 + \frac{T_{HOT} - Y\,T_{COLD}}{(Y-1)T_O} = 1 + T_e/T_O \qquad (15.11)$$

$$T_e = \frac{T_{HOT} - Y\,T_{COLD}}{(Y-1)} \qquad (15.12)$$

Equation (15.12) can be rearranged to define another useful parameter known as the equivalent noise ratio, or ENR, of a noise source as shown in Eq. (15.13) [1].

$$F = \frac{\left(T_{HOT}/T_O - 1\right) + Y\left(1 - T_{COLD}/T_O\right)}{(Y-1)} = \frac{ENR + Y\left(1 - T_{COLD}/T_O\right)}{(Y-1)} = \left.\frac{ENR}{(Y-1)}\right|_{T_{COLD} = T_O} \qquad (15.13)$$

Note that when T_{COLD} is set to the reference temperature for F, which the IEEE gives as $T_O = 290°$ Kelvin, then the device noise factor has a simple relationship to both ENR and Y [1,3].

Practically speaking, the noise factor is usually given in decibels and called the noise figure, $NF = 10 \log F$. While the most accurate noise sources use variable temperature loads, the most convenient variable noise sources use avalanche diodes with calibrated noise power versus bias current [6]. The diode noise sources usually contain an internal pad to reduce the impedance variation between on (hot) and off (cold) states. Also, the diodes come with an ENR versus frequency calibration curve.

15.4 Phase Noise and Jitter

15.4.1 Introduction

The noise we have been discussing was broadband noise. When noise is referenced to a carrier frequency it appears to modulate the carrier and so causes amplitude and phase variations in the carrier [7]. Because

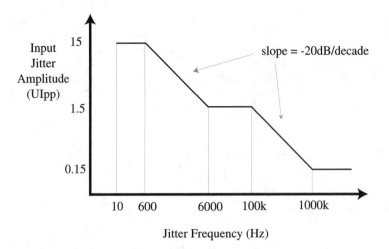

FIGURE 15.6 SONET Category II jitter tolerance mask for OC-48.

of the amplitude-limiting mechanism in an oscillator, oscillator phase modulation noise (PM) is much larger than amplitude modulation noise (AM) close enough to the carrier to be within the oscillator loop bandwidth. The phase variations, caused by the noise at different offset, or modulation frequencies create a variance in the zero crossing time of the oscillator. This zero crossing variance in the time domain is called jitter and is critical in digital communication systems. Paradoxically, even though jitter is easily measured in the time domain and often defined in picoseconds it turns out to be better to specify jitter in the frequency domain as demonstrated by the jitter tolerance mask for an OC-48 SONET signal [8]. The jitter plot shown in Fig. 15.6 can be translated into script L versus frequency using the equations in the following section [9].

Jitter is best specified in the frequency domain because systems are more sensitive to some jitter frequencies than others. Also, jitter attenuators, which are simply narrowband phase-locked loops (PLLs), have well-defined frequency domain transfer functions that can be cascaded with the measured input jitter to derive the output jitter.

$$V_O(t) = V_C \{1 + m(t)\} \cos[\omega_C t + \beta(t)] \tag{15.14}$$

15.4.2 Mathematical Basics

Consider the time domain voltage given in Eq. (15.14). This signal contains both amplitude modulation, $m(t)$, and phase modulation, $\beta(t)$ [10]. If we let $m(t) = m_1(t) \cos(\omega_m t)$, $\beta(t) = \beta_1(t) \sin(\omega_m t)$, and we define $|\beta_1(t)| \ll 1$, then Eq. (15.14) can be expanded into AM and PM sidebands as shown in Eq. (15.15).

$$
\begin{aligned}
V_O(t) \cong\ & V_C \cos[\omega_C t] \\
& + V_C m_1(t)/2 \{\cos[\omega_C t + \omega_m t] + \cos[\omega_C t - \omega_m t]\} \\
& + V_C \beta_1(t)/2 \{\cos[\omega_C t + \omega_m t] - \cos[\omega_C t - \omega_m t]\}
\end{aligned}
\tag{15.15}
$$

We can let $m_1(t)$ and $\beta_1(t)$ be fixed amplitudes as when sinusoidal test signals are used to characterize an oscillator, or we can let $m_1(t)$ and $\beta_1(t)$ be slowly varying, with respect to ω_m, noise signals. The latter case gives us the narrowband Gaussian noise approximation, which can represent an oscillator spectrum when the noise signals are summed over all modulation frequencies, ω_m [7].

Several notes should be made here. First, as $\beta_1(t)$ becomes large the single sidebands of Eq. (15.15) expand into a Bessel series that ultimately generates a flat-topped spectrum close into the average carrier frequency. This flat-topped spectrum is essentially the FM spectrum created by the large phase excursions that result from the $1/f^3$ increase in phase noise at low modulating frequencies. Second, if $|m_1(t)| = |\beta_1(t)|$, and they are fully correlated, then by altering the phases between $m_1(t)$ and $\beta_1(t)$ we can cancel the upper or lower sideband at will. This second point also shows that a single sideband contains equal amounts of AM and PM, which is useful for testing and calibration purposes [11].

Oscillator noise analysis uses several standard terms such as AM spectral density, PM spectral density, FM spectral density, script L, and jitter [10,12]. These terms are defined in Eqs. (15.16) through (15.20). The AM spectral density, or $S_{AM}(f_m)$, shown in Eq. (15.16) is derived by computing the power spectrum of $m(t)$, given in Eq. (15.14), with a 1-hertz-wide filter. S_{AM} is called a spectral density because it is on a 1-hertz basis. Similarly, the PM spectral density, or $S_\phi(f_m)$ in radians^2/Hz, is shown in Eq. (15.17) and is derived by computing the power spectrum of $\beta(t)$ in Eq. (15.14). The FM spectral density, or $S_{FM}(f_m)$ in Hz^2/Hz, is typically derived by using a frequency discriminator to measure the frequency deviations in a signal. Because frequency is simply the rate of change of phase, FM spectral density can be derived from PM spectral density as shown in Eq. (15.18). Script L is a measured quantity usually given in dBc/Hz and best described by Fig.15.7. It is important to remember that the definition of script L requires the sidebands to be due to phase noise. Because script L is defined as a measure of phase noise it can be related to the PM spectral density as shown in Eq. (15.19). Two complications arise in using script L. First, most spectrum analyzers do not determine if the sidebands are only due to phase noise. Second, the constant relating script L to S_ϕ is 2 if the sidebands are correlated and $\sqrt{2}$ if the sidebands are uncorrelated and the spectrum analyzer does not help in telling these two cases apart. Jitter is simply the rms value of the variation in zero crossing times of a signal compared with a reference of the same average frequency. Of course, the jitter of a signal can be derived by accumulating the phase noise as shown in Eq. (15.20). In Eqs. (15.16) through (15.20) $\Im\{x\}$ denotes the Fourier transform of x [7]. Most of these terms can also be defined from the Fourier transform of the autocorrelation of $m(t)$ or $\beta(t)$. The spectral densities are typically given in dB using a 1-Hz measurement bandwidth, abbreviated as dB/Hz.

$$S_{AM}\left(f_m\right) = \Im\{m(t)\}\,\Im\{m*(t)\} = m_1^2\big/2\ \delta\left(f_m - f_a\right)\Big|_{m(t) = m_1 \cos(\omega_a t)} \tag{15.16}$$

$$S_\phi\left(f_m\right) = \Im\{\beta(t)\}\,\Im\{\beta*(t)\} = \beta_1^2\big/2\ \delta\left(f_m - f_a\right)\Big|_{\beta(t) = \beta_1 \cos(\omega_a t)} \tag{15.17}$$

$$S_{FM}\left(f_m\right) = f_m^2\, S_\phi\left(f_m\right) \qquad\qquad \text{because } \omega_m = d\phi\big/dt \tag{15.18}$$

$$\text{script } L\left(f_m\right) = P_{SSB}\left(f_m\right)\big/\text{Hz}\big/P_C = S_\phi\left(f_m\right)\big/2\Big|_{\text{when phase noise has correlated sidebands}} \tag{15.19}$$

$$\text{jitter} = \sqrt{\left({}_0\!\int^\infty S_\phi\left(f_m\right)df_m\right)} = \beta_1\big/\sqrt{2}\,\Big|_{\beta(t) = \beta_1 \cos(\omega_m t)} \tag{15.20}$$

In the above equations f_m indicates the offset frequency from the carrier. In Eq. (15.19) P_{SSB} is defined as phase noise, but often is just the noise measured by a spectrum analyzer close to the carrier frequency, and P_C is the total oscillator power. The jitter given in Eq. (15.20) is the total jitter that results from a time domain measurement. Jitter as a function of frequency, f_m, is just the square root of $S_\phi(f_m)$. Jitter as a function of frequency can be translated to various other formats, such as degrees, radians, seconds, and unit intervals (UIs), using Eq. (15.21) [9].

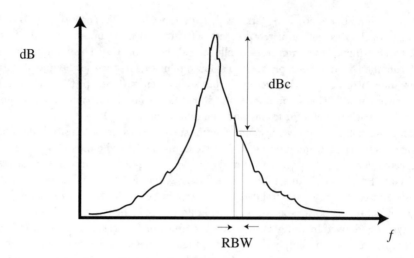

FIGURE 15.7   Typical measured spectrum on a spectrum analyzer.

$$\text{jitter} = \left\{ \begin{array}{l} \beta_{\text{RMS}} \text{ in radians} \\[4pt] \text{UI} = \beta_{\text{RMS}} \big/ 2\pi \text{ in unit intervals} \\[4pt] \text{UI} \big/ f_{\text{C}} = \beta_{\text{RMS}} \big/ \omega_{\text{C}} \text{ in seconds} \\[4pt] 360 \text{ UI} = 360\beta_{\text{RMS}} \big/ 2\pi \text{ in degrees} \end{array} \right\} \qquad (15.21)$$

For most free-running oscillators the $1/f^3$ region of the phase noise dominates the jitter so integrating the $1/f^3$ slope gives $\text{UI} \approx f_a/2 \; 10^{\text{scriptL}(f_a)/10}$ where f_a is any frequency on the $1/f^3$ slope and script L is in dBc/Hz. For PLL-based sources with large noise pedestals, a complete integration should be done.

15.4.3   Phase Noise Measurements

Phase noise is typically measured in one of three ways: spectrum analyzer, PLL, or transmission line discriminator [1,10,13-15]. The spectrum analyzer is the easiest method of measuring script $L(f_m)$ for any oscillator noisier than the spectrum analyzer reference source. Figure 15.7 shows a typical source spectrum. Care must be taken to make the resolution bandwidth, RBW, narrow enough to not cover a significant slope of the measured noise [1]. The spectrum analyzer cannot distinguish between phase and amplitude noise, so reporting the results as script L only holds where $S_\phi > S_{\text{AM}}$, which usually means within the $1/f^3$ region of the source. Spectrum analyzer measurements can be very tedious when the oscillator is noisy enough to wander significantly in frequency.

   PLL-based phase noise measurement is used in most commercial systems [14]. Figure 15.8 shows a PLL-based phase noise test set. The reference oscillator in Fig. 15.8 is phase locked to the device under test (DUT) through a low pass filter (LPF) with a cutoff frequency well below the lowest desired measurement frequency. This allows the reference oscillator to track the DUT and downconvert the phase noise sidebands without tracking the noise as well. The low frequency spectrum analyzer measures the noise sidebands and arrives at a phase noise spectral density by factoring in the mixer loss or using a calibration tone [11]. A PLL system requires the reference source to be at least as quiet as the DUT. Another DUT can be used as a reference with the resulting noise sidebands increasing by 3 dB, but usually the reference is much quieter than the DUT so fewer corrections have to be made.

   A transmission line frequency discriminator can provide accurate and high-resolution phase noise measurements without the need for a reference oscillator [13]. The discriminator resolution is

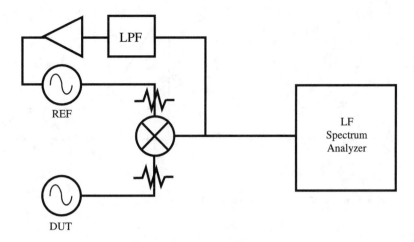

FIGURE 15.8   PLL phase noise measurement.

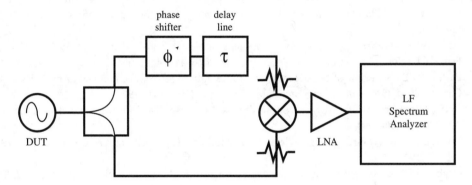

FIGURE 15.9   Transmission line discriminator.

proportional to the delay line delay, τ. The phase shifter is adjusted so that the mixer signals are in quadrature, which means the mixer DC output voltage is set to the internal offset voltage (approximately zero). Transmission line discriminators can be calibrated with an offset source of known amplitude, as discussed previously, or with a source of known modulation sensitivity [11] (see Fig. 15.9). The disadvantages of a transmission line discriminator are that high source output levels are required to drive the system (typically greater than 13 dBm), and the system must be retuned as the DUT drifts. Also, it is important to remember that the discriminator detects FM noise which is related to phase noise as given in Eq. (15.18) and shown in Fig. 15.10.

15.5 Summary

Accurate noise measurement and analysis must recognize that noise is a random process. While nonlinear devices will affect the noise statistics, linear networks will not change the noise statistics. Noise statistics are also important for analyzing multiple noise sources because the correlation between the noise sources must be considered. At very high frequencies it is easier to work with noise power flow than individual noise voltage and current sources, so methods such as the Y-Factor technique have been developed for amplifier noise figure measurement. Measuring oscillator noise mostly involves the phase variations of a source. These phase variations can be represented in the frequency domain as script L, or in the time domain as jitter. Several techniques of measuring source phase noise have been developed which trade off accuracy for cost and simplicity.

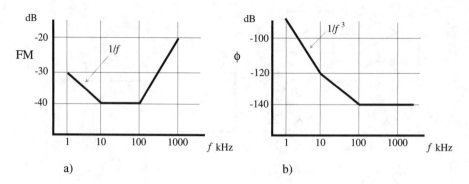

FIGURE 15.10   (a) FM and (b) phase noise spectral densities for the same device.

References

1. Ambrozy, A., *Electronic Noise*, McGraw-Hill, New York, 1982.
2. Papoulis, A., *Probability, Random Variables and Stochastic Processes*, McGraw-Hill, New York, 1965.
3. Haus, H.A., IRE Standards on Methods of Measuring Noise in Linear Twoports, *Proc. IRE*, 60–68, Jan. 1960.
4. Staff, *HP 8560 E-Series Spectrum Analyzer User's Guide*, Hewlett-Packard, Dec. 1997.
5. Friis, H.T., Noise Figures of Radio Receivers, *Proc. IRE*, 419–422, July 1944.
6. Pastori, W.E., A Review of Noise Figure Instrumentation, *Microwave J.*, 50–60, Apr. 1983.
7. Carlson, B.A., *Communication Systems*, McGraw-Hill, New York, 1975.
8. ANSI, Telecommunications-Synchronous Optical Network (SONET)- Jitter at Network Interfaces, *ANSI T1.105.03-1994*, 1994.
9. Adler, J.V., Clock Source Jitter: A Clear Understanding Aids Clock Source Selection, *EDN*, 79–86, Feb. 18, 1999.
10. Ondria, J.G., A Microwave System for Measurement of AM and FM Noise Spectra, *IEEE Trans. MTT*, 767–781, Sept. 1968.
11. Buck, J.R., and Healey, D.J. III, Calibration of Short-Term Frequency Stability Measuring Apparatus, *Proc. IEEE*, 305–306, Feb. 1966.
12. Blair, B.E., *Time and Frequency: Theory and Fundamentals*, U.S. Dept. Commerce, NBS Monograph 140, 1974.
13. Schielbold, C., Theory and Design of the Delay Line Discriminator for Phase Noise Measurement, *Microwave J.*, 103–120, Dec. 1983.
14. Harrison, D.M., Howes, M.J., and Pollard, R.D., The Evaluation of Phase Noise in Low Noise Oscillators, *IEEE MTT-S Digest*, 521–524, 1987.
15. Staff, *Noise Measurements Using the Spectrum Analyzer, Part One: Random Noise*, Tektronix, Beaverton, 1975.
16. Haus, H.A. and Adler, R.B., *Circuit Theory of Linear Noisy Networks*, MIT Press, Cambridge, MA, 1959.

16

Nonlinear Microwave Measurement and Characterization

J. Stevenson Kenney
Georgia Institute of Technology

16.1 Introduction

While powerful methods have been developed to analyze complex linear circuits, it is unfortunate that almost all physical systems exhibit some form of nonlinear behavior. Often the nonlinear behavior of a microwave circuit is detrimental to the signals that pass through it. Such is the case with distortion within a microwave power amplifier. In some cases nonlinearities may be exploited to realize useful circuit functions, such as frequency translation or detection. In either case, methods have been devised to characterize and measure nonlinear effects on various signals. These effects are treated in this chapter and include:

- Harmonic Distortion
- Gain Compression

- Intermodulation Distortion
- Phase Distortion
- Adjacent Channel Interference
- Error Vector Magnitude

Many of the above characterizations are different manifestations of nonlinear behavior for different types of signals. For instance, both analog and digital communication systems are affected by *intermodulation distortion*. However, these effects are usually measured in different ways. Nevertheless, some standard measurements are used as figures of merit for comparing the performance to different circuits. These include:

- Output Power at 1 *dB* Gain Compression
- Third Order Intercept Point
- Spurious Free Dynamic Range
- Noise Power Ratio
- Spectral Mask Measurements

This chapter treats the characterization and measurement of nonlinearities in microwave circuits. The concentration will be on standard techniques for analog and digital communication circuits. For more advanced techniques, the reader is advised to consult the references at the end of this section.

16.2 Mathematical Characterization of Nonlinear Circuits

To analyze the effects of nonlinearities in microwave circuits, one must be able to describe the input-output relationships of signals that pass through them. Nonlinear circuits are generally characterized by input-output relationships called *transfer characteristics*. In general, any memoryless circuit described by transfer characteristics that does not satisfy the following definition of a *linear* memoryless circuit is said to be *nonlinear*.

$$v_{out}(t) = Av_{in}(t),\tag{16.1}$$

where v_{in} and v_{out} are the input and output time-domain waveforms and A is a constant independent of time. Thus, one form of a nonlinear circuit has a transfer characteristic of the form

$$v_{out}(t) = g\big[v_{in}(t)\big].\tag{16.2}$$

The form of $g(v)$ will determine all measurable distortion characteristics of a nonlinear circuit. Special cases of nonlinear transfer characteristics include:

- Time Invariant: g does not depend on t
- Memoryless: g is evaluated at time t using only values of v_{in} at time t

16.2.1 Nonlinear Memoryless Circuits

If a transfer characteristic includes no integrals, differentials, or finite time differences, then the instantaneous value at a time t depends only on the input values at time t. Such a transfer characteristic is said to be *memoryless*, and may be expressed in the form of a power series

$$g(v) = g_0 + g_1 v + g_2 v^2 + g_3 v^3 + \dots\tag{16.3a}$$

where g_n are real-valued, time-invariant coefficients. Frequency domain analysis of the output signal $v_{out}(t)$ where $g(v)$ is expressed by Eq. (16.3a) yields a Fourier series, whereby the harmonic components are governed by the coefficients G_n. If $v_{in}(t)$ is a sinusoidal function at frequency f_c with amplitude V_{in}, then the output signal is a harmonic series of the form

$$v_{out}(t) = G_0 + G_1 V_{in} \cos(2\pi f_c t) + G_2 V_{in}^2 \cos(4\pi f_c t) + G_3 V_{in}^3 \cos(6\pi f_c t) + \dots \qquad (16.3b)$$

The coefficients, G_n are functions of the coefficients g_n, and are all real. The extent that the coefficients g_n are nonzero is called the *order* of the nonlinearity. Thus, from Eq. (4.24b), it is seen that an nth order system will produce harmonics of nth order of amplitude $G_n V_{in}^n$.

16.2.2 Nonlinear Circuits with Memory

As described in Eq. (16.3a), $g(v)$ is said to be *memoryless* because the output signal at a time t depends only on the input signal at time t. If the output depends on the input at times different from time t, the nonlinearity is said to have *memory*. A nonlinear function with a finite memory (i.e., a *finite impulse response*) may be described as

$$v_{out}(t) = g\left[v_{in}(t), v_{in}(t - \tau_1), v_{in}(t - \tau_2), \dots v_{in}(t - \tau_n) \right]. \qquad (16.4)$$

The largest time delay, τ_n, determines the length of the memory of the circuit. *Infinite impulse response* nonlinear systems may be represented as functions of integrals and differentials of the input signal

$$v_{out}(t) = g\left[v_{in}(t), \int_{-\infty}^{t} v_{in}(\tau)d\tau, \frac{\partial^n v_{in}}{\partial t^n} \right]. \qquad (16.5)$$

The most general characterization of a nonlinear system is the *Volterra Series*.[1] Consider a linear circuit that is stimulated by an input signal $v_{in}(t)$. The output signal $v_{out}(t)$ is then given by the convolution with the input signal $v_{in}(t)$ and the *impulse response* $h(t)$. Unless the impulse response takes the form of the *delta function* $\delta(t)$, the output $v_{out}(t)$ depends on values of the input $v_{in}(t)$ at times other than t, i.e., the circuit is said to have memory.

$$v_{out}(t) = \int_{-\infty}^{\infty} v_{in}(\tau)h(t - \tau)d\tau. \qquad (16.6a)$$

Equivalently, in the frequency domain,

$$V_{out}(f) = V_{in}(f)H(f). \qquad (16.6b)$$

In the most general case, a nonlinear circuit with reactive elements can be described using a Volterra series, which is said to be a power series with *memory*.

$$v_{out}(t) = g_0 + \int_{-\infty}^{\infty} v_{in}(\tau)g_1(t - \tau)d\tau + \int_{-\infty}^{\infty}\int_{-\infty}^{\infty} v_{in}(t - \tau_1)v_{in}(t - \tau_2)g_2(\tau_1, \tau_2)d\tau_1 d\tau_2 + \dots \qquad (16.7a)$$

An equivalent representation is obtained by taking the *n*-fold Fourier transform of Eq. (16.7a)

$$V_{out}\left(f_1, f_2, \ldots\right) = G_0 \delta\left(f_1\right) + G_1\left(f_1\right) V_{in}\left(f_1\right) + G_2\left(f_1, f_2\right) + \cdots \qquad (16.7b)$$

Notice that the Volterra series is applicable to nonlinear effects on signals with discrete spectra (i.e., a signal consisting of a sum of sinusoids). For instance, the DC component of the output signal is given by $g_0 = G_0$, while the fundamental component is given by $G_1(f_1) V_{in}(f_1)$, where G_1 and V_{in} are the Fourier transforms of the impulse response g_1 and v_{in}, respectively, evaluated at frequency f_1. The higher order terms in the Volterra series represent the harmonic responses and intermodulation response of the circuit.

Fortunately, extraction of high order Volterra series representations of nonlinear microwave circuits is rarely required to gain useful information on the deleterious and/or useful effects of distortion on common signals. Such simplifications often involve considering the circuit to be memoryless, as in Eq. (16.3a,b), or having finite order, or having integral representations, as in Eq. (16.5).

16.3 Harmonic Distortion

A fundamental result of the distortion of nonlinear circuits is that they generate frequency components in the output signal that are not present in the input signal. For sinusoidal inputs, the salient characteristic is *harmonic distortion*, whereby signal outputs consist of integer multiples of the input frequency.

16.3.1 Harmonic Generation in Nonlinear Circuits

As far as microwave circuits are concerned, the major characteristic of a nonlinear circuit is that the frequency components of the output signal differ from those of the input signal. This is readily seen by examining the output of a sinusoidal input from Eq. (16.8).

$$v_{out}\left(t\right) = g_0 + g_1 A \cos\left(2\pi f_c t\right) + g_2 A^2 \cos^2\left(2\pi f_c t\right) + g_3 A^3 \cos^3\left(2\pi f_c t\right) + \cdots$$

$$= g_0 + \frac{g_2 A^2}{2} + \left(g_1 A + \frac{3 g_3 A^3}{4}\right) \cos\left(2\pi f_c t\right) + \frac{a g_2 A^2}{2} \cos\left(4\pi f_c t\right) + \frac{g_3 A^3}{4} \cos\left(6\pi f_c t\right) + \cdots \qquad (16.8)$$

It is readily seen that, along with the *fundamental* component at a frequency of f_c, there exists a DC component, and *harmonic* components at integer multiples of f_c. The output signal is said to have acquired *harmonic distortion* as a result of the nonlinear transfer characteristic. This is illustrated in Fig. 16.1. The function represented by $g(v)$ is that of an ideal limiting amplifier. The net effect of the terms are summarized in Table 16.1.

16.3.2 Measurement of Harmonic Distortion

While instruments are available at low frequencies to measure the *total harmonic distortion* (*THD*), the level of each harmonic is generally measured individually using a spectrum analyzer. Such a setup is shown in Fig. 16.2.

Harmonic levels are usually measured in a relative manner by placing a marker on the fundamental signal and a delta marker at the nth harmonic frequency. When measured in this mode, the harmonic level is expressed in *dBc*, which designates *dB* relative to carrier (i.e., the fundamental frequency) level. While it is convenient to set the spectrum analyzer sweep to include all harmonics of interest, it may be necessary to center a narrow span at the harmonic frequency in order to reduce the *noise floor* on the spectrum analyzer. An attenuator may be needed to protect the spectrum analyzer from overload. Note that the power level present at the spectrum analyzer input includes all harmonics, not just the ones displayed on the screen. Finally, it is important to note that spectrum analyzers have their own nonlinear characteristics that depend on the level input to the instrument. It is sometimes difficult to ascertain

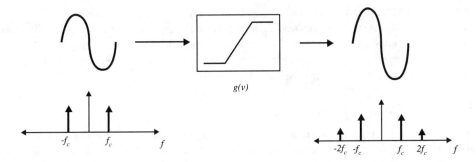

FIGURE 16.1   Effects of a nonlinear transfer characteristic on a sinusoidal input: harmonic distortion.

TABLE 16.1   Effect of Nonlinearities on Carrier Term by Term

Term	Amplitude	Qualitative Effect
DC	$g_0 + g_2A^2/2$	Small offset added due to RF detection
Fundamental	$20\log(g_1A + 3\ g_3A^3/4)$	Amplitude changed due to compression
2nd Harmonic	$40\log(g_2A^2/2)$	2:1 slope on P_{in}/P_{out} curve
3rd Harmonic	$60\log(g_3A^3/4)$	3:1 slope on P_{in}/P_{out} curve

whether measured harmonic distortion is being generated within the device or with the test instrument. One method to do this is to use a step attenuator at the output of the device and step up and down. If distortion is being generated with the spectrum analyzer, the harmonic levels will change with different attenuator settings.

16.4 Gain Compression and Phase Distortion

A major result of changing impedances in microwave circuits is signal gain and phase shift that depend on input amplitude level. A change in signal gain between input and output may result from signal

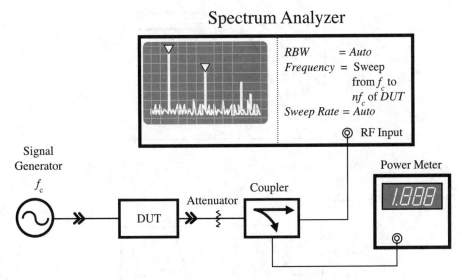

FIGURE 16.2   Setup used to measure harmonic distortion. Because harmonic levels are a function of output amplitude, a power meter is needed to accurately characterize the harmonic distortion properties.

clipping due to device current saturation or cutoff. Insertion phase may change because of nonlinear resistances in combination with a reactance. Though there are exceptions, signal gain generally decreases with increasing amplitude or power level. For this reason, the *gain compression* characteristics of microwave components are often characterized. Phase distortion may change either way, so it is often described as *phase deviation* as a function of amplitude or power level.

16.4.1 Gain Compression

Referring back to Eq. (16.8), it is seen that, in addition to harmonic distortion, the level fundamental signal has been modified beyond that dictated by the linear term, g_1. This effect is described as *gain compression* in that the gain of the circuit becomes a function of the input amplitude A. Figure 16.3 illustrates this result. For small values of A, the g_1 term will dominate, giving a 1:1 slope when the output power is plotted against the input power on a *log* (i.e., *dB*) scale. Note that the power level of the nth harmonic plotted in like fashion will have an n:1 slope.

Gain compression is normally measured on a *bandpass* nonlinear circuit.[2] Such a circuit is illustrated in Fig.16.4. It is interesting to note that an ideal limiting amplifier described by Eq. (16.9) when heavily overdriven at the input will eventually produce a square wave at the output, which is filtered by the bandpass filter. Note that the amplitude of the fundamental component of a square wave is at a level of $4/\pi$ times, or 2.1 dB greater than the amplitude of the square wave set by the clipping level.

$$v_{out}(t) = \begin{cases} g_1 v_{in}(t) & v_{out} < v_{\lim} \\ v_{\lim} & otherwise \end{cases} \qquad (16.9)$$

For a general third-order nonlinear transfer characteristic driven by a sinusoidal input, the bandpass output is given by

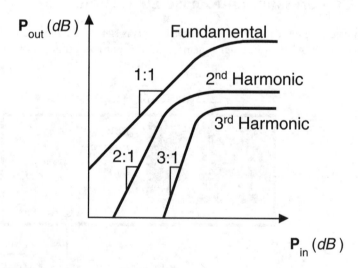

FIGURE 16.3   Output power vs. input power for a nonlinear circuit.

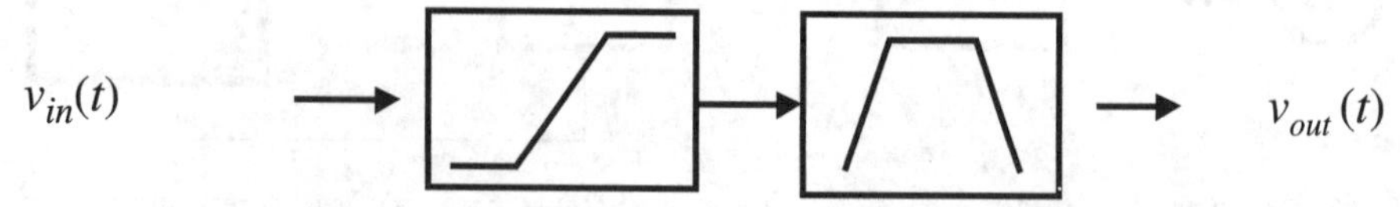

FIGURE 16.4   Bandpass nonlinear circuit.

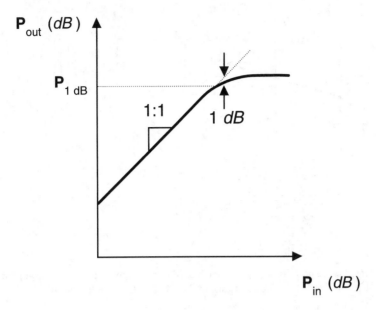

FIGURE 16.5 Gain compression of a bandpass nonlinear circuit. A figure of merit P_{1dB} is the output power at which the gain has been reduced by 1 dB.

$$v_{out}(t) = \left(g_1 A + \frac{3g_3 A^3}{4} \right) \cos\left(2\pi f_c t \right) \qquad (16.10)$$

A bandpass nonlinear circuit may be characterized by the power output at 1 dB gain compression, P_{1dB} as illustrated in Fig. 16.5.

16.4.2 Phase Distortion

Nonlinear circuits may also contain reactive elements that give rise to *memory* effects. It is usually unnecessary to extract the entire Volterra representation of a nonlinear circuit with reactive elements if a few assumptions can be made. For bandpass nonlinear circuits with memory effects of time duration of the order of the period of the carrier waveform, a simple model is often used to describe the phase deviation versus amplitude:

$$v_{out}(t) = A(t)\cos\left\{ 2\pi f_c t + \Phi\left[A(t)\right] \right\}. \qquad (16.11)$$

Equation (16.11) represents the *AM-PM* distortion caused by short-term memory effects (i.e., small capacitances and inductances in microwave circuits). The *effects* of *AM-PM* on an amplitude-modulated signal is illustrated in Fig. 16.6.

For the case of input signals with small deviations of amplitude ΔA, the phase deviation may be considered linear, with a proportionality constant k_ϕ as seen in Fig. 16.6. For a sinusoidally modulated input signal, an approximation for small modulation index FM signals may be utilized. One obtains the following expression for the output signal:

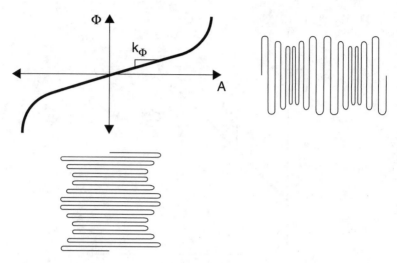

FIGURE 16.6 Effect of *AM-PM* distortion on a modulated signal. Input signal has *AM* component only. Output signal has interrelated *AM* and *FM* components due to the *AM-PM* distortion of the circuit.

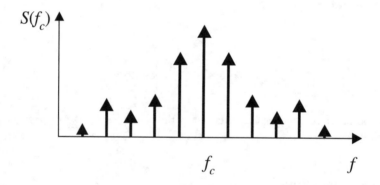

FIGURE 16.7 Output components of an amplitude-modulated signal distorted by *AM-PM* effects.

$$v_{out}(t) = \left[A + \Delta A \sin\left(2\pi f_m t\right) \right] \cos\left\{ 2\pi f_c t + k_\phi A + k_\phi \Delta A \sin\left(2\pi f_m t\right) \right\}$$

$$\approx A \cos\left(2\pi f_c t + k_\phi A\right) \sum_{n=0}^{\infty} J_n\left(k_\phi \Delta A\right) \cos\left(2n\pi f_m t\right) \tag{16.12}$$

where J_n is the nth order Bessel function of the first kind.[3] Thus, like amplitude distortion, *AM-PM* distortion creates sidebands at the harmonics of the modulating signal. Unlike amplitude distortion, these sidebands are not limited to the first sideband. Thus, *AM-PM* distortion effects often dominate the out-of-band interference beyond $f_c \pm f_m$ as seen in Fig. 16.7.

The *FM* modulation index k_ϕ may be used as a figure of merit to assess the impact of *AM-PM* on signal with small amplitude deviations. The relative level of the sidebands may be calculated from Eq. (16.12). It must be noted that two sidebands nearest to the carrier may be masked from the *AM* components of the signal, but the out-of-band components are readily identified.

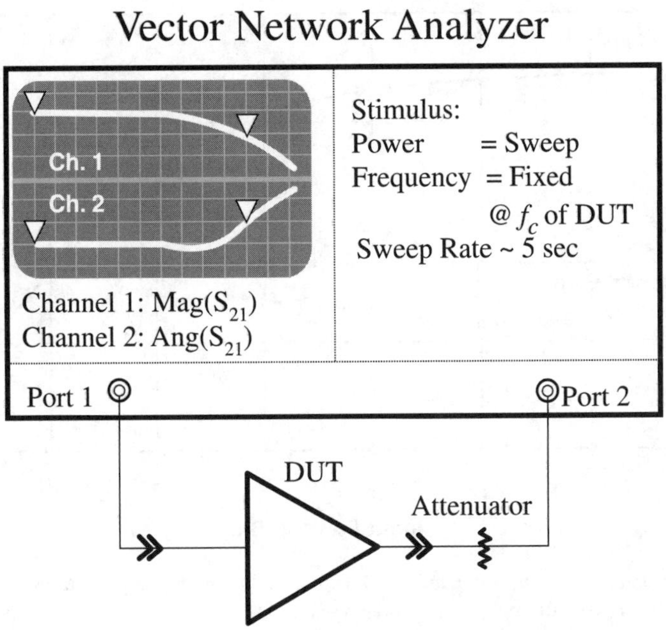

Vector Network Analyzer

FIGURE 16.8 Setup used to measure gain compression and *AM-PM*.

16.4.3 Measurement of Gain Compression and Phase Deviation

For bandpass components where the input frequency is equal to the output frequency, such as amplifiers, gain compression and phase deviation of a nonlinear circuit are readily measured with a *network analyzer* in power sweep mode. Such a setup is shown in Fig. 16.8. P_{1dB} is easily measured using delta markers by placing the reference marker at the beginning of the sweep (i.e., where the DUT is not compressed), and moving the measurement marker where $\Delta Mag(S_{21}) = -1$ dB. Sweeping at too high a rate may affect the readings. The sweep must be slow enough so that steady-state conditions exist in both the thermal case and the DC bias network within the circuit. Sweeper retrace may also affect the first few points on the trace. These points must be neglected when setting the reference marker.

The *FM* modulation index is often estimated by measuring the phase deviation at 1 dB gain compression $\Delta\Phi(P_{1dB})$

$$k_\phi \approx \frac{\Delta\Phi\left(P_{1dB}\right)}{2Z_0\sqrt{P_{1dB}}}. \tag{16.13}$$

For circuits such as mixers, where the input frequency is not equal to the output frequency, gain compression may be measured using the network analyzer with the measurement mode setup for frequency translation. The operation in this mode is essentially that of a *scalar network analyzer*, and all phase information is lost. *AM-PM* effects may be measured using a *spectrum analyzer* and fitting the sideband levels to Eq. (16.12).

The gain compression and phase deviation of a GaAs power amplifier is shown in Fig. 16.9. P_{1dB} for this amplifier is approximately 23 *dBm* or 0.5 W. The phase deviation $\Delta\Phi$ is not constant from low power to P_{1dB}. Nevertheless, as a figure of merit, the modulation index k_ϕ may be calculated from Eq. (16.13) to be 0.14°/*V*. Notice that for higher power levels, the amplifier is well into compression, and the phase deviation occurs at a much higher slope than k_ϕ would indicate.

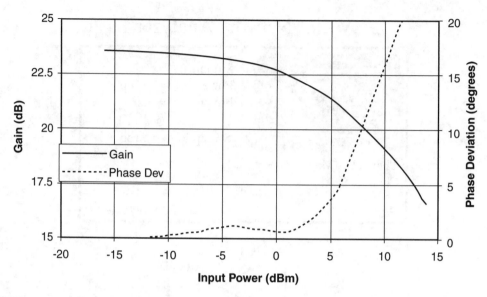

FIGURE 16.9   Measured gain compression and *AM-PM* of a 0.5 W 1960 MHz GaAs MESFET power amplifier IC using an HP8753C Vector Network Analyzer in power sweep mode.

16.5 Intermodulation Distortion

When more than one frequency component is present in a signal, the distortion from a nonlinear circuit is manifested as *intermodulation distortion* (*IMD*).[4] The *IMD* performance of microwave circuits is important because it can create unwanted interference in adjacent channels. While *bandpass* filtering can eliminate much of the effects of harmonic distortion, intermodulation distortion is difficult to filter out because the IMD components may be very close to the carrier frequency. A common figure of merit is *two-tone* intermodulation distortion.

16.5.1   Two-Tone Intermodulation Distortion

Consider a signal consisting of two sinusoids

$$v_{in}(t) = A\cos(2\pi f_1) + A\cos(2\pi f_2).$$ (16.14)

Such a signal may be represented in a different fashion by invoking well-known trigonometric identities.

$$v_{in}(t) = A\cos(2\pi f_c)\cos(2\pi f_m),$$ (16.15a)

where

$$f_c = \frac{f_1 + f_2}{2} \quad \text{and} \quad f_m = \left|\frac{f_1 - f_2}{2}\right|$$ (16.15b)

Applying such a signal to a memoryless nonlinearity as defined in Eq. (4.24), one obtains the following result:

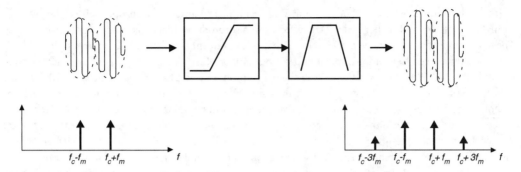

FIGURE 16.10 Intermodulation distortion of a two-tone signal. The output bandpass signal contains the original input signal as well as the harmonics of the envelope at the sum and difference frequencies.

$$v_{out} = \left[\left(g_1 A + \frac{3g_3 A^3}{4} \right) \cos\left(2\pi f_m t\right) + \frac{3g_3 A^3}{4} \cos\left(6\pi f_m t\right) \right] \cos\left(2\pi f_c t\right). \qquad (16.16)$$

Thus, it is seen that the IMD products near the input carrier frequency are simply the odd-order harmonic distortion products of the modulating *envelope*. This is illustrated in Fig. 16.10.

16.5.2   Third Order Intercept Point

Referring to Fig. 16.11, note that the output signal varies at a 1:1 slope on a *log-log* scale with the input signal, while the IMD products vary at a 3:1 slope. Though both the fundamental and the IMD products saturate at some output power level, if one were to extrapolate the level of each and find the intercept point, the corresponding output power level is called the *third order intercept point* (IP_3). Thus, if the IP_3 of a nonlinear circuit is known, the IMD level relative to the output signal level may be found from

$$IMD_{dBc} = 2\left(P_{out,dBm} - IP_{3,dBm} \right). \qquad (16.17)$$

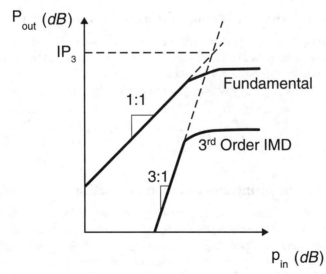

FIGURE 16.11 Relationship between signal output power and intermodulation distortion product power levels. Extrapolating the trends, a figure of merit called the *third order intercept point* (IP_3) is obtained.

It must be noted that 3rd order *IMD* is only dominant for low levels of distortion (<10 *dB* below P_{1dB}). At higher levels, 5th and higher order *IMD* effects can also produce sidebands at the 3rd order frequency. The net result is that the relative *IMD* level will change at a rate greater than 2:1 compared to carrier level. Care should be taken to avoid extrapolating IP_3 from points where this may be occurring. Another point of caution is *AM-PM* effects. In theory, the sidebands produced by phase modulation are in quadrature with those produced by *AM* distortion, and thus should add directly to the *IMD* power. However, the author's experience has shown that these *AM-PM* products can be rotated in phase and thus vector added to the *AM* sidebands. Since the *FM* sidebands are antiphase, one *FM* sideband adds constructively to the *AM* sidebands, while the other adds destructively. The net effect is an imbalance in the *IMD* levels from lower to higher sideband frequencies. Most specifications of *IMD* level will measure the worst case of the two.

For a limiting amplifier, an often used rule of thumb may be derived that predicts a relationship between P_{1dB} and IP_3.[4]

$$IP_3 = P_{1dB} + 9.6 \ dB. \tag{16.18}$$

While this may not be rigorously relied upon for every situation, it is often accurate within ±2 *dB* for small-signal amplifiers and class-A power amplifiers.

16.5.3 Dynamic Range

Because intermodulation distortion generally increases with increasing signal levels, IP_3 may be used to establish the *dynamic range* of a system. The signal level at which the *IMD* level meets the noise floor is defined at the *spurious free dynamic range (SFDR)*.[5] This is illustrated in Fig. 16.12.

The *SFDR* of a system with gain G may be derived from IP_3 and the *noise figure NF*

$$SFDR = \frac{2IP_3 - 2\left[10\log\left(kT_{eq}B\right) + NF + G\right]}{3}, \tag{16.19}$$

where k is Boltzman's constant, T_{eq} is the equivalent input noise temperature, and B is the bandwidth of the system.

16.5.4 Intermodulation Distortion of Cascaded Components

The question often arises when two components are cascaded of what effect the driving stage *IMD* has on the total *IMD*. This is shown in Fig. 16.13. To the degree that the *IMD* products produced by the nth stage are uncorrelated with those of the $n + 1$ stage, the output *IMD* may be calculated as the power addition of 3rd order *IMD* levels (P_3) with levels adjusted accordingly for gain.

$$\left(P_3\right)_{n+1} = 10\log\left[10^{\left[\left(P_3\right)_n + G_1\right]/10} + 10^{\left[3\left(P_1\right)_{n+1} - 2\left(IP_3\right)_{n+1}/10\right]}\right]. \tag{16.20}$$

16.5.5 Measurement of Intermodulation Distortion

IMD is normally measured with two-signal generators and a *spectrum analyzer*. Such a setup is shown in Fig. 16.14. Care must be taken to isolate the signal generators, as *IMD* may result from one output mixing with signal from the opposing generator. The carrier levels should be within 0.5 *dB* of each other for accurate *IMD* measurements. Also, it is usually recommended that a power meter be used to get an accurate reading of output power level from the DUT. Relative *IMD* level is measured by placing a reference marker on one of the two carrier signals, and placing a delta marker at either sideband. Finally, the input level must be maintained well below the input IP_3 of the spectrum analyzer to insure error-free reading of the DUT.

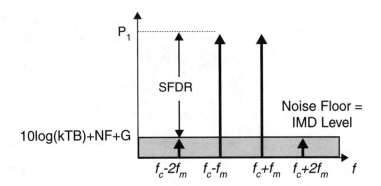

FIGURE 16.12 An illustration of spurious free dynamic range, which defines the range of signal levels where the worst case signal-to-noise ratio is defined by the noise floor of the system, rather than the *IMD* level.

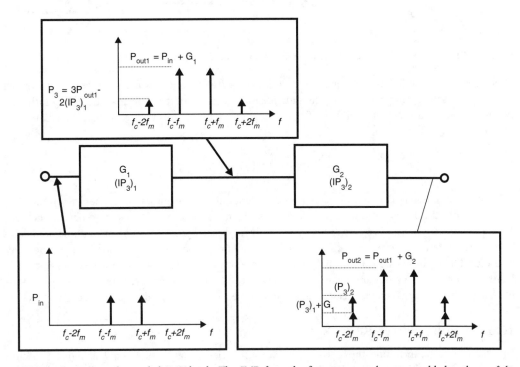

FIGURE 16.13 Effect of cascaded *IMD* levels. The *IMD* from the first stage may be power added to those of the second stage with levels adjusted for the gain of the stage.

16.6 Multicarrier Intermodulation Distortion and Noise Power Ratio

While two-tone intermodulation distortion serves to compare the linearity of one component to another, in many applications, a component will see more than two carriers in the normal operation of a microwave system. Thus direct measurement of multitone *IMD* is often necessary to insure adequate carrier-to-interference level within a communication system.

16.6.1 Peak-to-Average Ratio of Multicarrier Signals

The major difference between two-tone signals and multitone signals is the *peak-to-average (pk/avg)* power ratio.[6] From Eq. (16.14), it is clear that the average power of a two-tone signal is equal to the sum

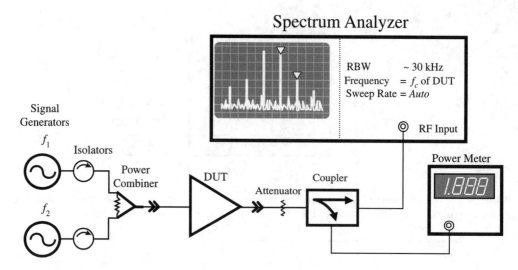

FIGURE 16.14   Setup used to measure two-tone intermodulation distortion (*IMD*).

of powers from the individual carriers. However, from Eq. (16.15), one may derive that the *peak envelope power* (*PEP*) is four times the level of the individual carriers. Thus, it is said that the *pk/avg* ratio of a two-tone signal is a factor of 2, or 3 *dB*. From inductive reasoning, it is then clear that the *pk/avg* ratio of an *n*-tone signal is

$$pk/avg = 10\log(n). \qquad (16.21)$$

While the absolute peak of a multicarrier is dependent only on the number of carriers, the probability distribution of the *pk/avg* ratio depends on the modulation. Figure 16.15 shows the difference between 16 phase-aligned tones, and 16 carriers with randomly modulated phases. In general, multiple modulated signals encountered in communication systems will mimic the behavior of random phase modulated sinusoids. Phase aligned sinusoids may be considered a worst case condition. As the number of carriers increases, and if their phases are uncorrelated, the Central Limit Theorem predicts that the distribution of *pk/avg* approaches that of white Gaussian random noise.[7] The latter signal is treated in the next section.

16.6.2   Noise Power Ratio

For many systems, including those that process multicarrier signals, white Gaussian noise is a close approximation for the real-world signals. This is a result of the *Central Limit Theorem,* which states that the probability distribution of a sum of a large number of random variables will approach the Gaussian distribution, regardless of the distributions of the individual signals.[7] One metric that has been employed to describe the *IMD* level one would expect in a dense multicarrier environment is the *noise power ratio*. This concept is illustrated in Fig. 16.16.

16.6.3   Measurement of Multitone IMD and Noise Power Ratio

Thus it is clear that power ratings for components must be increased for peak power levels given by Eq. (16.21). Furthermore, two-tone intermodulation distortion may not be indicative of *IMD* of multitone signals. Measurement over various power levels is the only way to accurately predict multitone IMD. Figure 16.17 illustrates a setup that may be used to measure multitone intermodulation.

The challenge in measuring *NPR* is creating the signal. It is clear that, to get an accurate indication of *IMD* performance, the signal bandwidth must not exceed the bandwidth of the device under test. Furthermore, to measure *NPR*, one must notch out the noise power over a bandwidth approximating

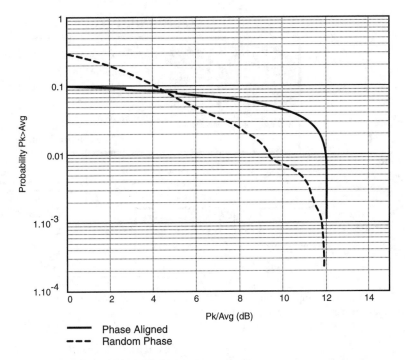

FIGURE 16.15 Distribution of *peak-to-average* ratio of a phase-aligned 16 carrier signal and a random-phase 16 carrier signal. The *y*-axis shows the probability that the signal exceeds a power level above average on the *x*-axis. While both signals ultimately have the same *pk/avg* ratio, their distributions are much different.

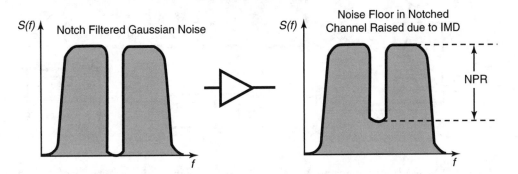

FIGURE 16.16 An illustration of *noise power ratio*. *NPR* is essentially a measure of the carrier-to-interference level experienced by multiple carriers passing through a nonlinear component.

one channel *BW*. As an example, an *NPR* measurement on a component designed for North American Digital Cellular System (IS-136) ideally would produce a 25-MHz wide noise source with one channel of bandwidth equal to 30 *kHz*. The *Q* of a notch filter to produce such a signal would be in excess of 25,000. Practical measurements employ filters with *Q*s around 1000, and are able to achieve more than 50 *dB* of measurement range. Such a setup is shown in Fig. 16.18.

16.7 Distortion of Digitally Modulated Signals

While standard test signals such as a two-tone or band-limited Gaussian noise provide relative figures of merit of the linearity of a nonlinear component, they cannot generally insure compliance with government or industry system-compatibility standards. For this reason, methods have been developed to measure and

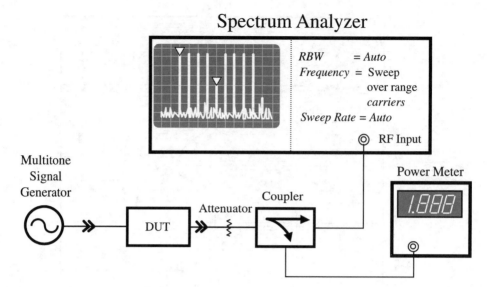

FIGURE 16.17   Measurement setup for multitone *IMD*. Tones are usually spaced equally, with the middle tone deleted to allow measurement of the worst-case *IMD*.

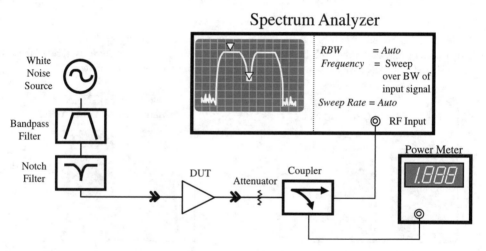

FIGURE 16.18   Noise Power Ratio measurement setup. The rejection of the notch filter should be at least 10 *dB* below the *NPR* level to avoid erroneous measurement.

characterize the intermodulation distortion of the specific digitally modulated signals used in various systems. Table 16.2 summarizes the modulation formats for North American digital cellular telephone systems.[8,9]

16.7.1   Intermodulation Distortion of Digitally Modulated Signals

Amplitude and phase distortion affect digitally modulated signals the same way they affect analog modulated signals: gain compression and phase deviation. This is readily seen in Fig. 16.19. Because both amplitude and phase modulation are used to generate digitally modulated signals, they are often expressed as a *constellation* plot, with the in-phase component $I = A\cos\phi$ envelope plotted against the quadrature component $Q = A\sin\phi$. The instantaneous power envelope is given by

$$P(t) = I(t)^2 + Q(t)^2 = A(t)^2. \qquad (16.22)$$

TABLE 16.2   Modulation Formats for North American Digital Cellular Telephone Systems

Standard	Multiple Access Mode	Channel Power Output	Modulation	Channel Bandwidth
IS-136[8]	TDMA	+28 dBm	$\pi/4$-DQPSK	30 kHz
IS-95[9]	CDMA	+28 dBm	OQPSK	1.23 MHz

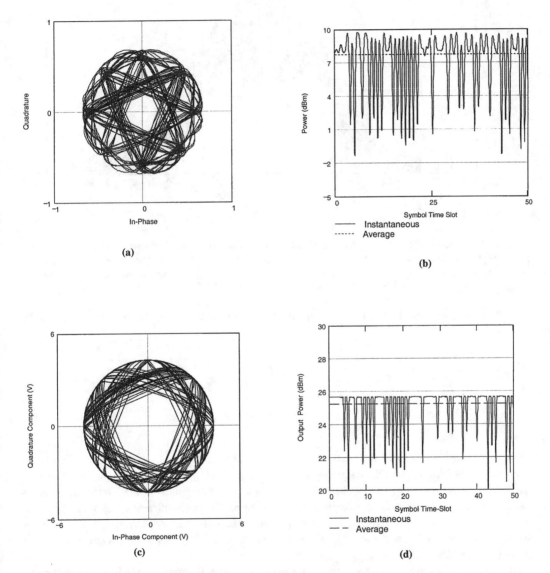

FIGURE 16.19   Effect of amplitude and phase distortion on digitally modulated signals. (a) Shows a $\pi/4$ DQPSK signal constellation, and its associated power envelope in (b). When such a signal is passed through a nonlinear amplifier, the resulting envelope is clipped (d), and portions of the constellation are rotated (c).

When the envelope is clipped and/or phase rotated, the resulting *IMD* is referred to as *spectral regrowth*. Figure 16.20 shows the effect of nonlinear distortion on a digitally modulated signal. The out-of-band products may lie in adjacent channels, thus causing interference to other users of the system. For this reason, the *IMD* of digitally modulated signals are often specified as *adjacent channel power ratio (ACPR)*.[10]

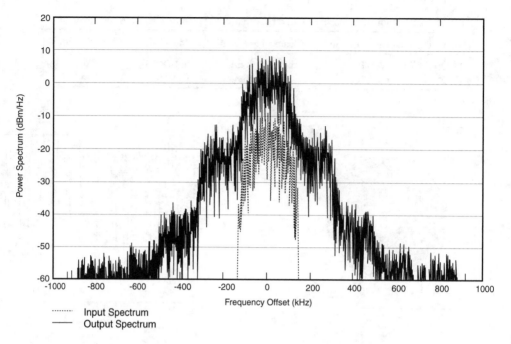

FIGURE 16.20   Effect of nonlinear distortion on a digitally modulated signal. The lower power input signal to a power amplifier has a frequency spectrum that is well contained within a specified channel bandwidth. IMD due to nonlinear distortion creates out-of-band products that may fall within the adjacent channels, causing interference to other users of the system.

ACPR may be specified in a number of ways, depending on the system architecture. In general, *ACPR* is given by

$$ACPR = \frac{I_{adj}}{C_{ch}} = \frac{\int_{f_o - B_{adj}}^{f_o - B_{adj}} S(f)df}{\int_{-B_{ch}/2}^{B_{ch}/2} S(f)df}, \qquad (16.23)$$

where I_{adj} is the total interference power in a specified adjacent channel bandwidth, B_{adj} at a given frequency offset f_o from the carrier frequency, and C_{ch} is the channel carrier power in the specified channel bandwidth B_{ch}. Note that the carrier channel bandwidth may be different from the interference channel bandwidth because of regulations enforcing interference limits between different types of systems. Furthermore, the interference level may be specified in more than one adjacent channel. In this case, the specification is referred to as the *alternate channel power ratio*. Table 16.3 shows ACPR specifications for various digital cellular standards.

In addition to the out-of-band interference due to the intermodulation distortion in-band interference will also result from nonlinear distortion. The level of the in-band interference is difficult to measure directly because it is superimposed on the channel spectrum. However, when the signal is demodulated, errors in the output *I-Q* constellation occur at the sample points. This is shown in Fig. 16.21. Because the demodulator must make a decision as to which symbol (i.e., which constellation point) was sent, the resulting errors in the *I-Q* vectors may produce a false decision, and hence cause *bit errors*.

There are two methods to characterize the level *I-Q* vector error: *error vector magnitude* (*EVM*), and a quality factor called the ρ-*factor*.[11] Both *EVM* and ρ-*factor* provide an indication of signal distortion, but they are calculated differently. *EVM* is the *rms* sum of vector errors divided by the number of samples.

TABLE 16.3 ACPR and EVM Specifications for Digital Cellular Subscriber Equipment

Standard	ADJ. CH. PWR	ALT. CH. PWR	EVM
IS-136	−26 dBc/30 kHz @ ± 30 kHz	−45 dBc/30 kHz @ ± 60 kHz	12.5%
IS-95	−42 dBc/30 kHz @ > ± 885 kHz	−54 dBc/30 kHz @ > ± 1.98 MHz	23.7%

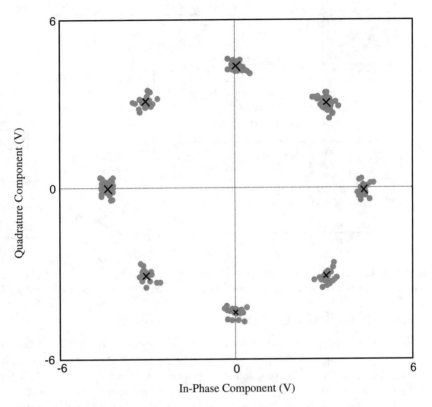

FIGURE 16.21 Errors in the demodulated *I-Q* constellation may result from the in-band *IMD* products. The *rms* summation of errors from the desired location (given by the × markers) give the *error vector magnitude* of the signal distortion.

$$EVM = \frac{1}{n}\sqrt{\sum_{n}\left[\left|I(t_n) - S_{In}\right|^2 + \left|Q(t_n) - S_{Qn}\right|^2\right]}, \qquad (16.24)$$

where the *I-Q* sample points at the nth sample windows are given by $I(t_n)$ and $Q(t_n)$, and the nth symbol location point in-phase and quadrature components are given by S_{In} and S_{Qn} respectively.

Whereas *EVM* provides an indication of *rms* % error of the signal envelope at the sample points, ρ-*factor* is related to the waveform quality of a signal. It is related to *EVM* by

$$\rho = \frac{1}{1 - EVM^2} \qquad (16.25)$$

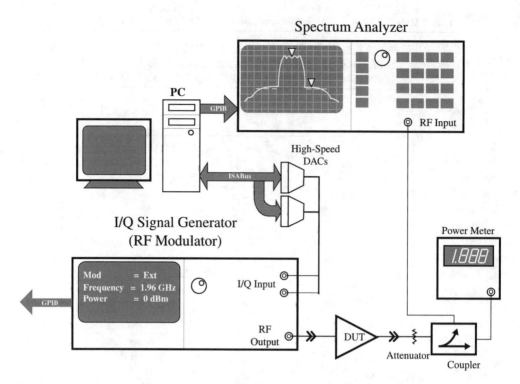

FIGURE 16.22   Measurement setup for ACPR. Waveforms are created using PCs or specialized arbitrary waveform generators. In either case, the baseband waveform must be upconverted to the center frequency of the DUT.

16.7.2   Measurement of ACPR, EVM, and Rho-Factor

ACPR may be measured using a setup similar to those for measuring *IMD* (Fig. 16.22). The major difference involves generating the test signal. Test signals for digitally modulated signals must be synthesized according to system standards using an *arbitrary waveform generator* (*AWG*), which generates *I*- and *Q*-baseband envelopes. In the most basic form, these are high speed digital-to-analog converters (*DACs*). The files used to generate the envelope waveforms may be created using commonly available mathematics software, and are built in many commercially available *AWGs*. The *I*- and *Q*-baseband envelopes are fed to an *RF* modulator to produce a modulated carrier at the proper center frequency.

In the case of *CDMA* standards, deviations between test setups can arise from different selections of Walsh codes for the traffic channels. While a typical *CDMA* downlink (base station to mobile) signal has a *pk/avg* of approximately 9.5 *dB*, it has been shown that some selections of Walsh codes can result in peak-to-average ratios in excess of 13 *dB*.[12] Measurement of *EVM* is usually done with a *vector signal analyzer* (*VSA*) (Fig. 16.23). This instrument is essentially a receiver that is flexible enough to handle a variety of frequencies and modulation formats. Specialized software is often included to directly measure *EVM* or rho-factor for well-known standards used in microwave radio systems.

16.8 Summary

This section has treated characterization and measurement techniques for nonlinear microwave components. Figures of merit were developed for such nonlinear effects as harmonic level, gain compression, and intermodulation distortion. While these offer a basis for comparison of the linearity performance between like components, direct measurement of adjacent channel power and error vector magnitude are preferred for newer wireless systems. Measurement setups for the above parameters were suggested in each section. For more advanced treatment, the reader is referred to the references at the end of this section.

Vector Signal Analyzer

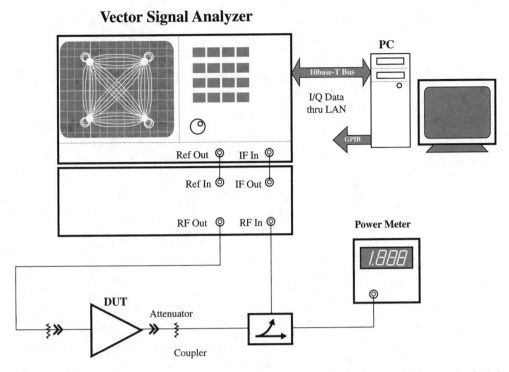

FIGURE 16.23 Setup for measuring *EVM*. The *VSA* demodulates the *I-Q* waveform and calculates the deviation from ideal to calculate *EVM* and ρ-*factor* as given in Eq. (16.24) and Eq. (16.25), respectively.

References

1. Maas, S.A., *Nonlinear Microwave Circuits*, Artech House, Boston, 1988.
2. Blachman, N.M., Band-pass nonlinearities, *IEEE Trans. Information Theory*, IT-10, 162–64, April, 1964.
3. Andrews, L.C., *Special Functions of Mathematics for Engineers*, 2nd ed., McGraw-Hill, New York, 1992, chap. 6.
4. Cripps, S.C., *RF Power Amplifiers for Wireless Communications*, Artech House, Boston, 1999, chap. 7.
5. Carson, R.S., *Radio Concepts: Analog*, John Wiley & Sons, New York, 1990, chap. 10.
6. Kenney, J.S., and Leke, A., Design considerations for multicarrier CDMA base station power amplifiers, *Microwave J.*, 42, 2, 76–86, February, 1999.
7. Papoulis, A., *Probability, Random Variables, and Stochastic Processes*, 3rd ed., McGraw-Hill, New York, 1991, chap. 8.
8. IS-136 Interim Standard, Cellular System Dual-Mode Mobile Station — Base Station Compatibility Standards, Telecommunications Industry Assoc.
9. IS-95 Interim Standard, Mobile Station — Base Station Compatibility Standard for Dual-Mode Wideband Spread Spectrum Cellular Systems, Telecommunications Industry Assoc.
10. Kenney, J.S. and Leke, A., Power amplifier spectral regrowth for digital cellular and PCS applications, *Microwave J.*, 38, 10, 74–92, October 1995.
11. Lindsay, S.A., Equations derive error-vector magnitude, *Microwaves & RF*, April, 1995, 158–67.
12. Braithwaite, R.N., Nonlinear amplification of CDMA waveforms: an analysis of power amplifier gain errors and spectral regrowth, *Proc. 48th Annual IEEE Vehicular Techn. Conf.*, 2160–66, 1998.

17

Theory of High-Power Load-Pull Characterization for RF and Microwave Transistors

John F. Sevic
Ultra RF, Inc.

17.1 Introduction

In both portable and infrastructure wireless systems the power amplifier often represents the largest single source of power consumption in the radio. While the implications of this are obvious for portable applications, manifested as talk-time, it is also important for infrastructure applications due to thermal management, locatability limitations, and main power limitations. Significant effort is devoted toward developing high-performance RF and microwave transistors and circuits to improve power amplifier efficiency. In the former case, an accurate and repeatable characterization tool is necessary to evaluate the performance of the transistor. In the latter case, it is necessary to determine the source and load impedance for the best trade-off in overall performance. Load-pull is presently the most common technique, and arguably the most useful for carrying out these tasks. In addition, load-pull is also necessary for large-signal model development and verification.

Load-pull as a design tool is based on measuring the performance of a transistor at various source and/or load impedances and fitting contours, in the gamma-domain, to the resultant data; measurements at various bias and frequency conditions may also be done. Several parameters can be superimposed over each other on a Smith chart and trade-offs in performance established. From this analysis, optimal source and load impedances are determined.

Load-pull can be classified by the method in which source and load impedances are synthesized. Since the complex ratio of the reflected to incident wave on an arbitrary impedance completely characterizes the impedance, along with a known reference impedance, it is convenient to classify load-pull by how the reflected wave is generated.

The simplest method to synthesize an arbitrary impedance is to use a stub tuner. In contrast to early load-pull based on this method, contemporary systems fully characterize the stub tuner *a priori*, precluding the need for determining the impedance at each load-pull state [1]. This results in a significant reduction in time and increases the reliability of the system. This method of load-pull is defined as passive-mechanical. Passive-mechanical systems are capable of presenting approximately 50:1 VSWR, with respect to 50 Ω, and are capable of working in very high power environments. Repeatability is better than –60 dB. Maury Microwave and Focus Microwave each develop passive-mechanical load-pull systems [2,3]. For high-power applications, e.g., > 100 W, the primary limitation of passive-mechanical systems is self-heating of the transmission line within the tuner, with the resultant thermally induced expansion perturbing the line impedance.

Solid-state phase-shifting and attenuator networks can also be used to control the magnitude and phase of a reflected wave, thereby effecting an arbitrary impedance. This approach has been pioneered by ATN Microwave [4]. These systems can be based on a lookup table approach, similar to the passive-mechanical systems, or can use a vector network analyzer for real-time measurement of tuner impedance. Like all passive systems, the maximum VSWR is limited by intrinsic losses of the tuner network. Passive-solid-state systems, such as the ATN, typically exhibit a maximum VSWR of 20:1 with respect to 50 Ω. These systems are ideally suited for medium power applications and noise characterization (due to the considerable speed advantage over other types of architectures).

Tuner and fixture losses are the limiting factor in achieving a VSWR in excess of 50:1 with respect to 50 Ω. This would be necessary not only for characterization of high-power transistors, but also low-power transistors at millimeter-wave frequencies, where system losses can be significant. In these instances, it is possible to synthesize a reflected wave by sampling the wave generated by the transistor traveling toward the load, amplifying it, controlling its magnitude and phase, and reinjecting it toward the transistor. Systems based on this method are defined as active load-pull. Although in principle active load-pull can be used to create very low impedance, the power necessary usually limits the application of this method to millimeter-wave applications [5,6]. Because active load-pull systems are capable of placing any reflection coefficient on the port being pulled (including reflections greater than unity) these systems can be very unstable and difficult to control. Instability in a high-power load-pull system can lead to catastrophic failure of the part being tested.

The present chapter is devoted to discussing the operation, setup, and verification of load-pull systems used for characterization of high-power transistors used in wireless applications. While the presentation is general in that much of the discussion can be applied to any of the architectures described previously, the emphasis is on passive-mechanical systems. There are two reasons for limiting the scope. The first reason is that passive-solid-state systems are usually limited in the maximum power incident on the tuners, and to a lesser extent, the maximum VSWR the tuners are capable of presenting. The second reason is that currently there are no active load-pull systems commercially available. Further, it is unlikely that an active load-pull system would be capable of practically generating the sub 1 Ω impedances necessary for characterization of high-power transistors.

The architecture of the passive-mechanical system is discussed first, with a detailed description of the necessary components for advanced characterization of transistors, such as measuring input impedance and ACPR [7]. Vector network analyzer calibration, often overlooked, and the most important element of tuner characterization, is presented next. Following this, tuner, source, and load characterization methods are discussed. Fixture characterization methods are also presented, with emphasis on use of pre-matching fixtures to increase tuner VSWR. Finally, system performance verification is considered.

17.2 System Architecture for High-Power Load-Pull

Figure 17.1 shows a block diagram of a generalized high-power automated load-pull system, although the architecture can describe any of the systems discussed in the previous section. Sub-harmonic and

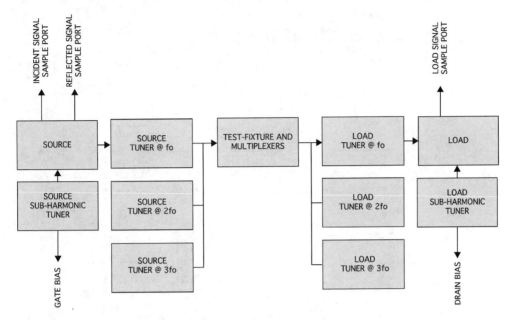

FIGURE 17.1 Block diagram of a generalized high-power load-pull system, illustrating the source, tuners, test-fixture, and load. The incident, reflected, and load signals are sampled at the three sampling points shown. Also shown, though not necessary, are harmonic and sub-harmonic tuners.

harmonic tuners are also included for characterization of out-of-band impedances [8]. The signal sample ports are used to measure the incident and reflected voltage waves at the source-tuner interface and the incident voltage wave at the load. The signals at each of these ports are applied to the equipment necessary to make the measurements the user desires. Each of these blocks is described subsequently.

The source block of Fig. 17.1 usually includes all of the components necessary for generating the signal, leveling its power, providing gate/base bias for the device under test, and providing robust sampling points for the measurement equipment. Figure 17.2 shows the details of a typical source block. For flexibility and expediency in applying arbitrarily modulated signals, an arbitrary waveform generator and vector signal source are shown. The signal is typically created using MATLAB, and can represent not only digitally modulated signals, but also the more conventional two-tone signal. The signal is applied to a reference PA, which must be characterized to ensure that it remains transparent to the DUT; for high-power applications this is often a 50 W to 100 W PA.

Following the reference PA is a low-pass filter to remove harmonics generated from the source and/or reference PA. Next are the sampling points for the incident and reflected waves, which is done with two distinct directional couplers. Since the source tuner may present a high reflection, a circulator to improve directivity separates each directional coupler; the circulator also protects the reference PA from reflected power. The circulator serves to present a power-invariant termination for the source tuner, the impedance of which is critical for sub 1 Ω load-pull. The bias-tee is the last element in the source block, which is connected to the gate/base bias source via a low-frequency tuner network for sub-harmonic impedance control. Since the current draw of the gate/base is typically small, remote sensing of the power supply can be done directly at the bias-tee.

Although components within the source block may have type-N or 3.5 mm connectors, interface to the source tuner is done with an adapter to an APC 7 mm connector. This is done to provide a robust connection and to aid in the VNA characterization of the source block. Depending on the measurements that are to be made during load-pull, a variety of instruments may be connected to the incident and reflected sample ports, including a power meter and VNA. The former is required for real-time leveling and the latter for measuring the input impedance to the DUT [9].

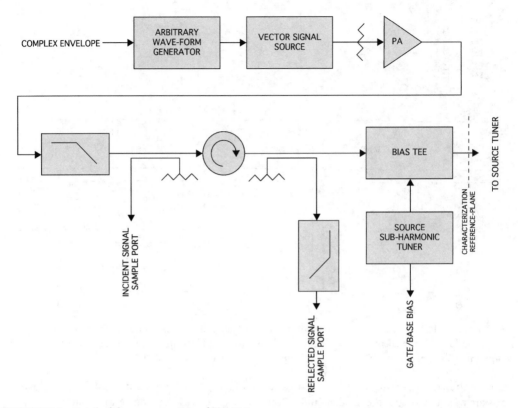

FIGURE 17.2   Detail of the source portion of Fig. 17.1.

The load block of Fig. 17.1 usually includes a port for sampling the load signal of the DUT and the padding and filtering necessary to interface the load signal to a power sensor. Figure 17.3 shows the details of a typical load block. The bias-tee comes first. Although remote-sense can be sampled here, in situations where significant current is required, the remote-sense should be sampled directly on the DUT test fixture. For a load-pull system capable of 100 W average power, the attenuator following the bias-tee should be appropriately rated and exhibit at least 30 dB attenuation.

The load signal is sampled at a directional coupler after the high-power pad. A spectrum analyzer is often connected at this port, and it may be useful to use a low coupling factor, e.g., −30 dB, to minimize the padding necessary in front of the spectrum analyzer. This results in an optimal dynamic range of the system for measuring ACPR. Following the directional coupler is a low-pass filter, to remove harmonics,[1] which is followed by another attenuator. This attenuator is used to improve the return loss of the filter with respect to the power sensor. As with the source block, interface to the load tuner and power sensor are done with APC 7 mm connectors to improve robustness and power-handling capability.

The DUT test-fixture is used to interface the source and load tuners to a package. For cost and package de-embedding reasons, it is useful to standardize on two or three laboratory evaluation packages. For hybrid circuit design, it is useful to design a test fixture with feeds and manifolds identical to those used in hybrid to mitigate de-embedding difficulties. The collector/drain side of the test fixture should also have a sampling port for remote sensing of the power supply.

---

[1]Although a filter is not necessary, characterization of a DUT in significant compression will result in the average power detected by the power sensor including fundamental and harmonic power terms. When the DUT is embedded into a matching network, the matching network will usually attenuate the harmonics; thus, inclusion of the low-pass filter more closely approximates the performance that will be observed in practice.

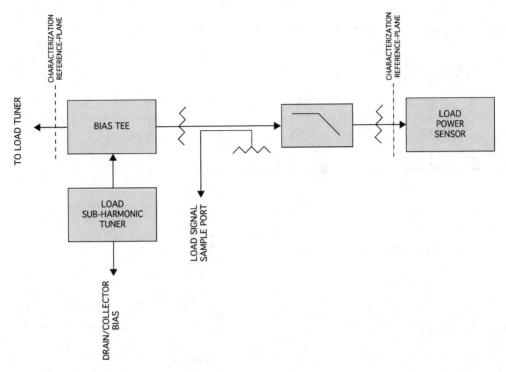

FIGURE 17.3   Detail of the load portion of Fig. 17.1.

After the load-pull system has been assembled, it is recommended that the maximum expected power be applied to the system and changes in impedance be measured due to tuner self-heating. This may be significant where average powers exceed 100 W or peak powers exceed several hundred watts. Any impedance change will establish the upper power limit of the system with respect to impedance accuracy.

17.3 Characterization of System Components

Each of the blocks described in the previous section must be characterized using s-parameters in order for a load-pull system to function properly. In this section, the characterization procedure for each of the sections of Fig. 17.1 is described, with emphasis on calibration of the vector network analyzer and the characterization of the transistor test fixture. Two-tier calibration and impedance re-normalization are considered for characterizing quarter-wave pre-matching test fixtures.

17.3.1   Vector Network Analyzer Calibration Theory

Due to the extremely low impedances synthesized in high-power load-pull, the vector network analyzer (VNA) calibration is the single most important element of the characterization process. Any errors in the measurement or calibration, use of low quality connectors, e.g., SMA or type-N, or adoption of low-performance calibration methods, e.g., SOLT, will result in a significant reduction in accuracy and repeatability. Only TRL calibration should be used, particularly for tuner and fixture characterization. Use of high-performance connectors is preferred, particularly APC 7 mm, due to its repeatability, power handling capability, and the fact that it has a hermaphroditic interface, simplifying the calibration process.

Vector network analysis derives its usefulness from its ability to characterize impedance based on ratio measurements, instead of absolute power and phase measurements, and from its ability to characterize and remove systematic errors due to nonidealities of the hardware. For a complete review of VNA architecture and calibration theory, the reader is encouraged to review notes from the annual ARFTG Short-Course given in November of each year [10,11].

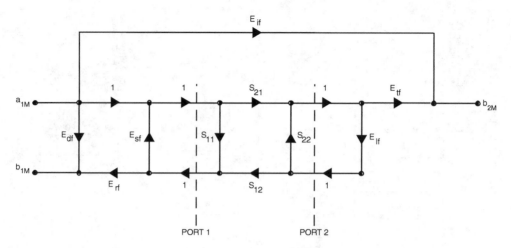

FIGURE 17.4   Signal-flow graph of the forward direction of a typical VNA.

Figure 17.4 shows a signal-flow graph of the forward direction of a common VNA architecture, where six systematic error terms are identified. An identical flow-graph exists for the reverse direction, with six additional error terms. Consider the situation where it is required to measure an impedance that exhibits a near total reflection, such as a load tuner set for 1 Ω. Assuming a 50 Ω reference impedance, nearly all of the incident power is reflected back toward the VNA, along with a phase shift of 180°. Consider what happens when the reflected wave is sampled at the VNA, denoted as b_{1M} in Fig. 17.4. If there is any re-reflection of the reflected wave incident at the VNA, an error will occur in measuring the actual impedance of the load. The ability of a VNA to minimize this reflected power is characterized by its residual source match, which is the corrected source impedance looking into the VNA. The uncorrected source impedance looking into the VNA is characterized by the E_{sf} term in the flow graph of Fig. 17.4.

Continuing with this example, Fig. 17.5 shows a plot of the upper bound on apparent load impedance versus the residual source match (with respect to a reference impedance of 50 Ω and an actual impedance of 1 Ω). For simplicity, it is assumed that the residual source match is in phase with the reflected signal. Also shown are typical residual source match performance numbers for an HP 8510C using an HP 8514B test set. From this graph it is clear that use of low-performance calibration techniques will result in latent errors in any characterization performed using a DUT with reflection VSWR near 50:1. Using a 3.5 mm SOLT calibration can result in nearly 20% uncertainty in measuring impedance. Note that TRL*, the calibration method available on low-cost VNAs, offers similar performance to 3.5 mm SOLT, due to its inability to uniquely resolve the test-set port impedances. This limitation is due to the presence of only three samplers instead of four, and does not allow switch terms to be measured directly. For this reason, it is recommended that three-sampler architectures not be used for the characterization process.

Similar arguments can be made for the load reflection term of Fig. 17.4, which is characterized by the residual load match error term. Identical error terms exist for the reverse direction too, so that there are a total of four error terms that are significant for low impedance VNA calibration.

TRL calibration requires a thru line, a reflect standard (known only within λ/4), and a delay-line. The system reference impedances will assume the value of the characteristic impedance of the delay-line, which if different from 50 Ω, must be appropriately re-normalized back to 50 Ω [12–15]. TRL calibration can be done in a variety of media, including APC 7 mm coaxial waveguide, rectangular/cylindrical waveguide, microstrip, and stripline. Calibration verification standards, which must be used to extract the residual error terms described above, are also easily fabricated. Figure 17.6 shows the residual forward source and load match response of an APC 7 mm calibration using an HP 8510C with an HP 8514B test set. These were obtained with a 30 cm offset-short airline and 30 cm delay-line, respectively [16,17,18]. The effective source match is computed from the peak-peak ripple using

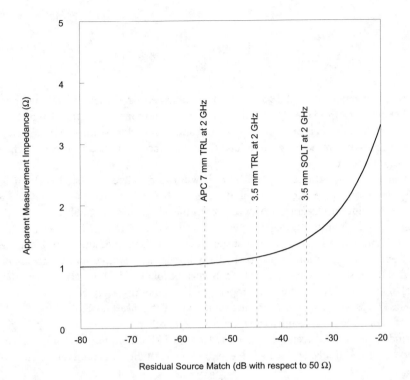

FIGURE 17.5 The influence of residual source match on the ability of a VNA to resolve a 1 Ω impedance with a 50 Ω reference impedance. The calibration performance numbers are typical for an HP 8510C with an 8514B test-set operating a 2 GHz.

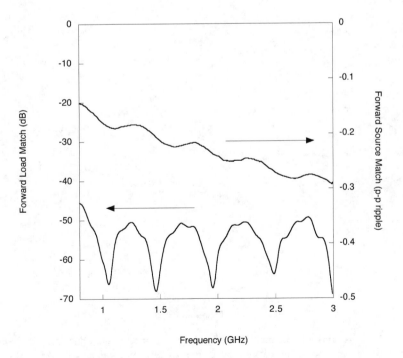

FIGURE 17.6 Typical response of an APC 7 mm TRL calibration using an offset-short and delay-line to extract source match and load match, respectively. This data was taken from an HP 8510C with an HP 8514B test set.

$$E_{sf} = 10 * \log_{10} \left[\frac{1 - 10^{-\frac{p-p\ ripple}{20}}}{1 + 10^{-\frac{p-p\ ripple}{20}}} \right] \tag{17.1}$$

where it is seen that better than −53 dB source match is obtained across the band. Due to finite directivity, 6 dB must be subtracted from the plot showing the delay-line response, indicating that better than −56 dB load match is obtained except near the low end of the band. Calibration performance such as that obtained in Fig. 17.6 is necessary for accurate tuner and fixture characterization, and is easily achievable using standard TRL calibration.

For comparison purposes, Figs. 17.7 and 17.8 show forward source and load match for 3.5 mm TRL and SOLT calibration, respectively. Here it is observed that the source match of the 3.5 mm TRL calibration has significantly degraded with respect to the APC 7 mm TRL calibration and the 3.5 mm SOLT calibration has significantly degraded with respect to the 3.5 mm TRL calibration.

Proper VNA calibration is an essential first step in characterization of any component used for high-power load-pull characterization, and is particularly important for tuner and fixture characterization. All VNA calibrations should be based on TRL and must be followed by calibration verification to ensure that the calibration has been performed properly and is exhibiting acceptable performance, using the results of Fig. 17.6 as a benchmark. Averaging should be set to at least 64. Smoothing should in general be turned off in order to observe any resonances that might otherwise be obscured. Although APC 7 mm is recommended, 3.5 mm is acceptable when used with a TRL calibration kit. Under no circumstances should type-N or SMA connectors be used, due to phase repeatability limitations and connector reliability limitations.

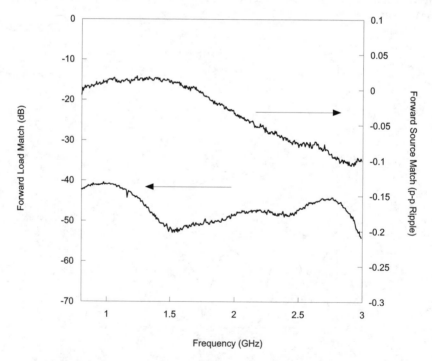

FIGURE 17.7   Typical response of a 3.5 mm TRL calibration using an offset-short and delay-line to extract source match and load match, respectively. This data was taken from an HP 8510C with an HP 8514B test set.

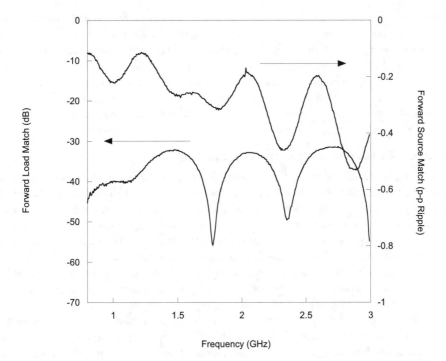

FIGURE 17.8   Typical response of a 3.5 mm SOLT calibration using an offset-short and delay-line to extract source match and load match, respectively. This data was taken from an HP 8510C with an HP 8514B test set.

17.3.2   S-Parameter Characterization of Tuners

Tuner characterization begins with proper calibration of the VNA, as described in the previous section. It is suggested at this point that any adapters on the tuner be serialized and alignment marks made to ensure that in the event of removal, they can be replaced in their original positions. Replacement of an adapter, for any reason, will require a new tuner characterization. Tuners should be leveled using a bubble-level and should be positioned such that the VNA test-port cables are not flexed. Proper torquing of all connector interfaces is essential. Since the tuner files usually consist of a small number of frequencies with respect to the number of frequencies present in a typical VNA calibration, it is appropriate to increase the number of averages to 128 or 256.

It is generally most useful to characterize a tuner without any additional components attached, such as a bias-tee, in order to maintain maximum flexibility in the use of the tuner subsequent to the characterization. For tuners that are being characterized for the first time, it is recommended that they be fully evaluated for insertion loss, minimum and maximum VSWR, and frequency response to ensure they are compliant with the manufacturer's specifications.

After characterization the tuner file should be verified by setting the tuner for arbitrary impedances near the center and edge of the Smith Chart over 2π radians. The error should be less than 0.2% for magnitude and 0.1° for phase. Anything worse than this may indicate a problem with either the calibration (verify it again) or the tuner.

17.3.3   S-Parameter Characterization of System Components

Characterization of system components consists of creating one-port and two-port s-parameter files of the source block and load block, as shown in Figs. 17.1 and 17.2, respectively. Each of these figures show suggested reference-planes for characterization of the network. Since the reflection coefficient of each port of the source and load blocks is in general small with respect to that exhibited by tuners, the VNA calibration is not as critical[2] as it is for tuner characterization. Nevertheless, it is recommended to use the same calibration as used for the tuner characterization and to sweep a broad range of frequencies to eliminate the possibility of characterization in the future at new frequencies.

If possible, each component of the source and load blocks should be individually characterized prior to integration into their respective block. This is particularly so for circulators and high-current bias-tees, which tend to have limited bandwidth. The response of the source and load block should be stored for future reference and/or troubleshooting.

17.3.4 Fixture Characterization to Increase System VSWR

In the beginning of this section it was indicated that high-power load-pull may require source and load impedances in the neighborhood of 0.1 Ω. This does not mean that the DUT may require such an impedance as much as it is necessary for generating closed contours, which are useful for evaluation of performance gradients in the gamma domain. A very robust and simple method of synthesizing sub 1 Ω impedances is to use a quarter-wave pre-matching network characterized using numerically well-defined two-tier calibration methods. To date, use of quarter-wave pre-matching offers the lowest impedance, though it is limited in flexibility due to bandwidth restrictions. Recently, commercially available passive mechanical systems cascading two tuners together have been made available offering octave bandwidths, though they are not able to generate impedances as low as narrowband quarter-wave pre-matching. In this section, a robust methodology for designing and characterizing a quarter-wave pre-matching network capable of presenting 0.1 Ω at 2 GHz is described [16,18]. It is based on a two-tier calibration with thin-film gold on alumina substrates (quarter-wave pre-matching networks on soft substrates are not recommended due to substrate variations and repeatability issues over time).

The theory of quarter-wave pre-matching begins with the mismatch invariance property of lossless networks [19]. Consider the quarter-wave line of characteristic impedance Z_{ref} shown in Fig. 17.9. This line is terminated in a mismatch of $VSWR_{load}$ with an arbitrary phase. The reference impedance of $VSWR_{load}$ is Z_L. The mismatch invariance property of lossless networks shows that the input VSWR is identical to the load VSWR, but it is with respect to the quarter-wave transformed impedance of Z_L. Thus, the minimum achievable impedance, which is real valued, is the impedance looking into the quarter-wave line when it is terminated in Z_L divided by $VSWR_{load}$. This is expressed as

$$R_{in,min} = \frac{\dfrac{Z_{ref}^2}{Z_L}}{VSWR_{load}} \tag{17.2}$$

Suppose it is desired to synthesize a minimum impedance of 0.1 Ω, which might be required for characterizing high power PCS and UMTS LDMOS transistors. If a typical passive-mechanical tuner is capable of conservatively generating a 40:1 VSWR, then the input impedance of the quarter-wave line must be approximately 4 Ω, requiring the characteristic impedance of the quarter-wave line to be approximately 14 Ω, assuming a Z_L of 50 Ω. To the extent that the minimum impedance deviates from the ideal is directly related to fixture losses. Thus, the importance of using a low-loss substrate and metal system is apparent.

Full two-port characterization of each fixture side is necessary to reset the reference plane of each associated tuner. Several methods are available to do this, including analytical methods based on approximate closed-form expressions, full-wave analysis using numerical techniques, and employment of VNA error correction techniques [20,21,22]. The first method is based on approximations that have built-in uncertainty, as does the second method, in the form of material parameter uncertainty. The third method

---

[2]If the magnitude of the reflection coefficient approaches the residual directivity of the VNA calibration, then errors may occur.

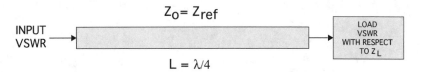

FIGURE 17.9 Network to describe the mismatch invariance property of lossless networks.

is entirely measurement based, and relies on well-behaved TRL error correction mathematics to extract a two-port characterization of each fixture half from a two-tier calibration. More importantly, using verification standards, it is possible to quantify the accuracy of the de-embedding, as described in the section on VNA calibration.

Using the error-box formulation of the TRL calibration it is possible to extract the two-port characteristics of an arbitrary element inserted between two reference planes of two different calibrations [11]. The first tier of the calibration is usually done at the test-port cables of the VNA. The second tier of the calibration is done in the media that matches the implementation of the test fixture, which is usually microstrip. Figure 17.10 illustrates the reference-plane definitions thus described. The second tier of the calibration will have its reference impedance set to the impedance of the delay standard, which is the impedance of the quarter-wave line. Although there are many methods of determining the characteristic impedance of a transmission line, methods based on estimating the capacitance per unit length and phase velocity are well suited for microstrip lines [12,15]. The capacitance per unit length and phase velocity uniquely describe the quasi-TEM characteristic impedance as

$$Z_O = \frac{1}{v_p C} \tag{17.3}.$$

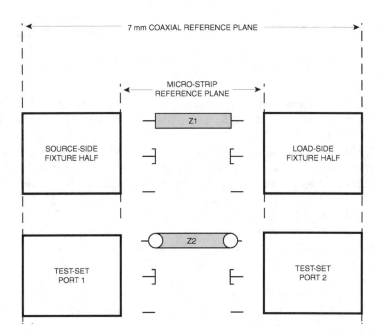

FIGURE 17.10 Reference-plane definitions for a two-tier calibration used for fixture characterization. The first tier is based on a TRL APC 7 mm calibration and the second tier is based on a microstrip TRL calibration.

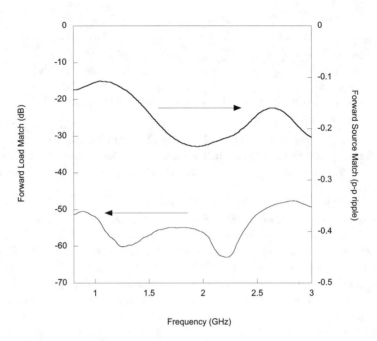

Frequency (GHz)

FIGURE 17.11   Microstrip TRL calibration using an offset-short and delay-line to extract source match and load match, respectively. This data was taken from an HP 8510C with an HP 8514B test set.

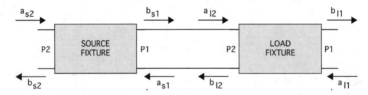

FIGURE 17.12   Port and traveling-wave definitions for cascading the source-fixture and load-fixture to examine the accuracy of the two-tier calibration fixture characterization.

Once the characteristic impedance of the delay-line is known, the s-parameters can be re-normalized to 50 Ω to make them compatible with the 50 Ω reference impedance that most automated load-pull systems use [2,3,15].

Figure 17.11 shows the forward source and load match of the second tier microstrip calibration used in the pre-matching fixture described in References 16 and 18. This fixture was intended to present 0.1 Ω at 2 GHz with extremely high accuracy. From the verification data, the resultant source match is better than −45 dB across the band and the resultant load match is better than −52 dB across the band. Comparing these results with Fig. 17.5 shows that the uncertainty is very low

A significant advantage of using a transforming network to increase system VSWR, whether it be a quarter-wave line or an additional cascaded tuner, is that the two-port characterization of each element is done at manageable impedance levels. Characterization of a tuner presenting a 50:1 VSWR in direct cascade of a quarter-wave pre-match network would result in a significant increase in measurement uncertainty since the VNA must resolve impedances near 0.1 Ω. Segregating the characterization process moves the impedances that must be resolved to the 1 Ω to 2 Ω range, where the calibration uncertainty is considerably smaller.

The final step of the fixture verification process is to verify that the two-tier calibration has provided the correct two-port s-parameter description of each fixture half. Figure 17.12 shows each fixture half cascaded using the port definitions adopted by NIST Multical™ [15]. With microstrip, an ideal thru can be approximated by butting each fixture half together and making top-metal contact with a thin conductive film. When this is not possible, it is necessary to extract a two-port characterization of the thru. The cascaded transmission matrix is expressed as

$$
\begin{bmatrix} A_{11} & B_{12} \\ C_{21} & D_{22} \end{bmatrix}_{cascade} = \begin{bmatrix} A_{11} & B_{12} \\ C_{21} & D_{22} \end{bmatrix}_{source} \begin{bmatrix} 1 & 0 \\ 0 & 1 \end{bmatrix}_{thru} \begin{bmatrix} A_{11} & B_{12} \\ C_{21} & D_{22} \end{bmatrix}_{load} \tag{17.4}
$$

where the middle matrix of the right-hand side is the transmission matrix of a lossless zero phase-shift thru network. Converting the cascade transmission matrix back to s-parameter form yields the predicted response of the cascaded test-fixture, which can then be compared to the measurements of the cascade provided by the VNA.

Figure 17.13 shows the measured and predicted cascade magnitude response of a typical PCS quarter-wave pre-matching fixture based on an 11 Ω quarter-wave line; the phase is shown in Fig. 17.14 [16,18]. The relative error across the band is less than 0.1%. This type of fixture characterization performance is necessary to minimize error for synthesizing sub 1 Ω impedances.

17.4 System Performance Verification

Just as verification of VNA calibration is essential, so too is verification of overall load-pull system performance essential. Performance verification can be done with respect to absolute power or with respect to power gain. The former is recommended only occasionally, for example when the system is assembled or when a major change is made. The latter is recommended subsequent to each power calibration. Each of the methods will be described in this section.

Absolute power calibration is done by applying a signal to the source tuner via the source block of Fig. 17.2. After appropriately padding a power sensor, it is then connected to DUT side of the source tuner and, with the tuners set for 1:1 transformation, the resultant power is compared to what the overall cascaded response is expected to be.

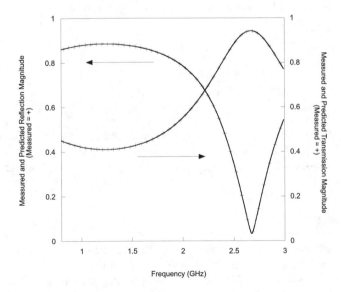

FIGURE 17.13   Forward reflection and transmission magnitude comparison of measured and cascaded fixture response. The error is so small the curves sit on top of each other.

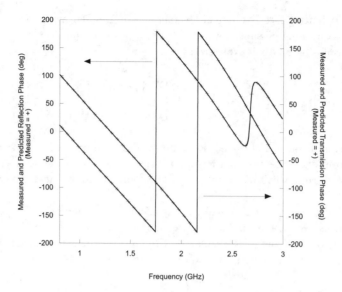

FIGURE 17.14   Forward reflection and transmission phase comparison of measured and cascaded fixture response. The error is so small the curves sit on top of each other.

This procedure is repeated for the load tuner except that the signal is injected at the DUT side of the load tuner and the power sensor is located as shown in Fig. 17.3. Splitting this verification in two steps assists in isolating any issues with either the source or load side. It is also possible to vary the impedance of each tuner and calculate what the associated available gain or power gain is, although this step is more easily implemented in the power gain verification.

Power gain verification starts with a two-port characterization of a known mismatch standard. The simplest way to implement this standard is to use one of the tuners, and then set the other tuner for the conjugate of this mismatch. In this case, the mismatch standard is an ideal thru, similar to the one used in fixture verification described in the previous section. Since it is unlikely that both the source and load tuners would have identical impedance domains, the measured loss must be compensated to arrive at actual loss. To compensate for this, the mismatch loss is computed as

$$G_{mm} = 10 \log_{10} \left[\frac{\left(1 - \left| \Gamma_s \right|^2 \right) \left(1 - \left| \Gamma_l \right|^2 \right)}{\left| 1 - \Gamma_s \Gamma_l \right|^2} \right] \tag{17.5}$$

where Γ_s and Γ_l are the source and load reflection coefficients, respectively, looking back into each tuner. Figure 17.15 shows a typical response of an entire cascade, including the quarter-wave pre-matching network. A transducer gain response boundary of ±0.1 dB is typical, and ±0.2 should be considered the maximum.

17.5 Summary

Load-pull is a valuable tool for evaluating high-power RF and microwave transistors, designing power amplifiers, and verifying large-signal model performance and validity domains. To enhance the reliability of the data that a load-pull system provides, it is essential that high performance VNA calibration techniques be adopted. Further, as emphasized in the present section, treating each section of the load-pull separately is useful from a measurement perspective and from a problem resolution perspective. In

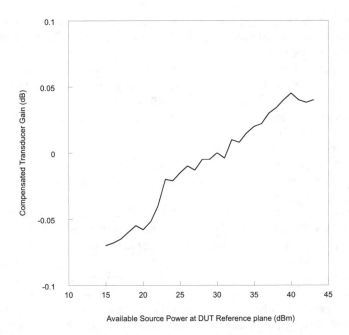

FIGURE 17.15   Measured transducer gain under the condition of conjugate match with mismatch loss compensation included.

the former case, it was shown that measuring quarter-wave pre-matching networks and tuners separately reduces the uncertainty of the calibration. In the latter case, it was shown that characterization of each section individually allows its performance to be verified prior to integrating it within the entire system.

The central theme of this section has been the VNA and its associated calibration. Due to the extremely low impedances synthesized in high-power load-pull, the VNA calibration is the single most important element of the characterization process. Any errors or uncertainty encountered in the VNA calibration will be propagated directly into the load-pull characterization files and may result in erroneous data, particularly if system performance verification is not performed.

To present the sub 1 Ω impedances necessary for evaluation of high-power transistors, transforming networks are required. These can be implemented using an impedance transforming network, such as a quarter-wave line, or by cascading two tuners together. The former offers the highest VSWR at the expense of narrow bandwidth, while the latter is in general more flexible. In either case, high performance and reliable characterization methods are necessary to attain the best possible results for using load-pull as a verification and design tool.

Acknowledgments

Kerry Burger (Philips), Mike Majerus (Motorola), and Gary Simpson and John King (Maury Microwave) have, in many ways, influenced the content of this section. Their support and friendship is happily acknowledged.

References

1. J. M. Cusak et al., Automatic load-pull contour mapping for microwave power transistors, *IEEE Transactions on Microwave Theory and Techniques*, 1146–1152, December 1974.
2. *Automated Tuner System User's Manual*, v.1.9, Maury Microwave Corporation, 1998.
3. *Computer Controlled Tuner System User's Manual*, v. 6.0, Focus Microwave Corporation, 1998.
4. *LP2 Automated Load-Pull System User's Manual*, ATN Microwave Corporation, 1997.

5. F. Larose, F. Ghannouchi, and R. Bosisio, A new multi-harmonic load-pull method for non-linear device characterization and modeling, *Digest of the IEEE International Microwave Symposium Digest*, 443–446, June 1990.

6. F. Blache, J. Nebus, P. Bouysse, and J. Villotte, A novel computerized multi-harmonic load-pull system for the optimization of high-efficiency operating classes in power transistors, *IEEE International Microwave Symposium Digest*, 1037–1040, June 1995.

7. J. Sevic, R. Baeten, G. Simpson, and M. Steer, Automated large-signal load-pull characterization of adjacent-channel power ratio for digital wireless communication system, *Proceedings of the 45th ARFTG Conference*, 64–70, November 1995.

8. J. Sevic, K. Burger, and M. Steer, A novel envelope-termination load-pull method for the ACPR optimization of RF/microwave power amplifiers, *Digest of the IEEE International Microwave Symposium Digest*, 723–726, June 1998.

9. G. Simpson and M. Majerus, Measurement of large-signal input impedance during load-pull, *Proceedings of the 50th ARFTG Conference*, 101–106, December 1997.

10. D. Rytting, ARFTG Short-Course: Network Analyzer Calibration Theory, 1997.

11. R. Marks, Formulation of the basic vector network analyzer error model including switch terms, *Proceedings of the 50th ARFTG Conference*, 115–126, December 1997.

12. R. Marks and D. Williams, Characteristic impedance measurement determination using propagation measurement, *IEEE Microwave and Guided Wave Letters*, 141–143, June 1991.

13. G. Engen and C. Hoer, Thru-reflect-line: an improved technique for calibrating the dual six-port automatic network analyzer, *IEEE Transactions on Microwave Theory and Techniques*, 987–993, December 1979.

14. R. Marks, A multi-line method of network analyzer calibration, *IEEE Transactions on Microwave Theory and Techniques*, 1205–1215, July 1990.

15. *MultiCal™ User's Manual*, v. 1.0, National Institute of Standards and Technology, 1997.

16. J. Sevic, A sub 1 Ω load-pull quarter-wave pre-matching network based on a two-tier TRL calibration, *Proceedings of the 52nd ARFTG Conference*, 73–81, December 1998.

17. D. Balo, Designing and calibrating RF fixtures for SMT devices, *Hewlett-Packard 1996 Device Test Seminar*, 1996.

18. J. Sevic, A sub 1 Ω load-pull quarter-wave pre-matching network based on a two-tier TRL calibration, *Microwave Journal*, 122–132, March 1999.

19. R. Collin, *Foundations for Microwave Engineering*, McGraw-Hill: New York, 1966.

20. B. Wadell, *Transmission Line Design Handbook*, Artech House: Boston, 1991.

21. *EM User's Manual*, v. 6.0, Sonnet Software, Inc., Liverpool, NY, 1999.

22. *HP 8510C User's Manual*, Hewlett-Packard Company, 1992.

18

Pulsed Measurements

Anthony E. Parker
Macquarie University

James G. Rathmell
The University of Sydney

Jonathan B. Scott
Agilent Technologies

18.1 Introduction

Pulsed measurements ascertain the radio-frequency (RF) behavior of transistors or other devices at an unchanging bias condition. A pulsed measurement of a transistor begins with the application of a bias to its terminals. After the bias has settled to establish a **quiescent condition**, it is perturbed with pulsed stimuli during which the change in terminal conditions, voltage and current, is recorded. Sometimes a RF measurement occurs during the pulse. The responses to the pulse stimuli quantify the behavior of the device at the established quiescent point. **Characteristic curves**, which show the relationship between terminal currents or RF parameters and the instantaneous terminal potentials, portray the behavior of the device.

Pulsed measurement of the characteristic curves is done using short pulses with a relatively long time between pulses to maintain a constant quiescent condition. The characteristic curves are then specific to the quiescent point used during the measurement. This is of increasing importance with the progression of microwave-transistor technology because there is much better consistency between characteristic curves measured with pulses and responses measured at high frequencies. When the behavior of the device is bias- or rate-dependent, pulsed measurements yield the correct high-frequency behavior because the bias point remains constant during the measurement. Pulse techniques are essential for characterizing devices used in large-signal applications or for testing equipment used in pulse-mode applications. When measurements at high potentials would otherwise be destructive, a pulsed measurement can safely explore breakdown or high-power points while maintaining a bias condition in the safe-operating area (SOA) of the device. When the device normally operates in pulse mode, a pulsed measurement ascertains its true operation.

The response of most microwave transistors to high-frequency stimuli depends on their operating conditions. If these conditions change, the characteristic curves vary accordingly. This causes **dispersion** in the characteristic curves when measured with traditional curve-tracers. The operating condition when sweeping up to a measurement point is different than that when sweeping down to the same point. The implication is that any change in the operating conditions during the measurement will produce ambiguous characteristic curves.

Mechanisms collectively called **dispersion effects** contribute to dispersion in characteristic curves. These mechanisms involve thermal, rate-dependent, and electron trapping phenomena. Usually they are slow acting, so while the operating conditions of the device affect them, RF stimuli do not. Even if the sequence of measurement precludes observation of dispersion, dispersion effects may still influence the resulting characteristic curves.

Pulsed measurements are used to acquire characteristic curves that are free of dispersion effects. The strategy is to maintain a constant operating condition while measuring the characteristic curves. The pulses are normally short enough to be a signal excursion rather than a change in bias, so dispersion effects are negligible. The period between pulses is normally long enough for the quiescent condition of the device to recover from any perturbation that may occur during each pulse.

Pulse techniques cause less strain, so are suitable for extending the range of measurement into regions of high power dissipation and electrical breakdown. Pulse techniques are also valuable for experiments in device physics and exploration of new devices and material systems at a fragile stage of development.

Stresses that occur when operating in regions of breakdown or overheating can alter the characteristic curves permanently. In many large-signal applications, there can be excursions into both of these regions for periods brief enough to avoid undue stress on the device. To analyze these applications, it is desirable to extend characteristic curves into the stress regions. That is, the measurements must extend as far as possible into regions that are outside the SOA of the device. This leads to another form of dispersion, where the characteristic curves change after a measurement at a point that stresses the device.

Pulsed measurements can extend to regions outside the SOA without stressing or damaging the device. If the pulses are sufficiently short, there is no permanent change in the characteristic curves. With pulses, the range of the measured characteristic curves can often extend to completely encompass the signal excursions experienced during the large-signal operation of devices.

In summary, pulsed measurements yield an extended range of characteristic curves for a device that, at specific operating conditions, corresponds to the high-frequency behavior of the device. The following sections present the main principles of the pulse technique and the pulse-domain paradigm, which is central to the technique. The pulse-domain paradigm considers the characteristic curves to be a function of quiescent operating conditions. Therefore, the basis for pulse techniques is the concept of measurements made in **isodynamic** conditions, which is effectively an invariable operating condition. A discussion is included of the requirements for an isodynamic condition, which vary with the transistor type and technology. There is also a review of pulsed measurement equipment and specifications in terms of cost and complexity, which vary with application. Finally, there is an examination of various pulsed measurement techniques.

18.2 Isothermal and Isodynamic Characteristics

For the analysis of circuit operation and the design of circuits, designers use transistor **characteristics**. The characteristics consist of **characteristic curves** derived from measurements or theoretical analysis. These give the relationship between the variable, but interdependent terminal conditions and other information that describes the behavior of the device. To be useful, the characteristics need to be applicable to the operating condition of the device in the circuit.

In all circuits, when there is no signal, the device operates in a **quiescent condition** established by bias networks and power supplies. The **DC characteristics** are characteristic curves obtained with slow curve tracers, conventional semiconductor analyzers, or variable power supplies and meters. They are essentially data from a set of measurements at different bias conditions. Consequently, the quiescent operating point of a device is predictable with DC characteristics derived from DC measurements. Figure 18.1 shows a set of DC characteristics for a typical microwave MESFET. This figure also shows the very different set

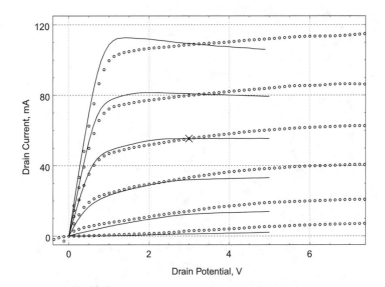

FIGURE 18.1   Characteristic curves for a MESFET. Shown are the DC characteristics (–) and the pulsed characteristics (O), with 300 ns pulses separated by 200 ms quiescent periods, for the quiescent point $V_{DS} = 3.0$ V, $I_D = 55.4$ mA (×). Gate-source potential from –2.0 to +0.5 V in 0.5 V steps is the parameter.

of **pulsed characteristics** for the same device made at the indicated quiescent point. The pulsed characteristics give the high-frequency behavior of the MESFET when biased at that quiescent point.

A clear example of a dispersion effect that causes the observed difference between the DC and pulsed characteristics is heating due to power dissipation. When the characteristics are measured at a slow rate (≈10 ms per point), the temperature of the device at each data point changes to the extent that it is heated by the power being dissipated at that point. Pulsed characteristics are determined at the constant temperature corresponding to the power dissipation of a single bias point. This measurement at constant temperature is one made in **isothermal** conditions.

In general, device RF characteristics should be measured in a constant bias condition that avoids the dynamics of thermal effects and any other dispersion effects that are not invoked by a RF signal. Such a measurement is one made in **isodynamic** conditions.

18.2.1   Small-Signal Conditions

Devices operating in **small-signal** conditions give a nearly linear response, which can be determined by **steady-state RF** measurements made at the quiescent point. A network analyzer, operated in conjunction with a bias network, performs such a measurement in isodynamic conditions. Once the quiescent condition is established, RF measurements characterize the terminal response in terms of small-signal parameters, such as Y-parameters. A different set of small-signal parameters is required for each quiescent condition.

It is not possible to correlate the small-signal parameters with the DC characteristics when there are dispersion effects. For example, the output conductance (drain-source admittance) of a typical MESFET varies with frequency as shown in Fig. 18.2. For this device, the small-signal conductance varies little with frequency above about 1 MHz. The conductance is easily determined from the real part of Y_{22} measured with a network analyzer. The conductance can also be determined from the slope of the pulsed characteristics at the quiescent point. The data from short pulses, shown in Fig. 18.1 in the regime of 1 to 10 MHz, give an **isodynamic characteristic** for this typical device because the calculated conductance is the same as that measured at higher frequencies. With longer pulses, corresponding to lower frequencies, dispersion effects influence the conductance significantly. The characteristics measured at rates below

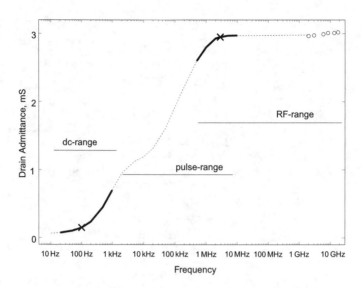

FIGURE 18.2 Frequency variation of drain-source admittance for the typical MESFET of Fig. 18.1 at the quiescent point $V_{DS} = 3.0$ V, $I_D = 55.4$ mA. An indicative response (- -) connects measured $\Re(Y_{22})$ from a RF network analyzer (O) and calculation from the pulsed and DC data in Fig. 18.1 ($\times$). Also indicated are the typical frequency ranges applicable to DC, pulsed, and RF measurements.

1 MHz can vary with the type of measurement because each point affects the subsequent point. The dispersion effects are prominent at the slow 10 to 1000 Hz rate of curve-tracer operation, which is why dispersion is observed in curve-tracer measurements. True DC measurements usually require slower rates.

18.2.2 Thermal Model

Thermal dispersion has a significant influence on the output conductance. To quantify this, consider the relationship between the terminal current i_T [A] and voltage v_T [V]. The small-signal terminal conductance is $g = d\, i_T/d\, v_T$ [S]. To explore the influence of thermal dispersion on this parameter, assume that the terminal current is a linear function of temperature rise ΔT [K] with thermal-coefficient λ [1/K], so that

$$i_T = i_O\left(1 - \lambda\,\Delta T\right). \tag{18.1}$$

The thermodynamic rate equation relates the temperature rise of the device to time t and heat flow due to power dissipation $Q = i_T\, v_T$ [W]:

$$mC\,R_T\,d\Delta T\big/dt + \Delta T = R_T\,Q. \tag{18.2}$$

The term mC [J/K] is the product of mass and heat capacity of the thermal path to ambient temperature, and R_T [K/W] is the thermal resistance of the path. There is a time constant $\tau = mC\,R_T$ [s] associated with this rate equation.

With **isothermal** conditions, the temperature rise remains constant during operation of the device. This occurs when the rate of change of power dissipation, due to signal components, is either much faster or much slower than the thermal time constant. With high-frequency signals, it is the quiescent power dissipation at the quiescent terminal current, I_T, and voltage, V_T, that sets the temperature rise. The rate Eq. (18.2) reduces to $\Delta T = R_T\,Q$ where $Q = I_T\,V_T$ is constant. The isothermal terminal current [Eq. (18.1)] is then:

$$i_T = i_O\left(1 - \lambda R_T I_T V_T\right).\tag{18.3}$$

The terminal conductance determined by small-signal RF measurement is then:

$$g = di_T/dv_T = di_O/dv_T\left(1 - \lambda R_T I_T V_T\right).\tag{18.4}$$

During measurement of DC characteristics, which are made at rates slower than the thermal time constant, the rate Eq. (18.2) reduces to $\Delta T = R_T\, i_T\, v_T$. This is different at each measurement point, so the DC terminal current [Eq. (18.1)] becomes:

$$i_T = i_O\left(1 - \lambda R_T\, i_T v_T\right).\tag{18.5}$$

An estimate of the terminal conductance from DC characteristics would be

$$G = di_O/dv_T\left(1 - \lambda R_T\, i_T\, v_T\right) - \lambda R_T\, i_O\left(i_T + Gv_T\right).\tag{18.6}$$

The difference between the small-signal conductance g in Eq. (18.4) and the DC conductance G in Eq. (18.6) is due to thermal dependence. If $\lambda = 0$, then $g = G$. Without knowing λR_T it is not possible to determine the small-signal conductance from the DC characteristics.

Figure 18.3 shows an attempt to determine the pulsed characteristics from the DC characteristics. The thermal effect is removed from the DC characteristics with a value of $\lambda R_T = 0.3\ \text{W}^{-1}$ determined from a complete set of pulsed characteristics made over many quiescent points. Multiplying the drain current of each point (v_{DS}, i_D) in the DC characteristics by $(1 - \lambda R_T\, I_D\, V_{DS})/(1 - \lambda R_T\, i_D\, v_{DS})$ normalizes it to the temperature of the quiescent point (V_{DS}, I_D) used in the pulsed measurement. Figure 18.3 demonstrates that although temperature, explained by the simple model above, is a dominant effect, other dispersion effects also affect the characteristics. The ambient-temperature DC characteristics exhibit changes in threshold potential, transconductance, and other anomalous characteristics, which occur because electron trapping, breakdown potentials, and leakage currents also vary with bias. The pulsed measurements made in isodynamic conditions are more successful at obtaining characteristic curves that are free of these dispersion effects.

18.2.3 Large-Signal Conditions

Transistors operating in **large-signal** conditions operate with large signal excursions that can extend to limiting regions of the device. Large-signal limits, such as clipping due to breakdown or due to excessive input currents, can be determined from an extended range of characteristic curves. Steady-state DC measurements are confined to operating regions in the SOA of the device. It is necessary to extrapolate to breakdown and high-power conditions, which may prompt pushing the limits of measurements to regions that cause permanent, even if non-destructive, damage. The stress of these measurements can alter the characteristics and occurs early in the cycle of step-and-sweep curve tracers, which leads to incorrect characteristic curves in the normal operating region. The observed dispersion occurs in the comparison of the characteristics measured over moderate potentials, measured before and after a stress.

Pulsed measurements extend the range of measurement without undue stress. Figure 18.4 shows characteristic curves of a HEMT that encompasses regions of breakdown and substantial forward gate potential. The diagram highlights points in these regions, which are those with significant gate current. The extended characteristics are essential for large-signal applications to identify the limits of signal excursion. The pulsed characteristics in the stress regions are those that would be experienced during a

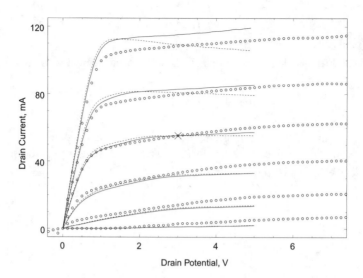

FIGURE 18.3 The MESFET characteristic curves shown in Fig. 18.1 with a set of DC characteristics (–) normalized to the temperature due to power dissipation at the quiescent point, $V_{DS} = 3.0$ V, $I_D = 55.4$ mA (×). Also shown are the raw DC characteristics (- -) and the pulsed characteristics (O) for the quiescent point V_{DS}, I_D. Gate-source potential from –2.0 to +0.5 V in 0.5 V steps is the parameter.

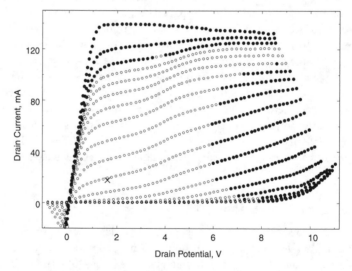

FIGURE 18.4 An example of the extended characteristic curves for an HEMT obtained with 500 ns pulses at 10 ms intervals for the quiescent point $V_{DS} = 1.6$ V, $I_D = 17.7$ mA (×). The solid points • are those for which the magnitude of gate current is greater than 1 mA. Gate-source potential from –3.0 to +1.5 V in 250 mV steps is the parameter.

large-signal excursion because the measurement is made in isodynamic conditions set by the operating point. There is little correlation between these and an extrapolation from DC characteristics because the stress regions are significantly affected by bias conditions and temperature.

18.2.4 Pulsed Measurements

Dispersion effects in microwave devices generate a rich dynamic response to large signals and changing operating conditions. The dynamic behavior affects the DC and high-frequency characteristics but is not observable in either. Thus, pulsed measurement techniques are required to quantify various aspects of the dynamic behavior.

The pulsed current/voltage (**pulsed-I/V**) characteristics are characteristic curves determined from an isodynamic measurement with short pulses separated by long relaxation periods at a specific quiescent point. Each quiescent point has its own pulsed-I/V characteristics, so a complete characterization of a device requires pulsed-I/V measurements over various quiescent points. Dispersion effects do not affect each pulsed-I/V characteristic but do affect the variation between characteristics measured in different quiescent conditions.

The pulsed characteristics vary with pulse length. Short pulses produce isodynamic pulsed-I/V characteristics, and very long pulses produce DC characteristics. A **time domain** pulsed measurement, performed by recording the variation of terminal conditions during a measurement pulse, can trace the transition from isodynamic to DC behavior. The time constants of the dispersion effects are present in the time domain characteristic. Note that the range of time domain measurements is limited to the SOA for the long pulses used.

Isodynamic small-signal parameters are determined from **pulsed-RF** measurements. During the measurement pulse, a RF signal is applied and a pulsed vector network analyzer determines the scattering parameters. The terminal potentials during each pulse are the **pulsed bias** for the RF measurement. Each operating point, at which the device relaxes between pulses, has its own set of pulsed-bias points and corresponding RF parameters. Pulsed-RF characteristics give small-signal parameters, such as reactance and conductance, as a surface function of terminal potentials. There is a small-signal parameter surface for each quiescent operating point and the dispersion effects only affect the variation of each surface with quiescent condition. Pulsed-RF measurements are also required for pulse-operated equipment, such as pulsed-radar transmitters, that have off-state quiescent conditions and pulse to an on-state condition that may be outside the SOA of the device.

Pulse timing and potentials vary with the measurement type. The periods required for isodynamic conditions and safe-operating potentials for various types of devices are discussed in the next section. The complexity and cost of pulse equipment, which also varies with application, is discussed in the subsequent section.

18.3 Relevant Properties of Devices

Three phenomena present in active devices that cause measurement problems best addressed with pulsed measurements. These are the SOA constraint, thermal dependency of device characteristics, and dependency of device characteristics upon charge trapped in and around the device. The following discusses these phenomena and identifies devices in which they can be significant.

18.3.1 Safe-Operating Area

The idea of a safe operating area is simply that operating limits exist beyond which the device may be damaged. The SOA limits are generally bounds set by the following four mechanisms:

- A maximum voltage, above which a mechanism such as avalanche breakdown can lead to loss of electrical control or direct physical alteration of the device structure.

- A maximum power dissipation, above which the active part of the device becomes so hot that it is altered physically or chemically.

- A maximum current, above which some part of the device like a bond wire or contact region can be locally heated to destruction.

- A maximum current-time product, operation beyond which can cause physical destruction at local regions where adiabatic power dissipation is not homogeneous.

It is important to realize that damage to a device need not be catastrophic. The above mechanisms may change the chemical or physical layout of the device enough to alter the characteristics of the device without disabling it.

Pulsed-I/V measurements offer a way to investigate the characteristics of a device in areas where damage or deterioration can occur, because it is possible to extend the range of measurements under pulsed conditions, without harm. This is not a new idea — pulsed capability has been available in curve tracers for decades. These pulsed systems typically have pulses no shorter than a few milliseconds or a few hundred microseconds. However, shorter pulses allow further extension, and for modern microwave devices, true signal response may require sub-microsecond stimuli.

There are time constants associated with SOA limitations. For example, the time constant for temperature rise can allow very high power levels to be achieved for short periods. After that time, the device must be returned to a low-power condition to cool down. The SOA is therefore much larger for short periods than it is for steady DC conditions. Figure 18.5 shows successful measurement of a 140 μm² HBT well beyond the device SOA. The example shows a sequence of measurement sweeps with successively increasing maximum collector potential. There is no deterioration up to 7.5 V, which is an order of magnitude above that which would rapidly destroy the device under static conditions. The sweeps to a collector potential greater than 7.5 V alter the device so its characteristics have a lower collector current in subsequent sweeps. Shorter pulses may allow extension of this limit.

Different active devices are constrained by different SOA limits. For instance, GaN FETs are not usually limited by breakdown, whereas certain III-V HBTs are primarily limited by breakdown; silicon devices suffer more from a current-time product limit than do devices in the GaAs system. Pulsed I/V measurements provide a way for device designers to identify failure mechanisms, and for circuit designers to obtain information about device characteristics in regions where signal excursions occur, which are outside the SOA.

18.3.2   Thermal Dispersion

GaAs devices, both FETs and HBTs, have greater thermal resistance than do their silicon counterparts. They tend to suffer larger changes in characteristics per unit change in junction temperature. Perhaps the first need for pulsed-I/V measurements arose with GaAs MESFETs because of the heating that occurs

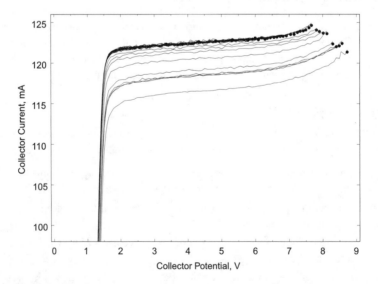

FIGURE 18.5   A single collector characteristic measured on a 140 μm² III-V HBT with sequentially increasing maximum voltage (shown by •) applied in 1 μs pulses. Note the progressive deterioration above a certain instantaneous dissipation level.

in simple DC measurement of these devices. Such a measurement turns out to be useless in high-frequency terms because each part of the measurement is at a vastly different temperature. This does not represent device characteristics in a RF situation where the temperature does not perceptibly change in each signal period. The sole utility of DC characteristics is to help predict quiescent circuit conditions.

A pulsed-I/V measurement can approach isothermal conditions, and can circumvent this problem. Figure 18.1, showing the DC and pulsed characteristics of a simple MESFET, exemplifies the difference. It is remarkable that the characteristics are for the same device.

Silicon devices, both FET and BJT, are relatively free of thermal dispersion effects, as are GaN FETs. The susceptibility of any given device, and the pulse duration and duty cycle required to obtain isothermal data, must be assessed on a case-by-case basis. Methods for achieving this are explored in the later discussion of measurement techniques.

18.3.3 Charge Trapping

Temperature is not the only property of device operation that can give rise to dispersion. Charge trapped in substrate or defects is particularly troublesome in FETs. Rather than power dissipation, currents or junction potentials can control slow-moving changes in the device structure. These phenomena are not as well understood as their thermal counterparts.

Exposing charge-trapping phenomena that may be influencing device performance is more difficult, but is still possible with an advanced pulsed-I/V system. One method is to vary the quiescent conditions between fast pulses, observing changes in the pulsed characteristic as quiescent fields and currents are varied independently, while holding power dissipation constant. Figure 18.6 shows two pulsed characteristics measured with identical pulse-stimulus regimes, but with different quiescent conditions. Since the power dissipation in the quiescent interval is unchanged, temperature does not vary between the two experiments, yet the characteristics do. The difference is attributed to trapped charge exhibiting a relatively long time constant.

Charge-trapping dispersion is most prevalent in HEMTs, less so in HFETs and MESFETs, and has yet to be reported in bipolar devices such as HBTs.

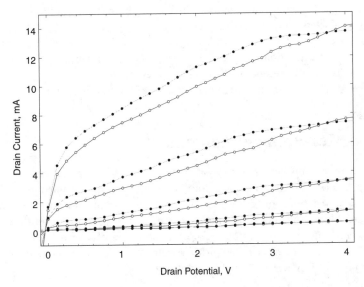

FIGURE 18.6 Two pulsed-I/V characteristics for the same GaAs FET measured at different quiescent conditions, $V_{DS} = 1.1$ V, $I_D = 0.4$ mA (O) and $V_{DS} = 2.2$ V, $I_D = 0.2$ mA (•). They have identical power dissipation. The measurements used 300 ns pulses separated by 200 ms quiescent periods. Gate-source potential from 0.0 to +0.5 V in 125 mV steps is the parameter.

18.3.4   Time Constants

Avalanche effects can operate extremely quickly — much faster than any pulse system — so SOA measurements exhibit the thermal and charge trapping time constants. Thermal effects typically have several time constants associated with them, each associated with the thermal capacity of some part of the system, from the active region to the external heat sink. Small devices typically have time constants of the order of one microsecond; larger devices may have their smallest significant time constant ten times larger than this. Package time constants tend to be of the order of milliseconds to tens or hundreds of milliseconds. External heat sinks add long time constants, though anything above a few seconds is disregarded or treated as environmental drift, since measurement or control of such external temperature is straightforward.

Charge trapping phenomena are more variable. Indeed, there are reports of devices susceptible to disruption from charge stored apparently permanently, after the fashion of flash memory. Values of the order of hundreds of microseconds are common, ranging up to milliseconds and longer.

Because of the wide variation of time constants, it is hard to know *a priori* what settings are appropriate for any measurement, let alone what capability ought to be specified in an instrument to make measurements. Values of less than 10 µs for pulse width and 1 ms for quiescent time might be marginally satisfactory, while 500 ns pulses with 10 ms quiescent periods would be recommended.

18.3.5   Pulsed-I/V and Pulsed-RF Characteristics

Pulsed-I/V measurement is sometimes accompanied by pulsed-RF measurements. The RF equipment acquires the raw data during the pulse stimulus part of the measurement. Given that pulsed-I/V systems characterize devices in isodynamic conditions, the need for measurement at microwave frequencies, simultaneously with pulse stimuli, might be questioned. The problem is that it may not be possible to infer the reactive parameters for a given quiescent point from static S-parameters that are measured over a range of DC bias conditions. This is because of significant changes in RF behavior linked to charge trapping or thermal dispersion effects.

Figure 18.7 compares S-parameters of an HBT measured at a typical operating point (well within the SOA) using a DC bias and using a 1µs pulsed bias at the same point with the device turned off between pulses. The differences, attributed to temperature, indicate the impact of dispersion effects on RF characteristics.

In addition, S-parameters cannot be gathered at bias points outside the SOA without pulse equipment. Pulse amplifiers often operate well beyond the SOA, so that a smaller, less expensive device can be used. This is possible when the duration of operation beyond SOA is brief, but again, it is not possible to characterize the device with DC techniques. For many of these applications, pulsed-RF network analyzers have been developed. These can measure the performance of the transistor during its pulsed operating condition.

18.4 Pulsed Measurement Equipment

Pulsed measurement systems comprise subsystems for applying bias, pulse, and RF stimuli, and for sampling current, voltage, and RF parameters. Ancillary subsystems are included to synchronize system operation, provide terminations for the device under test (DUT), and store and process data. A simple system can be assembled from individual pulse generators and data acquisition instruments. More sophisticated systems generate arbitrary pulse patterns and are capable of measurements over varying quiescent and pulse timing conditions. Pulsed-I/V systems can operate as stand-alone instruments or can operate in a pulsed-RF system to provide the pulsed bias.

18.4.1   System Architecture

The functional diagram of a pulsed measurement system, shown in Fig. 18.8, includes both pulsed-I/V and pulsed-RF subsystems. Pulse and bias sources, voltage and current sampling blocks, and associated

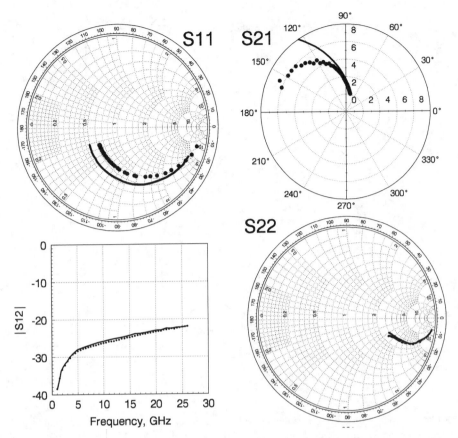

FIGURE 18.7 S-parameters measured at the same bias point with off-state and on-state quiescent conditions. The on-state parameters are from static, or DC measurements (–) and the off-state parameters are from measurements in a pulsed bias at the same point with off-state quiescent periods (·).

timing generators form the pulsed-I/V subsystem. A pulsed-RF source and mixer-based vector network analyzer form the pulsed-RF subsystem. The DUT is connected directly to the pulsed-I/V subsystem, or to bias networks that connect the pulsed-RF subsystem or RF terminations.

18.4.1.1 Pulsed-I/V System

Steady-state DC semiconductor parameter analyzers provide a source-monitor unit for each terminal of the DUT. The unit sources one of voltage or current while monitoring the other. In a pulsed measurement system, a pulsed voltage is added to a bias voltage and applied to the device. It is not practical to control the source potential within short pulse periods, so in order to ascertain the actual terminal conditions, both voltage and current are monitored. If a precise potential is required, then it is necessary to iterate over successive pulses, or to interpolate data from a range of pulsed measurements, or use longer pulse periods.

Simple systems use a pulse generator as the pulse source. Stand-alone pulse generators usually provide control of pulse and quiescent levels, so a single pulse point is measured during each test run. Such a system is easily assembled with pulse generators and is operated from their front panels. A single-point measurement mode is also employed by high-power pulsers that deliver high current pulses by dumping charge from capacitors, which are precharged during the quiescent period.

Systems that measure several pulse points in sequence use computer controlled arbitrary function generators to provide pulse and quiescent potentials. The function generators are essentially digital memory delivering values to a digital-to-analog converter. Pulse values are stored in every second

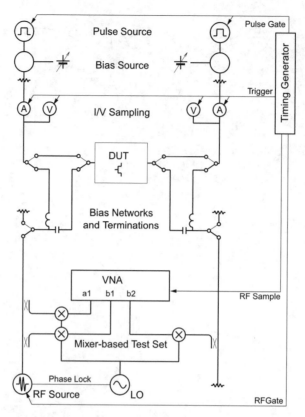

FIGURE 18.8   Simplified diagram of a generic pulsed measurement system. Alternative connections provide load terminations when there is no pulsed-RF test set or directly connect the pulsed-I/V subsystem to the DUT.

memory location and the quiescent value is stored in every other location. A timing generator then clocks through successive potentials at the desired pulse and quiescent time intervals. The quiescent potential is either simply delivered from the pulse generators or it is delivered from bench power supplies or other computer controlled digital-to-analog converters. In the latter cases, a summing amplifier adds the pulse and quiescent potentials and drives the DUT. This architecture extends the pulse power capability of the system. Whereas the continuous rating of the amplifier dictates the maximum quiescent current delivered to the device, the pulse range extends to the higher transient current rating of the amplifier.

In most systems, either data acquisition digitizers or digital oscilloscope channels sample current and voltage values. In a simple setup, an oscilloscope will display the terminal conditions throughout the pulse and the required data can be read on screen or downloaded for processing. Oscilloscope digitizers tend to have resolutions sufficient for displaying waveforms, but insufficient for linearity or wide dynamic range measurements. Data acquisition digitizers provide wider dynamic range and ability to sample at specific time points on each pulse or throughout a measurement sequence. When several pulse points are measured in sequence, the digitizers record pulse data from each pulse separately or time domain data from several points across each pulse. Either mode is synchronized by appropriate sampling triggers provided by a timing generator.

The position of the voltage and current sensors between the pulse source and the DUT is significant. There are transmission line effects associated with the cabling between the sensing points and the digitizers. The cable lengths and types of terminations will affect the transient response of, and hence the performance of, the pulse system. An additional complication is introduced when the DUT must be terminated for RF stability. A bias network is used but this introduces its own transient response to the measured pulses. For example, the initial 100 ns transient in Fig. 18.13 is generated by the bias network and is present when the DUT is replaced by a 50 Ω load.

Current is sensed by various methods that trade between convenience and pulse performance. With a floating pulse source, a sense resistor in the ground return will give the total current delivered by the source. There is no common-mode component in this current sensor, so a single-ended digitizer input is usable. The current reading will include, however, transient components from the charging of capacitances associated with cables between the pulser and the DUT. Low impedance cables can ameliorate this problem. Alternatively, hall-effect/induction probes placed near the DUT can sense terminal current. These probes have excellent common-mode immunity but tend to drift and add their own transient response to the data. A stable measurement of current is possible with a series sense resistor placed in line near the DUT. This eliminates the effect of cable capacitance currents, but requires a differential input with very good common-mode rejection. The latter presents a severe limitation for short pulses because common-mode rejection degrades at high frequency.

Data collection and processing in pulse systems is different than that of slow curve tracers or semiconductor parameter analyzers. The latter usually measure over a predefined grid of step-and-sweep values. If the voltage grid is defined, then only the current is recorded. The user relies on the instrument to deliver the specified grid value. In pulse systems, a precise grid point is rarely reached during the pulse period. The pulse data therefore includes measured voltage and current for each terminal. An important component in any pulse system is the interpretation process that recognizes that the pulse data do not lie on a regular grid of values. One consequence of this is that an interpolation process is required to obtain traditional characteristic curves.

18.4.1.2 Pulsed-RF System

Pulsed-RF test sets employ vector network analyzers with a wideband intermediate frequency (IF) receiver and an external sample trigger.[1] The system includes two RF sources and a mixer-based S-parameter test set. One source provides a continuous local oscillator signal for the mixers, while the other provides a gated RF output to the DUT. The local oscillator also provides a phase reference, so that a fast sample response is possible.

The pulsed bias must be delivered through bias networks, which are essential for the pulsed-RF measurement. During a pulsed-I/V measurement, the RF source is disabled and the RF test set provides terminations for the DUT. Pulsed-RF measurements are made one pulse point at a time. With the pulsed bias applied, the RF source is gated for a specified period during the pulse and the network analyzer is triggered to sample the RF signals. The same pulse point is measured often enough for the analyzer to work through its frequency list and averaging requirements.

18.4.2 Technical Considerations

A trade between cost, complexity, and technical performance arises in the specification and assembly of pulsed measurement systems. Important considerations are pulse timing capability, measurement resolution and range, total time required for a measurement task, and the flexibility of the pulse sequencing.

18.4.2.1 Pulse Events

Pulsed measurement systems produce a continuous, periodic sequence of **pulse events**. The generic timing of each part of a pulse event is shown in Fig. 18.9. Each pulse event provides a pulse stimulus and a quiescent period. The period of the pulse, T_{Pulse}, ranges from 10 ns to 1 s. Typically, pulsed-I/V measurements require 200 to 500 ns pulses, and true DC measurements require periods of 100 ms or more. To achieve sub-100 ns pulses, usually the DUT is directly connected to a pulse generator to avoid transmission-line effects. Quiescent periods, $T_{\text{Quiescent}}$, range from 10 µs to 1 s and often must be longer than 1 ms for isodynamic pulsed-I/V measurements.

One or both terminals of the DUT may be pulsed. In some systems, the pulse width on the second terminal is inset relative to the first, by τ_{inset}, which gives some control over the trajectory of the initial pulse transient to avoid possible damage to the DUT.

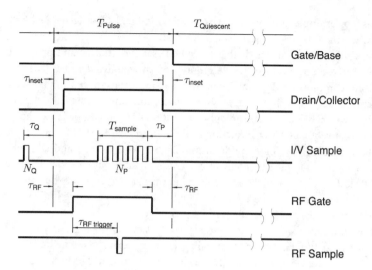

FIGURE 18.9   Generic timing diagram for each pulsed measurement pulse event.

Samples of current and voltage occur some time, τ_P, before the end of the pulse. Some systems gather a number, N_P, of samples over a period, T_{sample}, which may extend over the entire pulse if a time domain transient response is measured. The number of samples is limited by the sampling rate and memory of the digitizers. A measurement of the quiescent conditions some time, τ_Q, before the start of each pulse may also be made.

For pulsed-RF measurements, the RF source is applied for a period that is inset, by τ_{RF}, within the pulsed bias. A RF trigger sequences sampling by the network analyzer. The RF source is disabled during pulsed-I/V measurements.

18.4.2.2   Measurement Cycles

A pulsed **measurement cycle** is a periodic repetition of a sequence of pulse events. A set of pulse points, required to gather device characteristics, is measured in one or more measurement cycles. With single pulse-point measurements, there is only one pulse event in the sequence and a separate measurement cycle is required for each pulse point. This is the case with pulsed-RF measurements, with high-power pulsers, or with very-high-speed pulse generators. With arbitrary function generators, the measurement cycle is a sequence of pulse events at different pulse points; so one cycle can measure several pulse points.

Measurement cycles should be repeated for a **stabilizing period** to establish the bias condition of the measurement cycle, which is a steady-state repetition of pulse events. Then the cycle is continued while data are sampled. Typical stabilization periods can range from a few seconds to tens of seconds. These long times are required for initial establishment of stable operating conditions, whereas shorter quiescent periods are sufficient for recovery from pulse perturbations.

When several pulse points are measured in each cycle, the pulse stimulus is a steady-state repetition, so each pulse point has a well-known initial condition. Flexible pulse systems can provide an arbitrary initial condition within the cycle or use a pseudo-random sequencing of the pulse points. These can be used to assess the history dependence or isodynamic nature of the measurements. For example, it may be possible to precede a pulse point with an excursion into the breakdown region to assess short-term effects of stress on the characteristic.

18.4.2.3   Bias Networks

The most significant technical limitation to pulsed measurement timing is the bias network that connects the DUT to the pulse system. The network must perform the following:

- Provide RF termination for the DUT to prevent oscillations
- Pass pulsed-bias stimuli to the DUT

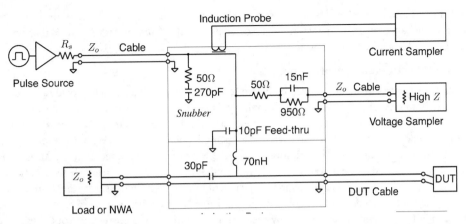

FIGURE 18.10 Schematic of a bias network that provides RF termination and pulsed bias feed with voltage and current measuring points.

- Provide current and voltage sample points
- Control transients and overshoots that may damage the DUT

These are contradictory requirements that must be traded to suit the specific application. In general, the minimum pulse period is dictated by the bias network.

For very-fast pulsed measurements, less than 100 ns, the pulse generator is usually connected directly to the DUT.[2] The generator provides the RF termination required for stability, and current and voltage are sensed with a ground-return sense resistor and a high impedance probe, respectively. Pulsed-RF measurements are not contemplated with this arrangement.

Systems that are more flexible use a modified bias network similar to that shown in Fig. 18.10. The DC-blocking capacitor must be small, so that it does not draw current for a significant portion of the pulsed bias, but must be large enough to provide adequate termination at RF frequencies. The isolating inductor must be small, so that it passes the pulsed bias, but must also be large enough to provide adequate RF isolation. In this example, the DUT is connected to a RF termination provided by a load or network analyzer. The DC-blocking capacitor, 30 pF, and isolating inductor, 70 nH, values are an order of magnitude smaller than are those in conventional bias networks. The network provides a good RF path for frequencies above 500 MHz and does not significantly disturb pulses longer than 100 ns. Modifying the network to providing a RF path at lower frequencies will disturb longer pulses.

The pulsed bias is fed to the bias network in Fig. 18.10 through a cable that will introduce transmission line transients. To control these, the source output impedance can provide line termination. Although this can provide significant protection from transients when fragile devices are being measured, it will limit the voltage and current range of the pulses. An alternative is to provide a termination at the bias network end of the cable with a series resistor-capacitor snubber. The values shown in this example are suitable for suppressing the 10 ns transients associated with a 1 m cable.

Voltage sampling in Fig. 18.10 is through a frequency-compensated network that provides isolation between the RF path and the cable connected to the voltage sampling digitizer. Without this isolation, the capacitance of the cable would load the pulsed bias waveform, significantly increasing its rise time. The voltage sample point should be as close as possible to the DUT to reduce the effect of the return pulse reflected from the DUT. The network in this example sets a practical limit of about 15 cm on the length of the cable connecting the DUT to the bias network.

In general, bias networks that provide RF terminations or pulsed-RF capability will limit the accuracy of measurements in the first 100 to 200 ns of a pulse. With such an arrangement, the pulse source need not produce rise times less than 50 ns. Rather, shaped rising edges would be beneficial in controlling transients at the DUT.

Current measurement with series sense resistors will add to the output impedance of the pulse source. Usually a capacitance of a few picofarads is associated with the sense or bias network that will limit resistance value for a specified rise time.

18.4.2.4   Measurement Resolution

Voltage and current ranges are determined by the pulse sources. Summing amplifiers provide a few hundred milliamps at 10 to 20 V. High-power, charge-dumping pulsers provide several amps and 50 V. Current pulses are achieved with series resistors and voltage sources. These limit the minimum pulse time. For example, a 1 kΩ resistor may be used to set a base current for testing bipolar transistors. With 10 pF of capacitance associated with the bias network, the minimum rise time would be of the order of 10 μs. An isodynamic measurement would need to use short collector-terminal pulses that are inset within long base-terminal pulses.

There is no practical method for implementing current limiting within the short time frame of pulses other than the degree of safety afforded by the output impedance of the pulse source.

Measurement resolution is determined by the sampling digitizers and current sensors. Oscilloscopes provide 8-bit resolution with up to 11-bit linearity, which provides only 100 μA resolution in a 100 mA range. The 12-bit resolution, with 14-bit linearity, of high-speed digitizers may therefore be desirable. To achieve the high resolutions, averaging is often required. Either the pulse system can repeat the measurement cycle to accumulate averages, or several samples in each pulse can be averaged.

18.4.2.5   Measurement Time

Measurement speed, in the context of production-line applications, is optimized with integrated systems that sequence several pulse points in each measurement cycle. As an example, acquiring 1000 pulse points with 1 ms quiescent periods, 500 ns pulse periods, and an averaging factor of 32 will necessarily require 32 s of pulsing. With a suitable stabilization period, and overhead in instrument setup and data downloading, this typical pulsed-I/V measurement can be completed in just less than one minute per quiescent point.

Single-point measurement systems have instrument setup and data downloading overhead at each pulse point. A typical 1000-point measurement usually requires substantially more than ten minutes to complete; especially when data communication is through GPIB controllers.

A pulsed-RF measurement is also slow because the network analyzer must step through its frequency list, and requires a hold-off time between RF sampling events. A typical pulsed-RF measurement with a 50-point frequency list, an averaging factor of 32, and only 100 pulse points, would take about half a minute to complete.

18.4.3   Commercial Measurement Systems

Figure 18.11 graphically portrays the areas covered in a frequency/signal level plane by various types of instruments used to characterize devices. The curve tracer, epitomized perhaps by the HP4145 and numerous analog predecessors made by companies such as Tektronix, cover the most basic measurement range. Beyond this range, instruments with some pulse capability, such as the HP4142 or HP4155/56, offer very wide capability, but this is still at speeds below that required for isodynamic .characterization. Network analyzers reach millimeter-wave frequencies but perform small-signal measurements by definition. Between these, pulsed-I/V systems such as those described below have the advantage of large-signal capability and speeds sufficient to give isodynamic characteristics.

The majority of pulsed measurements reported in the literature to date have been made with experimental equipment, or with systems under development. Three sub-microsecond systems are commercially available. These come with a range of options that require some assessment before purchase. This is partly a consequence of the immature nature of pulsed-I/V instrumentation (in comparison to conventional curve tracers), and partly a result of pulsed-I/V measurement being a more complicated problem.

Before reviewing the available systems, it is useful to identify an intrinsic problem for pulsed measurements. The performance limit on pulsed-I/V systems is frequently the DUT connection network and the form of the stimulus, not the measurement system itself.

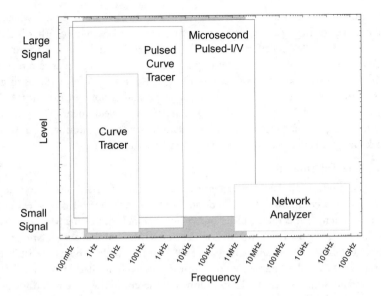

FIGURE 18.11 Relative position of various types of measurement equipment, including pulsed-I/V systems, in terms of measurement frequency and signal level. The shaded area indicates the frequency range of dispersion effects.

Network analyzers achieve very high-frequency resolution with a narrowband stimulus and receiver, which allows them to minimize noise and apply vector calibration techniques to eliminate parasitic disturbances. They define a measurement plane, behind which any fixed error is identified and eliminated by postprocessing of the data. They can also allow the DUT to come to a steady state during the measurement. Pulse systems conversely use a stimulus that contains many frequency components from the slow pulse repetition rate up to many times the fundamental component in the fast pulse. The measurement is both of wide bandwidth, and therefore noisy, and at high frequencies. Viewed in the time domain, the pulse width is limited by the charging of the unknown capacitance in the bias network, which can be minimized but not eliminated. For example, bias networks may contribute sufficient parasitic capacitance to limit pulsed measurements to 500 ns, or slower, with a pulse source impedance of 50 Ω. The situation is worse for current drive, and may be worse still, because of transients, for a voltage drive that does not match transmission line impedance. Thus, the system is infrequently limited by the minimum width of the pulse from the instrument, and some judgment needs to be exercised in each measurement setup.

18.4.3.1 GaAs Code

GaAs Code Ltd., based in Cambridge, England, offers a low-end pulsed-I/V measurement system.[3] It is controlled by a PC via a serial interface. Hardware cost is on the order of US$20,000. Specifications range from ±10 V, 0.5 A, 2.5 W up to +25 V, 1 A, 6 W, with output impedance at or above 10 Ω. Pulse width is from 100 ns to 1 ms. A higher power model is under development. Software supplied by GaAs Code allows control, plus generation of graphs that can be printed or incorporated into documents under the Windows operating system. The instrument works with various modeling software programs supplied by GaAs Code. No provision is made for synchronization with a network analyzer for pulsed-RF measurements.

18.4.3.2 Macquarie Research

Macquarie Research Ltd. offers an Arbitrary Pulsed-I/V Semiconductor Parameter Analyzer (APSPA).[4] The hardware is largely commercial VXI modules. Control is via proprietary software running on an embedded controller. Each measurement cycle can cover up to 2048 pulse points, which, together with the integrated bus architecture, gives fast measurement turnaround. System cost (hardware and software) is on the order of US$100,000. Specifications start at ±20 V, ±0.5 A and rise to 3 A in the VXI rack or to 50 V and 10 A with an external Agilent K-series pulse source. Output impedance ranges from less than 1 Ω to 50 Ω in discrete steps, depending upon options. Pulse timing is from 100 ns to greater than 1 s in 25 ns steps, with pseudo-random, arbitrary sequencing, and scripting capability. A 50 V, 5 A high-speed module, and support for low-cost digitizers, are under development. The proprietary software produces data files but does not support data presentation. Synchronization with an Agilent HP85108A pulsed network analyzer is included for routine pulsed-RF measurements.

18.4.3.3 Agilent Technologies

Agilent Technologies offers a pulsed-I/V system as a subsection of their pulsed modeling system.[5] The pulsed-I/V subsystem is composed of rack-mounted instruments controlled by a workstation running IC-CAP software. System cost is on the order of US$500,000 inclusive of the RF and pulsed subsystems, and software. The DC and pulsed-I/V system is approximately half of that cost, the pulsed-I/V subsystem constituting about US$200,000. Specifications are ±100 V at 10 A with an output resistance of about 1 Ω, based exclusively on K49 Pulse Sources. Pulse width is effectively limited by a lower bound of 800 ns. Data presentation and S-parameter synchronization are inherent in the system. A difficulty of the use of GPIB and K49s driven by conventional pulse generators is the overall measurement time, which at best is about 2 orders of magnitude slower than integrated multipoint systems. Only one pulse point is possible in each measurement cycle.

18.5 Measurement Techniques

With flexible pulsed measurement systems, a wide range of measurements and techniques is possible. Consideration needs to be given to what is measured and the measurement procedures, in order to determine what the data gathered represents. The following sections discuss different aspects of the measurement process.

18.5.1 The Pulse-Domain Paradigm and Timing

A general pulsed-I/V plane can be defined as the grid of terminal voltages pulsed to and from a particular quiescent condition. For isodynamic pulsing, a separate pulsed characteristic would be measured for each quiescent condition.

At each pulse point on an I/V-plane, measurements can be characterized in terms of the following:

- The quiescent point pulsed from, defined by the established bias condition and the time this had been allowed to stabilize.
- The actual pulse voltages, relative to the quiescent voltage, the sequence of application of the terminal pulses, and possibly the voltage rise times, overshoot, and other transients.
- The position in time of sampling relative to the pulses.
- The type of measurements made; voltage and current at the terminals of the DUT, together with RF parameters at a range of frequencies.

Thus, if a number of quiescent conditions are to be considered, with a wide range of pulsed terminal voltages, a large amount of data will be generated. The time taken to gather this data can then be an important consideration. Techniques of overnight batch measurements may need to be considered, together with issues such as the stability of the measurement equipment. Equipment architecture can be categorized in terms of the applications to measurement over a generalized I/V-plane. Those that allow arbitrary pulse sequences within each measurement cycle enable an entire I/V-plane to be rapidly sampled.

Systems intended for single pulses from limited quiescent conditions may facilitate precise measurement of a small region in the I/V-plane, but this is at the expense of speed and flexibility.

In the context of isodynamic pulsing, the most important consideration in interpreting the measured data is the sample timing. This is the time of current and voltage sampling relative to the application of the voltage pulses. As it is often information on time-dependent dispersion effects that is gathered, it is important to understand the time placement of sampling relative to the time constants of these rate-dependent effects.

For an investigation of dispersion effects, time domain pulse-profile measurements are used. Terminal currents and voltages are repeatedly sampled, from before the onset of an extended pulse, until after dispersion effects have stabilized. This can involve sampling over six decades of time and hence produces large amounts of data. From such data, the time constants of dispersion effects can be extracted. From pulse-profile measurements of a range of pulse points, and from a range of initial conditions, the dependence of the dispersion effects upon initial and final conditions can be determined.

For isodynamic measurements unaffected by dispersion, sampling must be done quickly after the application of the pulse, so that dispersion effects do not become significant. Additionally, the relaxation time at the quiescent condition, since the application of the previous pulse, must be long enough that there are no residual effects from this previous pulse. The device can then be considered to have returned to the same quiescent state. Generally, sampling must be done at a time, relative to pulse application, at least two orders of magnitude less than the time constants of the dispersion effects (for a less than 1% effect). Similarly, the quiescent time should be at least an order of magnitude greater than these time constants.

Note that for hardware of specific pulse and sampling speed limitations, there may be some dispersion effects too fast for observation. Thus, this discussion refers to those dispersion time constants greater than the time resolution of the pulse equipment.

Quantification of suitable pulse width, sample time, and quiescent time can be achieved with reference to the time constants observed in a time domain pulse profile. For example, for dispersion time constants in the 10 to 100 μs range, a pulse width of 1 μs with a quiescent time of 10 ms might be used. Sampling might be done 250 ns after pulse application, to allow time for bias network and cable transients to settle.

In the absence of knowledge of the applicable dispersion time constants, suitable pulse and quiescent periods can be obtained from a series of pulsed measurements having a range of pulse, sample, and quiescent periods. Observation of sampled current as a function of these times will reflect the dispersion effects present in a manner similar to that achievable with a time domain pulse-profile measurement.[6]

A powerful technique for verifying isodynamic timing is possible with measurement equipment capable of pulsing to points on the I/V-plane in a random sequence. If the quiescent time of pulse relaxation is insufficient, then the current measurement of a particular pulsed voltage will be dependent upon the particular history of previous pulse points. In conventional measurement systems, employing step-and-sweep sequencing whereby pulse points are swept monotonically at one terminal for a stepping of the other terminal, dispersion effects vary smoothly and are not obvious immediately. This is because adjacent points in the I/V-plane are measured in succession and therefore have similar pulse histories.

If, however, points are pulsed in a random sequence, adjacent points in the I/V-plane each have a different history of previous pulses. If pulse timing does not give isodynamic conditions, then the dispersion effects resulting from the pulse history will be evident in the characteristic curves. Adjacent points, having different pulse histories, will have dispersion effects of differing magnitude and hence markedly different values of current. This is observed in Fig. 18.12, showing isodynamic and non-isodynamic measurement of the characteristics of a particular device. The non-isodynamic sets of characteristics were measured with the same pulse timing. One characteristic was measured by sweeping the drain-terminal pulse monotonically for different gate-terminal pulse settings. The other characteristic was measured as a random sequence of the same pulses. The smooth shape of the former does not suggest dispersion effects. The apparently noisy variation between adjacent points in the latter indicates history-dependent dispersion is in effect.

Thus, by random sequencing of the pulse points, isodynamic timing can be verified. To obtain isodynamic characteristics, shown in Fig. 18.12, the quiescent relaxation time was increased and the pulse

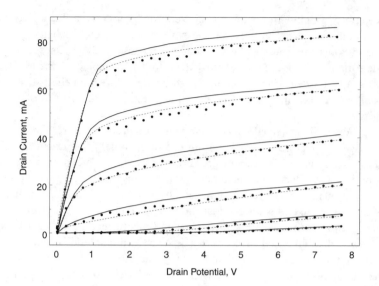

FIGURE 18.12 Characteristic curves for a MESFET measured with three different pulse sequences: a step-and-sweep with 1 μs pulses and 1 μs quiescent periods (- -), the same pulses sequenced in pseudo-random order (•), and an isodynamic measurement. The latter used 800 ns pulses with 1 ms quiescent periods.

time reduced, until both curves became smooth and identical. That is, until there is no observable history-dependent dispersion.

18.5.2 General Techniques

Within the context of the pulse-domain paradigm discussed in the previous section, and the available equipment, a number of specific measurement techniques and issues arise. These are affected by the equipment limitations and influence the data gathered. A number of these techniques and issues are discussed here.

18.5.2.1 Interpolation and Iteration

Often measurements are desired at a particular pulse point or specific grid of points. For a target pulse voltage, the actual voltage at the DUT at a certain time will usually be less. This results from various hardware effects such as amplifier output impedance and amplifier time constants, as well as cabling and bias network transients. Voltage drop across amplifier output impedance could be compensated for in advance with known current, but this current is being measured. This is why pulsed voltages need to be measured at the same time as the device currents.

If measurements are desired at specific voltage values, then one of two approaches can be used. Firstly, over successive pulses, the target voltage values can be adjusted to iterate to the desired value. This necessarily involves a measurement control overhead and can require considerable time for many points. If the thermal noise implicit in using wide-bandwidth digitizers is considered, it is of dubious value to iterate beyond a certain point.

Alternatively, if a grid of pulse points is sampled, covering the range of points of interest, then the device characteristics at these particular points can be interpolated from the measured points. Without iteration, these measured points can be obtained quickly. A least-squares fit to a suitable function can then be used to generate characteristics at as many points as desired. Thus, provided the sampled grid is dense enough to capture the regional variation in characteristics, the data gathering is faster. The main concept is that it is more efficient to rapidly gather an entire I/V-plane of data and then post-process the data to obtain specific intermediate points.

18.5.2.2 Averaging

The fast pulses generally required for isodynamic measurement necessitates the use of wide-bandwidth digitizers. Voltage and current samples will then contain significant thermal noise. A least-squares fit to an assumed Gaussian distribution to an I/V-grid can be employed to smooth data. Alternatively, or additionally, averaging can be used.

Two types of averaging processes present themselves. The first process is to average multiple samples within each pulse. This assumes a fast digitizer and that there is sufficient time within the pulse before dispersion becomes significant. If dispersion becomes significant over the intra-pulse period of sampling, then averaging cannot be employed unless some assumed model of dispersion is applied (a simple fitted time constant may suffice). An additional consideration with intra-pulse averaging is that voltage value within a pulse cannot be considered constant. The measurement equipment providing the voltage pulse has nonzero output impedance and time constants. Thus, the actual voltage applied to the DUT will vary (slightly) during the voltage pulse. Consecutive samples within this pulse will then represent the characteristics for different voltage values. These are valid isodynamic samples if the sample timing is still below the time constants of dispersion effects. However, they could not be averaged unless the device current could be modeled as a linear function of pulsed voltages (over the range of voltage variation).

The second averaging process is to repeat each pulse point for as many identical measurements as required and average the results. Unlike intra-pulse averaging, this inter-pulse averaging will result in a linear increase in measurement time, in that a measurement cycle is repeated for each averaging. Issues of equipment stability also need to be considered. Typically, both intra- and inter-pulse averaging might be employed. With careful application, averaging can provide considerable improvement in the resolution of the digitizers used, up to their limit of linearity.

18.5.2.3 Pseudo-Random Sequencing

As previously discussed, randomizing the order of the sequence of pulse points can provide a means of verifying that the quiescent relaxation time is sufficient. It can also provide information on the dispersion effects present. In this, a sequence of voltage pulse values is determined for the specified grid of terminal values. These are first considered a sweeping of one terminal for a stepping of the other. To this sequence, a standard pseudo-randomizing process is used to re-sequence the order of application of pulses. As this is deterministic for a known randomizing process, it is repeatable. This sequence is then applied to the DUT. Upon application of pulses, this random pulse sequence can help identify non-isodynamic measurement timing.

Additionally, if dispersion is present in the measured data, the known sequence of pulse points can provide information on history-dependent dispersion. With step-and-sweep sequencing of pulses, the prior history of each pulse is merely the similar adjacent pulse points. This represents an under-sampling of the dispersion effects. With random sequencing, consecutive pulse points have a wide range of earlier points, providing greater information on the dispersion effects.

Thus, for the known sequence of voltage pulses and the non-isodynamic pulse timing, a model of the dispersion effects can be fitted. These can then be subtracted to yield isodynamic device characteristics. This, however, only applies to the longer time-constant effects and requires that the timing be close to that of the time constants of the dispersion effects.

18.5.2.4 Pulse Profile

In normal isodynamic pulsing, pulse widths are kept shorter than the time constants of applicable dispersion effects. Relaxation periods between pulses, at the quiescent condition, are longer than these times. Typically, pulse widths of 1 μs and quiescent periods of 100 ms might be used.

In a pulse profile measurement, an extended pulse width of 0.1 to 1 s might be used, so that the dispersion effects can be observed. All dispersion time constants greater than the pulse rise and settling time are then observable. Quiescent periods between these extended pulses still need to be long, so that subsequent pulses can be considered as being from the same bias condition.

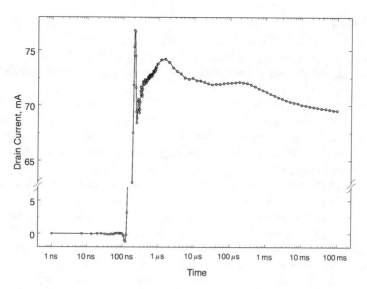

FIGURE 18.13 Transient response measured with eight repeated measurements at 50 ns intervals. Each repetition is shifted by 6.25 ns to give the composite response shown.

Plotted on a logarithmic time axis, the dispersion effects can be seen as a variation of device output current with time (see Fig. 18.13). Typically, output current might rise in the first 1 to 10 μs period due to junction heating and trapping effects, then fall due to channel heating. Time constants of the amplifier driving the pulses might need to be deconvolved before identifying those of the DUT alone. From such a plot, it can first be identified where isodynamic conditions apply. That is, how soon after pulse application sampling needs to be done before dispersion effects become significant. How long these dispersion effects take to stabilize will indicate how long the quiescent periods of isodynamic pulsing need to be. Secondly, values for dispersion time constants can be extracted from the data, together with other parameters applicable to a particular dispersion model.

Note that because the extended pulse widths of pulse profile measurements are intended to bring into effect heating and dispersion, the range of pulse points on the I/V-plane must be restricted. With isodynamic pulsing, it is possible to pulse to voltages well outside the SOA of the DUT. This is because the short pulses do not invoke the time-dependent thermal and current damage of static conditions. With pulse profile measurements, pulse widths extend to essentially static periods and so voltages must be restricted to the SOA for static measurements (although pulse profile techniques could be used to observe destruction outside the SOA).

Equipment issues influence pulse profile measurements in several ways. The first is pulse duration. Systems employing capacitor charge dumping for pulsing will be limited in the length of time that they can hold a particular output voltage. The second is output rise and settling times. Bias network and cable transients and the response time of data measurement will limit the earliest time after pulse application for which valid samples can be taken. This, typically, might be of the order of 100 ns, although with restrictions on application might extend down to 1 ns. This necessarily affects the range of dispersion effects observable to those having time constants greater than perhaps an order of magnitude more than this minimum time resolution.

Digitizer speed and bandwidth are another major issue in pulse profile measurements. A wide bandwidth is necessary so that sample values accurately reflect DUT conditions. In isodynamic pulsing, only one time point need be sampled, with a long time before the next pulse. With a pulse profile, it is desirable to repeatedly sample the pulse as fast as possible to observe variation with time. Sampling speed needs to be perhaps an order of magnitude faster than the time constant to be observed. Additionally, if bandwidth, jitter, and stability permit, an *equivalent time* sampling may be used. In this, repeated pulse profile measurements are performed, with sample times relative to pulse onset shifted slightly with each successive pulse. As an example, a 20 MHz digitizer, sampling at 50 ns intervals, might be applied to eight successive, identical pulses. Sampling is commenced 50 ns before the start of the first pulse, but offset an accumulating 6.25 ns on successive pulses. The sum of these then represents sampling at a rate of 160 MHz. This assumes the bandwidth of the digitizer input track-and-hold circuit is sufficient.

Sampling at a rate of 160 MHz generates a large amount of data when applied to a 1s long pulse. However, as the dispersion processes to be observed tend to be exponential in effect over time, then it is not necessary to continue sampling at this rate for the entire pulse profile. The sampling period needs to be less than 70% of the time constant to be observed, but typically sampling would be an order of magnitude faster for better amplitude resolution in noisy conditions. Thus, sampling may begin at 10 ns intervals for the first 100 ns, but then continue at every 100 ms toward the end of the 1 s pulse. Such logarithmic placement of sampling over the pulse is possible with digitizers that allow arbitrary triggering and systems that can generate arbitrary trigger signals. With such a system, sampling would be performed at a linear rate initially while requiring samples as fast as possible, reducing to a logarithmic spacing over time. For example, with a 20 MHz digitizer, sampling might be done every 50 ns for the first 1 μs, but then only ten samples per decade thereafter. This would give only 80 samples over a 1 s pulse, rather than the excessive 20 M samples from simple linear sampling. In this way, data can be kept to a manageable but adequate amount.

18.5.2.5 Output Impedance

In testing a device, whether the terminal current or voltage is the dependent variable or the independent variable is subjective and conditional upon the type of device (BJT or FET). However, pulsed measurement systems are usually implemented with sources of voltage pulses, for practical reasons. Thus, it is desirable to have negligible output impedance in the pulse generator or driving amplifier.

There exist, however, some situations where it is desirable to have significant output impedance in the pulse driver. For example, in testing FETs with very fast pulses, it is usually necessary to use a 50 Ω output impedance with the gate-terminal pulser to prevent RF oscillations.

When current is the more convenient independent variable, a large driver output impedance can simulate a current source. With bipolar devices (BJTs and HBTs), it is desirable to perform measurements at particular values of base current. This is a very strong function of base emitter voltage and hence difficult to control with a voltage source. With a large source resistance (e.g., 10 kΩ) in the base voltage driver, a reasonable current source can be approximated and base current controlled. This will necessarily severely limit the rise time of a base terminal pulse, so that typically this pulse would be first applied and allowed to stabilize before a fast pulse is applied to the collector terminal. This is fine for investigating isodynamic collector current in relation to dispersion effects due to collector voltage and power dissipation. However, the long base current pulse implies that base voltage and current-related dispersion effects are not isodynamic.

Output impedance is also used for current limiting and for safe exploration of the I/V-plane. The diode characteristic of the FET gate junction during forward conduction and breakdown means that gate current can become very large. Having 50 Ω in the gate-terminal pulser will limit this current to 20 mA typically. Similarly, 50 Ω in the drain-terminal pulser will limit drain current for a particular voltage pulse and constrain DUT output behavior to follow the load line determined by this 50 Ω load impedance and the applied voltage pulse. In this way, pulse voltage can be slowly increased to explore expanded regions of device operation safely. It will also curb transients.

18.5.2.6 Extending the Data Range

An important aspect of pulsed testing is that a wider range of data points can be tested. Beyond a certain range of terminal potentials or power, damage can be done to a device because of excessive temperature or current density. As the DUT temperature is a function of the time for which a given power level is applied, the shorter a pulse, the greater the voltage and/or instantaneous power that can be applied.

The conventional SOA of a device is that part of the I/V-plane for which the device can withstand static or continuous application of those voltage levels. Pulsed testing then extends this region, in particular to regions that are outside the static SOA, but are still encountered during normal RF operation of the device. This gives an extended range of data for use in modeling device operation, not only for isodynamic I/V characteristics, but also for RF parameters for extraction of parasitic resistances and capacitances. With a pulsed S-parameter system coupled with a pulsed-I/V system, the voltage pulses can take the DUT to an isothermal point outside the static SOA, where S-parameters can then be measured during this pulse.

18.5.2.7   Repetition

The characteristics of a device can change due to the manner in which it is used. For example, an excursion into a breakdown region can alter, although not damage, a device, permanently modifying its characteristics. To investigate such phenomena, an I/V-grid can be measured before and after such an excursion. Changes in the device characteristics can then be observed in the difference between the measurements.[7]

Of use in such investigations is the ability to specify an arbitrary list of pulse points. In this case, the list of points in the I/V-plane to be pulsed to would first list the regular grid, then the points of breakdown excursion, and then repeat the same regular grid points. Additionally, scripting capabilities might be used to create a series of such measurements.

18.5.2.8   Onion-Ring Destructive Testing

Often it is desired to test a device until destruction. An example of this might be breakdown measurements. Sometimes it is difficult not to destroy a *fragile* device during testing — especially devices fabricated with an immature technology. In either case, it is desirable to structure the sequence of pulse points from safe voltage and power levels to increasing levels up to destruction. It is essential in this that all data up to the point of device destruction is preserved.

Here again, scripting capabilities and the use of a list of pulse points allow measurements to be structured as a series of layers of pulse points, increasing in power and/or voltage level. In this way, the characteristics of a device can be explored as an extension, in layers, of the safe device operation or constant power level. Inter-pulse averaging and a waiting period for device stabilization would not normally be used in this form of measurement.

18.5.2.9   Quiescent Measurement

It is important to measure the bias point representing the isodynamic conditions of the DUT. This is the terminal voltage and current before each pulse and as such gives the quiescent thermal and trapping state of the device. This needs to be measured as part of the pulse exercise if the pulse sequence used is such that the average device temperature is raised.

The time spent at the quiescent point is usually quite long, affording opportunity for considerable averaging. Additionally, when pulsing too many points of the I/V-plane, the quiescent point can be measured many times. Thus, a comparatively noise-free measurement can be obtained.

Sample points for quiescent data would usually be placed immediately before a pulse. Several samples would be taken and averaged. It is assumed that the relaxation time at the quiescent condition, since the previous pulse, is very much greater than all relevant dispersion-effect time constants (unless these time constants are themselves being investigated). This is necessary if the samples are to be considered as representing a bias condition, rather than a transient condition.

Alternatively, or additionally, some samples might be taken immediately after a pulse. For these post-pulse samples to be considered to represent the bias condition, the pulse must be short enough for no significant dispersion effects to have occurred. Notwithstanding this, there may be useful information in observing relaxation after a pulse and in the change in device current immediately before and after a return from a pulse.

18.5.2.10 Timing

A number of different timing parameters can be defined within the paradigm of pulse testing. Referring to Fig. 18.9, a basic pulse cycle consists of an extended time at the quiescent bias point ($T_{\text{Quiescent}}$) and a (usually) short time at particular pulsed voltage levels (T_{Pulse}). In this diagram, T_{Pulse} refers to the time for which the gate or base voltage pulse is applied. The sum of these two times is then the **pulse event time** and the inverse of this sum would be the **pulse repetition frequency** for continuous pulsing.

A third timing parameter, τ_{inset}, reflects the relationship of the drain/collector pulse to the gate/base pulse. These voltage pulses need not be coincident, but will normally overlap. Often the gate pulse will be applied before the drain pulse is applied — an inset of 100 ns is typical. Sometimes it might be necessary for the drain pulse to lead the gate pulse in order to control the transition path over the I/V-plane. Thus, the parameter τ_{inset} might be positive or negative and might be different for leading and trailing pulse edges. In a simple system, it is most easily set to zero so that the terminal pulses are coincident.

These three parameters define pulse event timing — the times for which terminal voltage pulses are applied and the quiescent relaxation time. Note that actual voltage pulses will not be square shaped. For single-point pulsing, there might only be one pulse event, or a sequence of identical pulse events. For generalized pulsing over the I/V-plane, a measurement cycle may be an arbitrary sequence of different pulse points, all with the same cycle timing.

The number of sample points within a basic pulse event could be specified as both a number of samples within the pulse (N_P) and as a number of samples of the quiescent condition (N_Q). Typically these would be averaged, except in the case of a pulse profile measurement. The placement of these sample points within the pulse cycle need also be specified.

If the pulsed-I/V system is to be coupled with a pulsed-RF system, such as the Agilent Technologies HP85108, then relative timing for this needs to be specified. Figure 18.9 defines a time for application of the RF signal relative to the gate voltage pulse and a trigger point within this for RF sampling. These two signals can be supplied to the HP85108 for synchronization.

The above times would refer to the pulse event timing at the terminals of the DUT. Various instrument and cabling delays might require that these times be individually adjusted when referred to the pulse amplifiers and sample digitizers. Different signal paths for current and voltage digitizers might require separate triggers for these.

18.5.2.11 General Techniques

As well as the various measurement techniques just discussed, there exists a range of practical issues. For example, with combined pulsed-I/V and pulsed-RF systems, the RF must be turned off while measuring DUT current. This means that experiment times are longer than might be expected, as the pulsed-I/V and pulsed-RF data are gathered separately.

Another consideration is that the applied voltage pulses are not square shaped. Instrumentation and cable termination issues result in pulses having significant rise and fall times and in particular overshoot and settling. The devices being tested are generally fast enough to respond to the actual instantaneous voltages, rather than an averaged rectangular pulse. First, this means that sampling of both voltage and current must be performed, and that this must be at the same time. Second, as any pulse overshoot will be responded to, if this voltage represents a destructive level then damage may be done even when the target voltage settled to is safe. This particularly applies to gate voltage pulses approaching forward conduction or breakdown.

Also arising from the fact that the DUT is far faster in response than the pulse instrumentation, is the issue of pulsing trajectory. In pulsing from a bias point to the desired pulse point, the DUT will follow a path of voltage and current values across the I/V-plane, between the two points. Similarly, a path is followed in returning from the pulse point to the bias point. The actual trajectory followed between these two points will be determined by the pulse rise and fall times, overshoot and other transients, and by the relative inset of gate and drain pulses (Fig. 18.9).

A problem can arise if, in moving between two safe points on the I/V-plane, the trajectory passes through a destructive point. An example is pulsing to a point of low drain voltage and high current from a bias point of high drain voltage and low current. Here drain voltage is pulsing to a lower voltage while gate voltage is pulsing to a higher value. If the gate pulse is applied first, then the DUT will move through a path of high voltage and high current. This is a problem if it represents destructive levels and is dependent upon trajectory time. A similar problem exists in returning from the pulse point to the bias point. In general, because gate/drain coincidence cannot be sufficiently well controlled, consideration need be given to the trajectories that may be taken between two points on the I/V-plane and the suitability of these. With appropriate choice of leading and trailing overlaps between the gate and drain pulses, this trajectory can be controlled.

18.6 Data Processing

Having gathered data through pulsed measurements, various processing steps can follow. In this, reference need again be made to the pulse domain paradigm. In the simplest case, the data consists of a grid of pulse points for a fixed bias point, sampled free of dispersion effects. To this could be added further data of grids for multiple bias points. Rate dependence can be included with data from pulse profile measurements and grids with delayed sample times. In this way, the data can be considered as a sampling of a multidimensional space. The dimensions of this space are the terminal currents and voltages, both instantaneous and average, together with sample timing and temperature. RF parameters at a range of frequencies can also be added to this.

Processing of this data can be done in two ways. First, the data can be considered as raw and processed to clean and improve it. Examples of this form of processing are interpolation and gridding. Second, data can be interpreted against specific models. Model parameter extraction is the usual objective here. However, to fully use the information available in the pulsed data, such models need to incorporate the dispersion effects within the pulse domain paradigm.

18.6.1 Interpolation and Gridding

Data over the I/V-plane can be gathered rapidly about a grid of target pulse points. The grid of voltage values represents raw data points. Instrument output impedance and noise usually differentiate these from desired grid points. Interpolation and gridding can translate this data to the desired grid.

Data can be gathered rapidly if the precision of the target pulse-voltage values is relaxed. The data still represents accurate samples, however the actual voltage values will vary considerably. This variation is not a problem in model extraction, but can be a problem in the comparison of different characteristic curves (for different quiescent conditions) and the display of a single characteristic curve for a specified terminal voltage.

Gridding is performed as the simple two-dimensional interpolation of current values as a function of input and output pulse-voltage values. A second- or third-order function is usually used. The interpolated voltage values represent a regular grid of desired values, whereas the raw data values are scattered. A least-squares fit can be used if a noise model is assumed, such as thermal noise. Nothing is assumed about the underlying data, except for the noise model and the assumption that the data local variation can be adequately covered by the interpolation function used.

18.6.2 Intrinsic Characteristics

The simplest of models for data interpretation all assume series access resistances at each terminal. Fixed resistances can be used to model probe and contact resistances, as connecting external terminals to an idealized internal nonlinear device. For measured terminal current and assumed values of resistances, the voltage across the terminal access resistances is calculated and subtracted to give intrinsic voltages. These voltages can then be used in model interpretation.

For example, consider a FET with gate, drain, and source access resistances of R_G, R_D, and R_S respectively. If the measured terminal voltages and currents are v_{GS}, i_G, v_{DS}, and i_D respectively, then the intrinsic voltages can be obtained as:

$$v_{DS'} = v_{DS} - i_D R_D - \left(i_D + i_G\right)R_S,$$
$$v_{GS'} = v_{GS} - i_G R_G - \left(i_D + i_G\right)R_S.$$

(18.7)

If v_{GS}, i_G, v_{DS} and i_D are raw data, then a set of $v_{DS'}$, $v_{GS'}$ values can be used to obtain a grid of intrinsic data. This is easy to do with copious amounts of data gathered over the I/V-plane.

18.6.3 Interpretation

The data, raw or gridded, can be used to extract information on specific effects under investigation. In the simplest case, small-signal transconductance and conductance can be obtained as gradients, such as di_D/dv_{GS} and di_D/dv_{DS} in the case of a FET. These could then be used in circuit design where the device is being operated at a specific bias point. A second example is in the extrapolation of plots of voltage and current ratios to give estimates of terminal resistances for use in determining intrinsic values. The advantage of pulsed testing here is that an extended range of data can be obtained, extending outside the static SOA.

Another example of data interpretation is the use of measured history dependence to give information on dispersion effects. If, in pulsed testing, the quiescent relaxation time is insufficient, then pulse samples will be affected by dispersion. The use of shuffling of the pulse sequence enhances sampling of dispersion. Models of dispersion can then be fitted to this data to extract parameters for dispersion, as a function of terminal voltages and of pulse timing.

18.6.4 Modeling

The paradigm of pulsed testing assumes that DUT terminal currents are functions of both instantaneous and of average terminal voltages. This means that device response to RF stimuli will be different for different average or bias conditions. Pulsed testing allows separation and measurement of these effects.

A model of device behavior, for use in simulation and design, must then either incorporate this bias dependence or be limited to use at one particular bias condition. The latter is the usual case, where behavior is measured for a particular bias condition, for modeling and use at that bias condition.

If a model incorporates the bias-dependent components of device behavior, the wider sample space of pulsed testing can be utilized in model parameter extraction. From I/V-grids sampled for multiple bias conditions, the bias dependency of terminal current can be extracted as a function of both instantaneous and bias terminal voltages. From pulse profile measurements, dispersion effects can be modeled in terms of average terminal voltages, where this average moves from quiescent to pulse target voltage, over the pulse period, according to a difference equation and exponential time constants. The actual parameter extraction consists of a least-squares fit of model equations to the range of data available, starting from an initial guess and iterating to final parameter values. The data used would be I/V-grids, pulse profiles, and RF measurements over a range of frequencies, at a range of bias points, depending on the scope of the model being used. Important in all this is a proper understanding of what the sampled DUT data represents, in the context of the pulse domain paradigm, and of how the data is being utilized in modeling.

Empirical models that account for dispersion effects must calculate terminal currents in terms of the instantaneous and time-averaged potentials. In the case of a FET, the modeled drain current is a function of the instantaneous potentials v_{GS} and v_{DS}, the averaged potentials $\langle v_{GS} \rangle$, $\langle v_{DS} \rangle$ and average power $\langle i_{DS} v_{DS} \rangle$. The time averages are calculated over the time constants of the relevant dispersion effects. A model of thermal dispersion is:

$$i_{DS} = i_O \left(1 - \lambda R_T \left\langle i_{DS} v_{DS} \right\rangle \right),$$

(18.8)

where i_O includes other dispersion effects in a general form

$$i_O = I \left(v_{GS}, v_{DS}, \left\langle v_{GS} \right\rangle, \left\langle v_{DS} \right\rangle \right).$$

(18.9)

With a suitable value of λR_T, the thermal effects present in the characteristics of Fig. 18.1 can be modeled and the other dispersion effects can be modeled with the correct function for i_O in Eq. (18.9). The DC characteristics are given by the model when the instantaneous and time-averaged potentials track each other such that $\langle v_{GS}\rangle = v_{GS}$, $\langle v_{DS}\rangle = v_{DS}$, and $\langle i_{DS}\, v_{DS}\rangle = i_{DS}\, v_{DS}$. In this case, the model parameters can be fitted to the measured DC characteristics and would be able to predict the apparently negative drain conductance that they exhibit. In other words, the DC characteristics are implicitly described by

$$I_{DS} = I\left(V_{GS}, V_{DS}, V_{GS}, V_{DS}\right)\left(1 - \lambda R_T I_{DS} V_{DS}\right). \tag{18.10}$$

Of course, this would be grossly inadequate for modeling RF behavior, unless the model correctly treats the time-averaged quantities as constants with respect to high-frequency signals.

For each quiescent point $(\langle v_{GS}\rangle, \langle v_{DS}\rangle)$, there is a unique set of isodynamic characteristics, which relate the drain current i_{DS} to the instantaneous terminal-potentials v_{GS} and v_{DS}. Models that do not provide time-averaged bias dependence must be fitted to the isodynamic characteristics of each quiescent condition individually. Models in the form of Eqs. (18.8) and (18.9) simultaneously determine the quiescent conditions and the appropriate isodynamic characteristics.[8,9] Pulsed measurements facilitate this characterization and modeling of device RF behavior with bias dependency.

Defining Terms

Characteristic curves: For FETs/HBTs, a graph showing the relationship between drain/collector current (or RF parameters) as a function of drain/collector potential for step values of gate/base potential.
Bias condition: For a device, the average values of terminal potential and currents when the device is operating with signals applied.
Dispersion effects: Collective term for thermal, rate-dependent, electron trapping and other anomalous effects that alter the characteristic curves with the bias condition changes.
DC characteristics: Characteristic curves relating quiescent currents to quiescent terminal potentials.
Isodynamic characteristic: Characteristic curves relating instantaneous terminal currents and voltages for constant, and equal, bias and quiescent conditions.
Isothermal characteristic: Characteristic curves relating instantaneous terminal currents and voltages for constant operating temperature.
Pulsed bias: Pulsed stimulus that briefly biases a device during a pulsed-RF measurement.
Pulsed characteristics: Characteristic curves measured with pulsed-I/V or pulsed-RF measurements.
Pulsed-I/V measurement: Device terminal currents and voltages measured with pulse techniques.
Pulsed-RF measurement: Device RF parameters measured with pulse techniques.
Quiescent condition: For a device, the value of terminal potential and currents when the device is operating without any signals applied.

References

1. Teyssier, J.-P., ate al., 40-GHz/150-ns Versatile pulsed measurement system for microwave transistor isothermal characterization, *IEEE Trans. MTT*, 46, 12, 2043–2052, Dec. 1998.
2. Ernst, A.N., Somerville, M.H., and del Alamo, J.A., Dynamics of the kink effect in InAlAs/InGaAs HEMT's, *IEEE Electron Device Letters*, 18, 12, 613–615, Dec. 1997.
3. GaAs Code Ltd, Home page, 2000. [Online]. Available: URL: http://www.gaascode.com/.
4. Macquarie Research Ltd, Pulsed-bias semiconductor parameter analyzer, 2000. [Online]. Available: URL: http://www.elec.mq.edu.au/cnerf/apspa.
5. Agilent Technologies, HP85124 pulsed modeling system and HP85108 product information, 2000. [Online]. Available: URL: http://www.agilent.com.

6. Parker, A.E. and Scott, J.B., Method for determining correct timing for pulsed-I/V measurement of GaAs FETs, *IEE Electronics Letters*, 31, 19, 1697–1698, 14 Sept. 1995.
7. Scott, J.B., et al., Pulsed device measurements and applications, *IEEE Trans. MTT*, 44, 12, 2718–2723, Dec. 1996.
8. Parker, A.E. and Skellern, D.J., A realistic large-signal MESFET model for SPICE, *IEEE Trans. MTT*, 45, 9, 1563–1571, Sept. 1997.
9. Filicori, F., et al., Empirical modeling of low frequency dispersive effects due to traps and thermal phenomena in III-V FET's, *IEEE Trans. MTT*, 43, 12, 2972–2981, Dec. 1995

19

Microwave On-Wafer Test

Jean-Pierre Lanteri
M/A-COM TycoElectronics

Christopher Jones
M/A-COM TycoElectronics

John R. Mahon
M/A-COM TycoElectronics

19.1 On-Wafer Test Capabilities and Applications

19.1.1 Fixtured Test Limitations

Until 1985 the standard approach to characterize at microwave frequencies and qualify a semiconductor wafer before shipping was to dice it up, select a few devices, typically one in each quadrant, assemble them, and then test them in a fixture, recording s-parameters or power levels. Often, the parts were power transistors, the most common RF/microwave product then, and a part was used as a sample. For Gallium Arsenide (GaAs) Monolithic Microwave Integrated Circuits (MMICs), a transistor was similarly used for test coupon, or the MMIC itself. Typically, the parts were assembled in a leaded metal ceramic package, with epoxy or eutectic attach, and manually wedge bonded with gold wires for RF and bias connections. The package was then manually placed in a test fixture and held down by closing a clamp on the leads and body. The fixture was connected to the test equipment, typically a Vector Network Analyzer (VNA) or a scalar power meter, by Radio Frequency (RF) coaxial cables to present a 50 Ohms environment at the end of the coaxial cables. The sources of test uncertainty were numerous:

- Part placement in the package and bond wire loop profile, manually executed by an operator, lead to bond wire length differences and therefore matching variations for the Device Under Test (DUT).
- Package model inaccuracy and variability from package to package.
- RF and ground contacts through physical pressure of the clamp, applying force to the body of the package and the leads, with variable results for effective lead inductance and resistance, and potential oscillations especially at microwave frequencies.
- Fixture de-embedding empirical model for the connectors and transmission lines used on the RF ports.
- Calibration of the test equipment at the connectorized interface between the RF cables and the test fixture, not at the part or package test planes.

Most of these technical uncertainties arise because the calibration plane is removed from the product plane and the intermediate connection is not well characterized or not reproducible.

The main drawbacks of fixtured tests from a customer and business perspective were:

- Inability to test the very product shipped, only a "representative" sample is used due to the destructive nature of the approach. Especially for MMICs where the yield loss can be significant, this can lead to the rejection of many defective modules and products after assembly, at a large loss to the user.

- Cost of fixtured test; sacrificing parts and packages used for the test.

- Long cycle time; typically a day or two are needed for the parts to make it through assembly.

- Low rate production test; part insertion in a fixture is practically limited to a part per minute.

A first step was to develop test fixtures for bare die that could be precisely characterized. One solution was a modular fixture, where the die is mounted on an insert of identical length, which is sandwiched between two end pieces with transmission line and connector. The two end pieces can be fully characterized with a VNA to the end point of the transmission lines by Short-Open-Load-Thru (SOLT) or Thru-Reflect-Line (TRL) calibrations; wire bonding to preset inserts or between the two end-pieces butted together. Then the die is attached to the insert, assembled in between the end pieces, and wire bonded to the transmission lines. This approach became the dominant one for precise characterization and model extraction. The main advances were removal of die placement, package, lead contact and fixture as sources of variability, at the expense of a complex assembly and calibration process. The remaining limitations are bond loop variation, and destructiveness, and the length and cost of the approach, preventing its use in volume applications such as statistical model extraction or die acceptance tests.

19.1.2   On-Wafer Test Enabler: Coplanar Probes

The solution to accurate, high volume microwave testing of MMICs came from Cascade Microtech, the first company to make RF and microwave probes commercially available, along with extensive application support; their history and many useful application notes are provided on their Website (*www.cascademicrotech.com*). On-wafer test was common place for DC and digital applications, with high pin count probe cards available, based upon needles mounted on metal or ceramic blades. Although a few companies had developed RF frequency probes for their internal use, they relied on shortened standard DC probes, not the coplanar Ground-Signal-Ground (G-S-G) structure of Cascade Microtech's probes, and were difficult to characterize and use at microwave frequencies. The breakthrough idea to use a stable GSG configuration up to the probe tip enabled a reproducible 50 Ohms match to the DUT, leading to highly reproducible, nondestructive microwave measurements at the wafer level.[1,2] All intermediate interconnects were eliminated, along with their cost, delay, and uncertainty, provided that the DUT was laid out with the proper GSG inputs and outputs. Calibration patterns (Short, Open, Load, Thru, Line Stub) available on ceramic substrates or fabricated on the actual wafers provided standard calibration to the probe tips.[3,4] A few years later, PicoProbe (*www.picoprobe.com*) introduced a different mechanical embodiment of the same GSG concept.

About the same time, automatic probers with top plates fitted with probe manipulators for Cascade Microtech's probes became available. Agilent (then Hewlett Packard) introduced the 8510 Vector Network Analyzer, a much faster and easier way to calibrate microwave test equipment, and 50 Ohms matched MMICs dominated microwave applications. These events combined to completely change the characterization and die selection process in the industry. By the late 1980s, many MMIC suppliers were offering wafer qualification based upon RF test results on standard transistor cells in a Process Control Monitor (PCM) and providing RF tested Known Good Dies (KGD) to their customers.

TABLE 19.1 On-Wafer RF Test Capabilities Evolution

Year	Product	Configuration	Test Capability	Equipment
1985	Amplifier	2-Port	18 GHz s-Parameters	ANA
1987	Amplifier	Switched Multi-Port	26 GHz s-Parameters	ANA + Switch Matrix
1989	LNA	2-Port	Noise Figure	ANA + Noise System
1990	HPA	2-Port	Pulsed Power	Pulsed Power ANA
1991	Amplifier	2-Port	Intermodulation	Spectrum Analyzer
1991	LNA	2-Port, Zin Variable	Noise Parameters	Active Source Pull, ANA
1992	Mixer	3-Port	Conversion Parameters	ANA, Spectrum Analyzer
1993	HPA	2-Port, Zout Variable	Load Power Contours	Active Load Pull, ANA
1995	T/R Module	Switched Multi-Port	40 GHz s-Par, NF, Power	ANA, Noise, Spectrum
1998	Transceiver	Multi-Port	Modulation Parameters	Vector Signal Analyzer
1999	Amplifier	2-Port	110 GHz s-Parameters	ANA

19.1.3   On-Wafer Test Capabilities

At first, RF on-wafer testing was used only for the s-parameter test, for two port devices up to 18 GHz. Parameters of interest were gain, reflection coefficients, and isolation. Soon RF switching was introduced to test complex MMICs in one pass, switching the two ANA ports between multiple DUT ports. Next came noise figure test on-wafer, using noise source and figure meter combined with ANA. Power test on-wafer required a new generation of equipment, pulsed vector analyzers, to become reliable, and provided pulsed power, power droop, and phase droop.[5] Soon many traditional forms of microwave test equipment were connected to the DUT through complex switching matrixes for stimuli and responses, such as multiple sources, amplifiers, spectrum analyzers, yielding intermodulation distortion products. Next came active source pull equipment, and later on active load pull,[6] from companies such as ATN Microwave (*www.atnmicrowave.com*) and Cascade Microtech. The maximum s-Parameter test frequency kept increasing, to 26 GHz, then 40 GHz, 50 GHz, and 75 GHz. In the late 1990s new parameters such as Noise Power Ratio (NPR) and Adjacent Channel Power Ratio (ACPR) were required and could be accommodated by digitally modulated synthesizers and vector signal analyzers (Table 19.1). Today, virtually any microwave parameter can be measured on-wafer, including s-parameters up to 110 GHz.

19.1.4   On-Wafer RF Test Applications

On-wafer test ease of use, reasonable cost, and extensive parameter coverage has led to many applications in MMIC development and production, from device design and process development to high volume test for Known Good Die (KGD). The main applications are summarized in Table 19.2. Of course, all of the devices to test need to have been designed with one of the standard probe pad layouts (S-G-S, G-S, or S-G) to allow for RF probing.

TABLE 19.2 On-Wafer RF Test Applications

Application	DUT	Technique	Test	Test time/DUT	Volume/year
FET Model Development	Standard transistor	Transistor library	PCM transistor	MMIC or transistor	Assembly or package
Statistical Model Extraction	Source or load pull	s-par, NF, PP, set load	s-parameters, 50 Ohms	s-parameters, NF, PP	s-parameters, NF, PP
Process Monitoring	Noise parameters, load contours	Small and large signal models	Small signal model	Test specification	Test specification
Know Good Die Test	10 min	1 min	10 s	10–30 s	10–60 s
Module or Carrier Test	100s	1000s	10,000s	100,000s	100,000s

1. Model development and statistical model extraction is often performed on design libraries containing one type of element, generally Field Effect Transistors (FET), but sometimes inductors or capacitors, implemented in many variations that are characterized to derive a parametric model of the element.[7] The parts must be laid out with G-S-G (or G-S only for low microwave frequencies) in a coplanar and/or microstrip configuration. This test task would have taken months ten years ago, and is now accomplished in a few days. The ability to automatically perform all these measurements on significant sample sizes has considerably increased the statistical relevance of the device models. They are stored in a statistical database automatically used by the design and yield simulation tools. This allows first pass design success for complex MMICs.

2. Process monitoring is systematically performed on production wafers, sample testing a standard transistor in a Process Control Monitor (PCM) realized at a few places on each wafer. The layout is in a coplanar configuration that does not require back-side ground vias and therefore can be tested in process. Each time, a small signal model is extracted. Very good agreement between the tested s-parameters and the calculated ones from the extracted model can be seen in Fig. 19.1. The results are used during fabrication for pass/fail screening of wafers on RF parameters, and supplement the statistical model data.

3. On-wafer test is a production tool for dies, typically 100% RF tested when sold as is — as KGD — or used in expensive packages or modules. This is the norm for high power amplifiers in expensive metal ceramic packages, MMICs for Transmit/Receive (T/R) modules, bumped parts for flip-chip assembly, and military applications. The RF parameters of interest are measured at a few points across the DUT bandwidth, as seen in Fig. 19.2, and used to make the pass/fail decision. The rejected dies on the wafer are either marked with an ink dot, or saved in an electronic wafer map, as seen in Fig.19.3, which is used by the pick-and-place equipment to pick the passing devices. Final RF test on-wafer is usually not performed on high volume products. These achieve high yields and are all assembled in inexpensive packages, therefore it is easier and cheaper to plastic package all parts on the wafer to test them on automatic handlers and take the yield at this point.

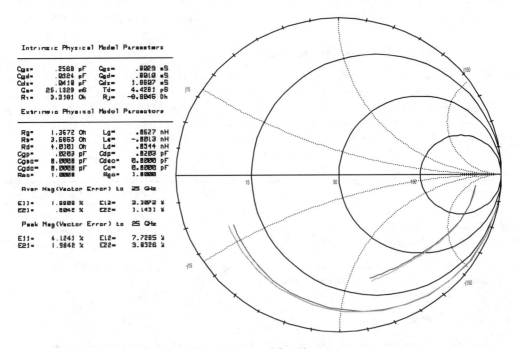

FIGURE 19.1   Equivalent circuit FET model extraction and fit with measurement.

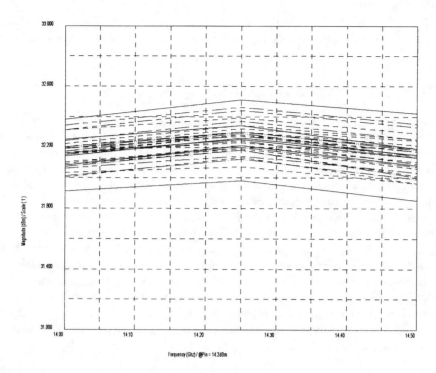

FIGURE 19.2   Pout response of Ku band PAs across a wafer.

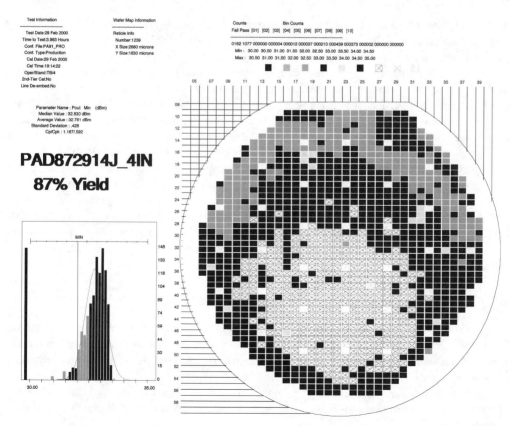

FIGURE 19.3   Wafer map of known good dies from on-wafer test.

4. The same "on-wafer" test application is used when testing packages, carriers, or modules manu-factured in array form on ceramic or laminate substrates, or leadless packages held in an array format by a test fixture.

19.2 Test Accuracy Considerations

In any test environment, three important variables to consider are accuracy, speed, and repeatability. The order of importance of these variables is based on price of the device, volume, and specification limits. High test speed is beneficial when it reduces the test cost-per-part and provides greater throughput without reaching an unacceptable level of accuracy and repeatability. Perfect accuracy would seem ideal, although in a high volume manufacturing environment "accuracy" is usually based on agreement between test results of two or more parties, primarily the vendor and end customer, for a specific product. The end customer, utilizing their available methods of measurement, usually defines most initial device specifications and sets the reference "accuracy," defining what parts work in the specific customer appli-cation. If due to methodology differences, a vendor's measurement is incompatible with that of a customer, yield and output can be affected without any benefit to the customer. It is not always beneficial, in this environment, to provide a more "accurate" method of measuring a product if the end customer is not testing it in the same fashion. Repeatability of the supplier measurement and correlation with the customer result are the more important criteria in that case.

Accuracy and repeatability considerations of any measurement system can be broken down into four primary parts, discussed in detail in the next sections.

19.2.1   Test Equipment Manufacturer

The manufacturer tolerances and supplied instrument error models are the first places to research when selecting the appropriate system. Most models will provide detail information on performance, dynamic range, and accuracy ratings of the individual instruments. Vendors like Agilent, Anritsu, Tektronix, and Boonton, to name a few, provide most hardware resources needed for automatic testing. There are many varieties of measurement instruments available on the market today. The largest single selection criterion of these is the frequency range. The options available diminish and the price increases dramatically as the upper frequency requirements increase. In the last decade many newer models with faster processors, countless menu levels, and more compact enclosures have come on the market making selections almost as difficult as buying a car. Most vendors will be competitive with each other in these matters. More important is support availability, access to resources when questions and problems arise, and software compatibility. Within the last decade many vendors have adopted a standard language structure for command programming of instru-ments known as SCPI (pronounced Skippy). This reduces software modification requirements when swapping instrumentation of one vendor with another. Some vendors have gone so far as to option the emulation of a more established competitor's model's instrument language to help inject their products into the market.

19.2.2   System Integration

Any system requiring full parametric measurement necessitates a complex RF matrix scheme to integrate all capabilities into a single function platform. Criteria such as frequency range, power levels, and device interface functionality drive the requirements of a RF matrix. Highly integrated matrices can easily exhibit high loss and poor matches that increase with frequency if care is not taken in the construction. These losses and mismatches can significantly degrade the accuracy of a system regardless of the calibration technique used. Assuming moderate power levels are to be used, frequency range is by far the most critical design consideration.

A system matrix must outperform the parts being tested on it. For complex systems requiring measurements such as intermodulation, harmonics, noise figure, or high port-to-port isolation, mechanical switches are the better alternative over solid state. Solid state switches would likely add their own performance limitations to the critical measurements being performed and cause erroneous results. Mechanical switches also have limitations to be considered. Although most mechanical switches have excellent transfer, isolation, and return loss characteristics, there is one issue that is sometimes overlooked. The return loss contact repeatability can easily vary by ± 5 milliunits and is additive based on the number of switches in series. To remove this error, directional couplers could be placed last in the matrix closest to the DUT and multiplexed to a common measurement channel within the network analyzer. This deviates from a conventional 2-port ANA configuration, but is worth consideration when measuring low VSWR devices.

19.2.3   Calibration Technique

Regardless of the environment, the level of system complexity and hardware resources can be minimized depending on the accuracy and speed requirements. Although the same criteria applies to both fixture and wafer environments, for optimum accuracy, errors can be minimized by focusing efforts on the physical limitations of the system integration, the most important being source and load matches presented to the DUT. By minimizing these parameter interactions, the accuracy of a scalar system can approach that of a full vector corrected measurement system.

The level of integration and hardware availability dictates the calibration requirements and capabilities of any test system. Simple systems designed for only one or two functions may necessitate assumptions in calibration and measurement errors. As an example, performing noise figure measurements on wafer using only a scalar noise figure system required scalar offsets be applied to attribute the loss of the probe environment, which cannot be dynamically ascertained through an automated calibration sequence. The same can also apply to a simple power measurement system consisting of only a RF source and a conventional power meter and assuming symmetry of input and output probes. These methods can and are used in many facilities, but can create large errors if care is not taken to minimize mismatch error terms that often come with contact degradation from repeated connections.

To obtain high accuracy up to the probe interface in a wafer environment requires a two-tier calibration method for certain measurements since it is usually difficult to provide a noise source or power sensor connection at the wafer plane. The most effective measurement tool for this second-tier calibration is a vector network analyzer. It not only provides full vector correction to the tips of the RF probes, but when the resulting vector measurements are used in conjunction with other measurement, such as noise figure and power, it can compensate for dynamic vector interactions between the measurement system and the device being tested. Equation (19.1), the vector relationship to the corrected input power (P_{A1}), and Eq. (19.2), the scalar offset normally applied in a simpler system, illustrate the relationship that would not be taken into account during a scalar power measurement when trying to set a specific input power level to the DUT. Usually a simple offset, P_{offset}, is added to the raw power measured at port A_0, (P_{A0}) to correct for the incident power at the device input A_1 (P_{A1}). This can create a large error when poor or moderate matches are present.

As an example, a device with a 10 dB return loss in a system with a 15 dB source match, not uncommon in a wafer environment, can create an error of close to ±0.5 dB in the input power setting when system interactions are ignored.

$$P_{A1} = \left| \frac{P_{A0}}{\left(1 - E_{sf} S_{11a}\right)^2} \right| \left(P_{offset}\right) \tag{19.1}$$

$$P_{A1} = P_{A0} \left(P_{offset}\right) \tag{19.2}$$

A similar comparison can be shown for the noise figure. Equations (19.3) and (19.4) illustrate the difference between the vector and scalar correction of the raw noise figure (R_{NF}) as measured by a standard noise figure meter. Depending on the system matches and the noise source gamma, the final corrected noise figure (C_{NF}) could vary considerably.

$$C_{NF} = R_{NF} + 10LOG\left(\frac{\left(\left|E_{10}^2\right|\right)\left(1-\left|G_{ns}^2\right|\right)}{\left(\left(\left|1-\left|E_{sf}+\left(E_{10}^2 G_{ns}/\left(1-G_{ns}E_{df}\right)\right)\right|^2\right|\right)\right)\left(\left|\left(1-E_{df}G_{ns}\right)^2\right|\right)}\right) \tag{19.3}$$

$$C_{NF} = R_{NF} + 10LOG\left(\left|E_{10}^2\right|\right) \tag{19.4}$$

For small signal correction, the forward path of the standard 12 Term, Full 2-Port Error model as given in Fig. 14.4 of Chapter 14 (Network Analyzer Calibration),[8] is applied. Equation (19.5) gives the derivation of the actual forward transmission (S_{21a}) from these error terms combined with raw measured data. By minimizing the mismatched terms E_{sf}, E_{lf}, E_{sr}, E_{lr}, E_{xf}, and E_{df}, detailed in section 4.2, Eq. (19.5) simplifies to Eq. (19.6). This simplified term is essentially the calculation used in standard scalar measurement systems and reflects an ideal environment. A further level of accuracy can be obtained when dealing with scalar systems that is very dependent on the type of device being tested. Looking at Eq. (19.5) it can be seen that in deriving S_{21a} many relationships between the error terms and measured values provide products that can further minimize errors based on the return loss components of the DUT as well as isolation in the reverse path. This makes an active device with good return losses and high reverse isolation a good candidate for a scalar measurement system when only concerned with gain as the functional pass/fail criteria. On the other hand, a switch or other control product has a potential for being a problem due to the symmetrical nature of the device if care is not taken to minimize the match terms. An even poorer candidate for a scalar system would be discrete transistors, which normally have not been tuned for optimum matching in the measurement environment. Figure 19.4 is an on-wafer measurement comparison of a discrete FET measurement using both full 2-port error correction as in Eq. (19.5) and the simplified scalar response Eq. (19.6) from 1 GHz to 25 GHz. The noticeable difference between these data sets is the "ripple" effect that is induced in the scalar corrected data, which stems from the vector sums of the error terms rotational relationship to the phase rotation of the measurement. Figure 19.5 shows the error terms E_{lf} and E_{lr} generated by multiple calibrations on the same vector test system used to measure the data in Fig. 19.4. Although the values seem reasonable, the error induced in the final measurement is significant.

This error is largely based on the poor input and output match of the discrete FET, as shown in Fig. 19.6, and their interaction with the system matches.

Figure 19.7, an example of better scalar-to-vector correlation, is an on-wafer measurement of a single pole double throw switch comparison using both full 2-port error correction as in Eq. (19.5) and the simplified scalar response Eq. (19.6) from 2 GHz to 20 GHz. Although the system matches are comparable to the discrete FET measurement, the device input and output return losses are both below 15 dB (Fig. 19.8). This product minimizes the errors induced by system to DUT interactions thus giving errors much smaller than that of the discrete FET measurement of Fig. 19.4.

$$\tag{19.5}$$

$$S_{21a} = \frac{\left(\left(S_{21m}-E_{xf}\right)/E_{tf}\right)\left(1+\left(S_{22m}-E_{dr}\right)\left(E_{sr}-E_{lf}\right)/E_{rr}\right)}{\left(1+\left(\left(S_{11m}-E_{df}\right)E_{sf}/E_{rf}\right)\right)\left(1+\left(\left(S_{22m}-E_{dr}\right)E_{sr}/E_{rr}\right)\right)-\left(\left(S_{21m}-E_{xf}\right)\left(S_{12m}-E_{xr}\right)E_{lf}E_{lr}/E_{tf}E_{tr}\right)}$$

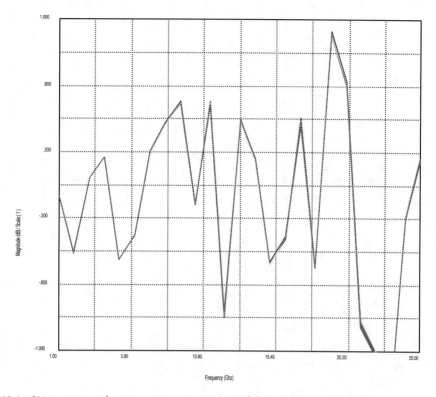

FIGURE 19.4   S21 vector to scalar measurement comparison of discrete FET (mismatched).

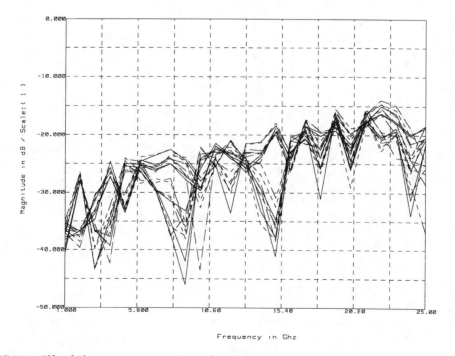

FIGURE 19.5   Elf and Elr error terms over a 5-month period.

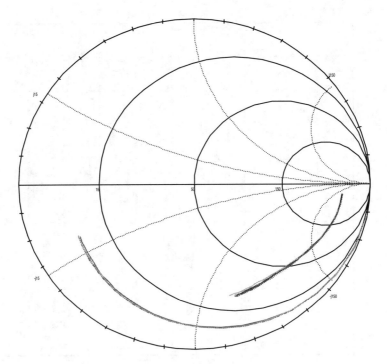

FIGURE 19.6   S11 and S22 of PCM FETs (mismatched) across a wafer.

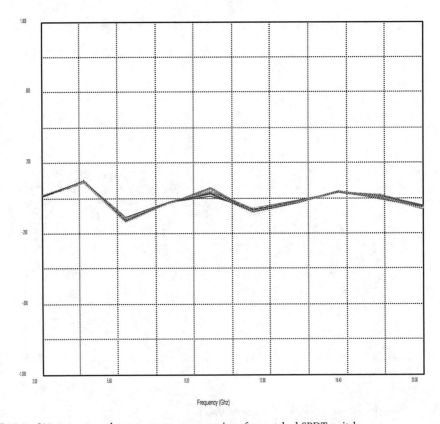

FIGURE 19.7   S21 vector to scalar measurement comparison for matched SPDT switch.

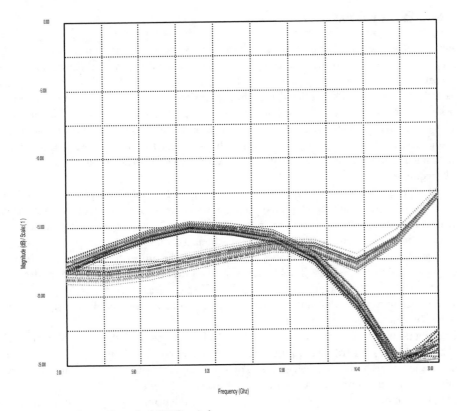

FIGURE 19.8   S11 and S22 of matched SPDT switch.

$$S_{21a} = \frac{S_{21m}}{E_{tf}} \bigg| E_{lf}, E_{sf}, E_{sr}, E_{lr}, E_{df} \rightarrow 0 \qquad (19.6)$$

19.2.4   Dynamic Range

Dynamic range is the final major consideration for accuracy of a measurement system. Dynamic range of any measurement instrument can be enhanced with changes in bandwidth or averaging. This usually degrades the speed of the test. A perfect example of this is a standard noise figure measurement of a medium gain LNA using an HP 8970 noise figure meter. Noise figure was measured on a single device one hundred times using 8 averages. The standard deviation is .02 dB, the cost for this is a 1.1-second measurement rate. By comparison, the same device measured with no averaging resulted in a standard deviation of .07 dB, but the measurement rate was less than 500 milliseconds.

Other methods can be applied to enhance the accuracy of the measurement without losing the speed. Placing a high gain 2nd stage LNA between the DUT and noise receiver will increase the dynamic range of the system and minimize the standard deviation obtained without losing the speed enhancement. These types of decisions should be made based on the parts performance and some experimentation.

Another obvious example is bandwidth and span setting on a spectrum analyzer. Sweep rates can vary from 50 milliseconds to seconds if optimization is not performed based on the requirements of the measurement. As in the noise measurement, this also should be evaluated based on the parts performance and some experimentation.

Highly customized systems that are optimized for one device type can overcome many dynamic range and mismatch error issues with additional components such as amplifiers, filters, and isolators. This can restrict or limit the capabilities of the system, but will provide speed enhancements and higher device output rates with minimal impact on accuracy.

19.3 On-Wafer Test Interface

On-wafer test of RF devices is almost an ideal measurement environment. Test interface technologies exist to support vector or scalar measurements. Common RF circuits requiring wafer test are: amplifiers, mixers, switches, attenuators, phase shifters, and coupling structures. The challenge is to select the interface technology or technologies that deliver the appropriate performance/cost relationship to support your product portfolio. Selection of test interface of wafer probes will be based on the measurements made and the desired product environment. It is common for high gain amplifiers to oscillate or for narrowband devices to shift frequency due to lack of bypass capacitors or other external components. It is recommended to consider wafer test during the circuit design stage to assure the circuit layout satisfies wafer test requirements.

A typical wafer probe system incorporates a test system, wafer prober, RF probes, and DC probes. Figure 19.9 shows a photograph of a typical production wafer prober. This prober has cassette feed, auto alignment, and is configured for a test system "test head." The test head connects to the test interface, which mounts in the hole on the left side of the machine. This prober uses a ring-type probe card as shown in Fig. 19.10. Conventional RF probes are mounted to the prober top plate using micro-manipulators arranged in quadrants. This allows access to each of the four sides of the integrated circuit. Figure 19.11 shows a two-port high frequency setup capable of vector measurements. Wafer prober manufacturers offer different top plates for different probe applications. Specification of top plate configuration is necessary for new equipment purchases.

Probe calibration standards are necessary to de-imbed the probe from the measurement. Calibrated open, short, and load standards are required for vector measurements. Probe suppliers offer calibrated standards designed specifically for their probes. For scalar measurements or when using complex probe assemblies, alternative calibration standards can be used, but with reduced measurement accuracy. Alternative calibration standards may be a custom test structure printed on a ceramic substrate or on a

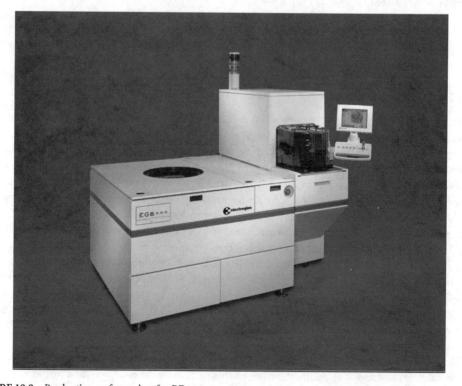

FIGURE 19.9   Production wafer prober for RF test.

FIGURE 19.10 Ring-type RF probe card.

FIGURE 19.11 RF probes mounted on manipulators.

wafer test structure. Scalar offsets can be applied for probe loss if you have a method of probe qualification prior to use. In general you have to decide if you are performing characterization or just a functionality screen of the device. This is important to consider early since measurement accuracy defines the appropriate probe technology, which places physical restrictions on the circuit layout.

When selecting the probe technology for any application you should consider the calibration approach, the maximum-usable frequency, the number of RF and DC connections required, the ability to support off-chip matching components, the cost of probes, and the cost of the calibration circuits. By understanding the advantages and limitations of each probe approach, an optimum technology/cost decision can be made. Remember that the prober top plate can be specified for ring frames or micro-manipulator type probes. Machine definition often dictates the types of probes to be used.

Traditional RF probes convert a coax transmission line into coplanar signal and ground probe points. This allows a coplanar or microstrip circuit with ground vias to be measured. These probes are offered as ground-signal and ground-signal-ground. They have been widely used for accurate high frequency measurements for many years. The ground-signal-ground probe offers improved performance above 12 GHz and can be used up to 100 GHz with proper construction. Probe spacing from signal to ground is referred to as the pitch. A common probe pitch is 0.006 in. Due to the small size, material selection significantly impacts RF performance and physical robustness. Many companies including Cascade Microtech and PicoProbe specialize in RF probes.

Cost considerations of probes are important. RF probes or membranes can cost anywhere from $300 to $3,000 each. This adds up quickly when you need multiple probes per circuit, plus spares, plus calibration circuits. When possible it is recommended to standardize the RF probe pitch. This will minimize setup time and the amount of hardware that has to be purchased and maintained. When custom probes are to be used, be prepared to incur the cost of probe and the calibration circuit development.

Wafer level RF testing using coplanar probing techniques can easily be accomplished provided the constraints of the RF probe design are incorporated into the circuit layout. This usually requires more wafer area be used for the required probe patterns and ground vias. These are standard and preferred design criteria for high frequency devices requiring on-wafer test. Devices without ground vias may require alternative interface techniques such as custom probes or membrane probes.

Although typical RF circuits have two or three RF ports and several DC, there are many that require increased port counts. Advanced probing techniques have been developed to support the need for increased RF and DC ports as well as the need for near chip matching and bypass elements. Probe manufacturers have responded by producing custom RF/DC probe cards allowing multiple functions per circuit edge. Figure 19.12 is an example of a single side four-port RF probe connected to a calibration substrate. Probe manufacturers have also secured the ability to mount surface mount capacitors on the end of probe tips to provide close bypass elements.

Another approach is Cascade Microtech's Pyramid Probe. It is a patented membrane probe technology that offers impedance lines, high RF and DC port count, and close location of external components. Figure 19.13 shows the pyramid probe with an off-chip bypass capacitor. One important aspect of the construction is that it incorporates an uninterrupted RF ground path throughout the membrane. This differs from the traditional coplanar probes that require the circuit to conduct the ground from one RF port to another. This allows for RF probing of lumped element circuits that do not utilize via holes and back side ground planes. This is becoming especially important to support developments such as chip scale packaging and multi-chip modules where the use of known good die is required for manufacturing.

For high volume devices where the circuit layout is optimized for the final package environment, considerations for on-wafer testing are secondary if not ignored. Products targeting the wireless market undergo aggressive die size reductions. Passive components such as capacitors, inductors, and resistors are often realized external to the integrated circuit. In this case the probes must be designed to simulate the packaged environment including the use of off chip components. Membrane technology is a good consideration for this. The membrane probe has the potential to emulate the package environment and external components that may be required at the final device level.

FIGURE 19.12   Four-part RF probe.

FIGURE 19.13   Cascade Microtech pyramid probe.

19.4 On-Wafer RF Test Benefits

The benefits of on-wafer RF testing are multiple and explain its success in the RF and microwave industry:

- Accuracy of RF test results with calibration performed at the probe tip, contact point to the DUT. The calibration techniques are now well established, supported by elaborate calibration standards, and easily implemented with software internally developed or purchased from the test equipment or probe vendors. This leads to accurate device models and higher first-pass design yields.

- Reproducibility of test results with stable impedance of the probe — be it 50 Ohms or a custom impedance — and automatic probe-to-pad alignment performed by modern wafer probers. Set probe placement on the pads during test and calibration is critical, especially above 10 GHz and for DUTs presenting a narrowband match.
- Nondestructive test of the DUT, allowing shipment of RF Known Good Dies to the user. This ability is key for multi-chip module or flip-chip onboard applications. The correlation between on-wafer and assembled device test results is excellent if the MMIC grounding is properly realized and the DC biasing networks are similar. For example, our experience producing 6 GHz power devices shows a maximum 0.2 dB difference in output power between wafer and module levels.
- Short cycle time for product test or statistical characterization and model extraction of library components, allowing for successful yield modeling and prediction.
- High throughput with complete automation of test and probing activities, and low cost, decreased by a factor of 10 in 10 years, to well below one dollar for a complex DUT today.

Wafer probing techniques are in fact gaining in importance today and are used for higher volume applications as Chip Scale Packages, Chip Size Packages (CSP), and flip chip formats become more common, bypassing the traditional plastic packaging step and test handler. Another increasing usage of on-wafer test is for parts built in array formats such as multi-chip modules or ball grid arrays. For these applications, robust probes are needed to overcome the low planarity of laminate boards. Higher speed test equipment such as that used with automatic handlers is likely to become more prevalent in wafer level test to meet volume needs. The probing process must now be designed to form a continuous flow, including assembly, test, separation, sorting, and packaging.

References

1. Strid, E.W., 26 GHz Wafer Probing for MMIC Development and Manufacture, *Microwave Journal*, August 1986.
2. Strid, E.W., On-Wafer Measurements with the HP 8510 Network Analyzer and Cascade Microtech Wafer Probes, RF & Microwave Measurement Symposium and Exhibition, 1987.
3. Cascade Microtech Application Note, On-Wafer Vector Network Analyzer Calibration and Measurements. (www.cascademicrotech.com)
4. Cascade Microtech Technical Brief TECHBRIEF4-0694, A Guide to Better Network Analyzer Calibrations for Probe-Tip Measurements. (www.cascademicrotech.com)
5. Mahon, J.R. et al., On-Wafer Pulse Power Testing, ARFTG Conference, May 1990.
6. Poulin, D.D. et al., A High Power On-Wafer Pulsed Active Load Pull System, *IEEE Trans. Microwave Theory and Tech.*, MTT-40, 2412–2417, Dec. 1992.
7. Dambrine, G. et al., A New Method for Determining the FET Small Signal Equivalent Circuit, *IEEE Trans. Microwave Theory and Tech.*, MTT-36, 1151–1159, July 1988.
8. Staudinger, J., Network Analyzer Calibration, *CRC Modern Microwave and RF Handbook*, CRC Press, Boca Raton, FL, chap. 4.2, 2000.

20

High Volume
Microwave Test

Jean-Pierre Lanteri
M/A-COM TycoElectronics

Christopher Jones
M/A-COM TycoElectronics

John R. Mahon
M/A-COM TycoElectronics

20.1 High Volume Microwave Component Needs

20.1.1 Cellular Phone Market Impact

High volume microwave test has emerged in the early 1990s to support the growing demand for GaAs RFICs used in cellular phones. Prior to that date, most microwave and RF applications were military and only required 10,000s of pieces a year of a certain MMIC type, easily probed or tested by hand in mechanical fixtures. For most companies in this industry, the turning point for high volume was around 1995 when some RFIC parts for wireless telephony passed the million per year mark. Cellular phones have grown to over 300 million units shipped in 1999 and represent 80% of the volume of microwave and RF ICs manufactured, driving the industry and its technology.

The cellular phone needs in terms of volume, test cost, and acceptable defect rate demanded new test solutions (Table 20.1) be developed that relied on the following key elements:

1. "Low" frequency ICs, first around 900 MHz and later on around 1.8 and 2.4 GHz, with limited bandwidth, allowing simpler device interfaces and fewer test points over frequency. Previously, MMICs were mostly military T/R module functions with frequencies ranging from 2 to 18 Ghz, with 30% or more bandwidths. They were tested at hundreds of frequencies, requiring specialized fast ramping Automatic Network Analyzers (ANA) such as Agilent's HP8510 or HP8530.
2. Standard plastic packages, based upon injection molding around a copper lead frame, to reach the low cost required in product assembly and test. Most early RFICs used large gull wing Dual In-line Packages (DIP), then Small Outline IC packages (SOIC), later Small Outline Transistor packages (SOT), and today's Micro Leadframe Flatpack (MLF).

TABLE 20.1 Microwave and RF IC Test Needs Evolution

Year	Product	Application	Package	Price	Volume	Test Time	Test Cost	Escape Rate
1991	T/R Module	Radar	Carrier	$200	10K/Y	1 min	$30	1%
1993	T/R Switch	Radar/Com	Ceramic	$40	100K/Y	30 sec	$4	0.5%
1995	RF Switch	Com	Plastic	$10	Mil/Y	10 sec	$1	0.1%
1997	RF MMIC	Com	SOIC	$3	Mil/M	3 sec	$0.30	0.05%
1999	RF MMIC	Com	SOT	$1	Mil/W	1 sec	$0.10	0.01%

3. Automatic handlers from the digital world, typically gravity fed, leveraging the plastic packages for full automation and avoiding human errors in bin selection. Previous metal or ceramic packages were mostly custom, bulky, and could only be handled automatically by pick-and-place type handlers, such as the one made by Intercontinental Devices in the early 1990s, barely reaching throughputs of a few hundred parts per hour.

4. Highly repeatable, accurate, and durable device contact interface and test board, creating the proper impedance environment for the device while allowing mechanized handling of the part. Most products before that were designed as matched to 50 Ohm impedance in and out, where cellular phone products will most often need to be matched in the user's system, and therefore on the test board. Adding to the difficulty, many handlers converted from digital applications hold the part in the test socket with a bulky mechanical clamp that creates ground discontinuities in the test board and spread the matching components further apart than designed in the part application.

5. Faster Automatic Network Analyzer (ANA) test equipment through hardware and software advances, later supplanted by specialized RFIC testers. The very high volumes reached by some parts, over a million pieces a week, allow dedication of a customized system to their testing to reduce measurement time and cost. Therefore the optimum test equipment first evolved from a powerful ANA-based system (HP8510, for example) with noise figure meter, spectrum analyzer, and multiport RF switch matrix, to an ad hoc set of bench-top equipment around an integrated ANA or ANA/spectrum analyzer. Next appeared products inspired from the digital world concept of the "electronic pin" tester, with RF functionality at multiple ports, such as the HP84000, widely used today.

6. Large databases on networked workstations and PCs for test results collection and analysis. The value of the information does not reside in the pass or fail outcome of a specific part, but in the statistical trends and operational performance measures available to company management. They provide feedback on employee training, equipment and calibration reproducibility, equipment maintenance schedules, handler supplier selection, and packaging supplier tolerances to name a few.

Although the high volume techniques described in this chapter would apply to most microwave and RF components, they are best fitted for products that do not require a broadband matched environment and that are packaged in a form that can be automatically tested in high-speed handlers.

20.1.2 High Volume RF Component Functions and Test Specifications

We will focus in this section on the different functions in the RF front end of a wireless phone to illustrate the typical products tested, their function, specification, and performance. The generic building blocks of a RF front end (Fig. 20.1) are switches (for antenna, Transmit/Receive (T/R), or band selection), input Low Noise Amplifiers (LNA), output Power Amplifiers (PA), up- and downconverters (typically comprising a mixer), Local Oscillator Amplifier (LOA), and Intermediate Frequency Amplifier (IFA). In most cases, these products are single band, either cellular or PCS, although new dual band components are appearing, requiring two similar tests in sequence, one for each band.

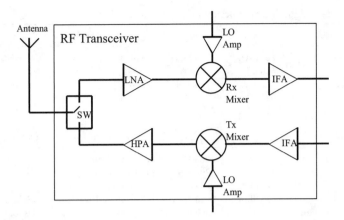

FIGURE 20.1   Typical RF transceiver building blocks.

The test equipment should therefore be capable of measuring DC parameters, network parameters such as gain or isolation, and spectral parameters such as IMD for most high volume products. Noise figure is required for LNAs and downconverters, and output power for HPAs. Typically, two types of RFIC testers will handle most parts, a general purpose RFIC for converters and eventually switches, and a specialized one for HPAs.

Typical specifications for the various parts are provided in Table 20.2. No specification is very demanding on the test instrument in absolute terms, but the narrow range of acceptance for each one requires outstanding reproducibility of the measurements, part after part. This will be the limiting factor in escape rate in our experience.

These specifications are dictated by the application and therefore largely independent of the technology used for fabrication of the RFIC. RFIC technology was predominantly GaAs Metal Semiconductor Field Effect Transistor (MESFET) until 1997, when GaAs Heterojunction Bipolar Transistor (HBT) appeared, soon followed by silicon products, in BiCMOS, SiGe BiCMOS, and CMOS technologies. The RF test is performed in a similar fashion for all implementation technologies of a given functionality.

20.1.3   High Volume Test Success Factors

The next sections will review in detail aspects of a successful back-end production of typical RF high volume parts; inexpensive, not too complex, packaged in plastic, produced at the rate of a million per week. The basic requirements addressed are:

- test equipment selection, balancing highest test speed with lowest test cost for the product mix
- automatic package handler keeping pace with the tester through parallel handling, and highly reliable
- part contactor performing at the required frequency, lasting for many contacts
- test software for fast set up of a new part with automatic revision control

Less obvious but key points for cost-effective high volume production are also discussed:

- tester, contactor, and test board calibration approach for reproducible measurements
- cost factors in a high volume test operation
- data analysis capabilities for relating yield to design or process
- test process monitoring tools, to ascertain the performance of the test operation itself

TABLE 20.2   Typical Product Specifications for High Volume Test

Switch Parameters	Min	Max	LNA Parameters	Min	Max
Frequency Range	800 MHz	1000 MHz	Frequency Range	800 MHz	1000 MHz
Control Leakage	–10 uA	10 uA	Current Consumption	8 mA	12 mA
Insertion Loss		0.5 dB	Linear Gain	15 dB	18 dB
Isolation	25 dB		Noise Figure		2 dB
Input IP3	60 dBm		Input IP3	–4 dBm	
PA Parameters	**Min**	**Max**	**Mixer Parameters**	**Min**	**Max**
Frequency Range	800 MHz	1000 MHz	Frequency Range	800 MHz	1000 MHz
Linear Current	160 mA	200 mA	IF Frequency Range	DC	100 MHz
Linear Gain	27 dB	35 dB	Conversion Loss		7.5 dB
Pout @ Pin = –1 dBm	25 dBm	30 dBm	LO to RF Leakage	38 dB	
Current @ Pin = –1 dBm		300 mA	1 dB Compression	21 dBm	
1 dB Compression	22.5 dBm		IMD @ Pin = –10 dBm	65 dBc	

20.2 Test System Overview

20.2.1   Hardware: Rack and Stack vs. High Speed IC Testers

Hardware considerations are based on the measurement requirements of your product set. To evaluate this, the necessary test dimensions should be determined. These dimensions can include but are not limited to swept frequency, swept spectrum, modulation schemes, swept power, and DC stimulus.

Commercially available hardware instruments can be combined to perform most RF/DC measurement requirements for manufacturing applications. These systems better known as "Rack and Stack" along with widely available third party instrument control software can provide a quick, coarse start-up for measurement and data collection, ideally suited for engineering evaluation. As the measurements become more integrated, the complexity required may exceed the generic capabilities of the third party software and may have to be supplemented with external software that can turn the original software into nothing more than a cosmetic interface.

To take the "Rack and Stack" system to a higher level requires a software expertise in test hardware communication and knowledge of the optimum sequencing of measurement events. Most hardware in a rack and stack system provides one dimension of competence, for example a network analyzer's optimum performance is achieved during a swept frequency measurement, a spectrum analyzer is optimized for frequency spectrum sweeps with fixed stimulus. Taking these instruments to a different dimension or repeating numerous cycles within their optimum dimension may not provide the speed required. Some instruments do provide multiple dimensions of measurement, but usually there is a setup or state change required that can add to the individual die test time. Another often-ignored aspect of these types of instruments is the overhead of the display processor, which is important in an engineering environment but an unnecessary time consumer in a manufacturing environment.

Commercially available high volume test systems usually provide equivalent speed in all dimensions by integrating one or two receivers with independently controlled stimulus hardware, unlike a network analyzer where the stimulus is usually linked to the receiver. These high-speed receivers combined with independently controlled downconverters, for IF processing, perform all the RF measurements that normally would take multiple instruments in a rack and stack system. Since these receivers are plug-in modules, whether for a PC back plane or a controlling chassis like a VXI card cage, they are also optimized for fast I/O performance and do not require a display processor, which can significantly impact the measurement speed. And since these receivers are usually based on DSP technology, complex modulation measurements such as ACPR can easily be made without additional hardware as would be required in most rack and stack systems.

TABLE 20.3   Speed Comparison of Rack and Stack and High Speed IC Tester

Repeat Count	Measurement/Stimulus	Rack and Stack		High Speed IC	
		Each	Total	Each	Total
3 Times	Set RF Source #1 Stimulus	100 mS	300 mS	50 mS	150 mS
3 Times	Set RF Source #2 Stimulus	100 mS	300 mS	50 mS	150 mS
12 Times	Set Analyzer to Span	250 mS	3000 mS	50 mS	600 mS
12 Times	Acquire Output Signal	50 mS	600 mS	40 mS	480 mS
	Total Time		4200 mS		1380 mS

In a normal measurement sequence of any complex device, the setting of individual stimulus far exceeds the time required to acquire the resulting output. A simple example of this would be a spectrum analyzer combined in a system with two synthesized sources to perform an intermodulation measurement at three RF frequencies. Accomplishing this requires extensive setting before any measurements can be made. Table 20.3 shows the measurement sequence and the corresponding times derived from a rack and stack system and a commercially available high speed IC measurement system for comparison. The measurement repeatability of these systems is equivalent for this example, therefore the bandwidth of the instrument setting is comparable.

As shown in the table, the acquisition of the output signal shows relatively no speed improvement with a difference of only 120 mS total. The most significant improvement is the setting of the acquisition span on the High Speed IC tester. This speed is the same as the setting of a RF stimulus since the only overhead is the setting of the LO source required for the measurement downconversion. The only optimization that could be performed with the rack and stack system would be higher speed RF sources having internal frequency list and power leveling capability. The change in span setting on a standard spectrum analyzer will always be a speed inhibitor since it is not its optimum dimension of performance.

From this type of table a point can be determined where the cost of a high-speed IC tester outweighs the speed increase it will yield. This criteria is based on complex multifunction devices that require frequent dimension hopping as described above. Other component types, such as filters requiring only broadband frequency sweeps in a single dimension, would show less speed improvement with an increase in frequency points since network analyzers are optimized for this measurement type.

Various vendors for high speed systems exist. Agilent Technologies (formerly Hewlett Packard), Roos Instruments, LTX, and Teradyne are just a few of the more well-known suppliers. The full system prices can range from a few hundred thousand dollars to well into the millions depending on the complexity/customization required.

A note of caution when purchasing a high speed IC tester: careful homework is warranted. Most IC testers are a three- to five-year commitment of capital dollars, and the one purchased should meet current and future product requirements. Close attention to measurement capabilities, hardware resources, available RF ports, DC pin count, and compatibility to existing test boards will avoid future upgrades, which are usually costly and delay time to market for new products if the required measurement capability is not immediately available.

20.2.2   System Software Integration

Software capabilities of third party systems require close examination, especially if it is necessary to integrate the outputs with existing resources on the manufacturing floor. Most high-speed IC testers focus on providing a test solution not a manufacturing solution. Network integration, software or test plan control, and data file organization is usually taken care of by the end customer. This software usually provides little operator input error checking or file name redundancy checking when dealing with multiple systems. The output file structure should have all the information required available in the file. Most third party systems provide an ASCII file output, which supports STDF (Standard Test Data Format), an industry standard data format

TABLE 20.4   Test Handler Manufacturers and Type

Manufacturer	Pick and Place	Gravity	Turret
Aetrium		X	X
Asseco	X	X	
Delta Daymark	X	X	
Exatron	X	X	
Intercontinental Microwave	X		
Ismeca	X		X
MultiTest	X	X	
Roos		X	

invented by Teradyne. As with the hardware, the software is fixed at a revision level. It is important to suggest improvements to the vendors to make the system more effective. Software revisions introduced by the vendor may not be available as fast as expected to correct observed deficiencies. It is still valuable to use the current revision level of the software to avoid known bugs and receive the best technical support.

20.2.3 RFIC Test Handlers

The primary function of the test handler is to move parts to the test site and then to sort them based on the test result. Package style and interface requirements will define what machines are available for consideration. The product will define the package and the handler is typically defined by the package. Common approaches include tube input — gravity handling, tray input — pick and place handling, and bulk input — turret handling. During the product design phase, selection of a package that works well with automation is highly recommended. The interface requirements are extremely critical for RF devices. Contact inductance, off chip matching components, and high frequency challenge our ability to realize true performance. The best approach is a vacuum pick up and plunge. This allows optimal RF circuit layout and continuous RF ground beneath the part.

Common test handler types and suppliers are listed in Table 20.4. Various options can be added to support production needs such as laser marking, vision inspection, and tape and reel. For specialized high volume applications, handlers are configured to accept lead frame input and tape and reel output providing complete reel-to-reel processing. When evaluating handlers for purchase, some extra time to identify process needs is very valuable. The machine should be configured for today's needs with the flexibility to address tomorrow's requirements. Base price, index time, jam rate, hard vs. soft tooling, conversion cost, tolerance to multiple package vendors, and vendor service should be considered. One additional quantitative rating is design elegance. An elegant design typically has the fewest transitions and fewest moving parts. Be cautious of machines that have afterthought solutions to hide their inherent limitations.

20.2.4 Contact Interface and Test Board

The test interface is comprised of a contactor and test board. The contactor provides compliance and surface penetration ensuring a low resistance connection is made to all device ports. Figure 20.2 shows a sectioned view of a pogo pin contactor. For RF applications the ideal contactor has zero electrical length and coupling capacitance. In the real-world contactors typically have 1 to 2 nH of series inductance and 0.2 to 0.4 pF of coupling capacitance. This can have significant impacts on electrical performance. Refer to Table 20.5 for a review of contactor manufacturers and parasitics. A more in-depth review of some available contactor approaches and suppliers is given in an article by Robert Crowley.[1] Parasitics of contactors can typically be compensated for in series ports using filter networks. Shunt ports however, such as an amplifier ground reference, challenge the use of contactors because the electrical length cannot be removed. The additional electrical length often shifts performance in magnitude or frequency beyond the range where scalar offsets can be used.

FIGURE 20.2 Pogo pin contactor.

TABLE 20.5 Test Contactor Manufacturers and Type

Manufacturer	Approach	Self Inductance	Mutual Inductance	Capacitance
Agilent	"YieldPro"	0.3 nH		0.17 pF
Aries	Microstrip Contact	0.01 pF	0.05 nH	0.04 pF
Exatron	Particle Interconnect	0.26 nH		0.024 pF
Johnstech International	"S" Contact	1.0 nH	0.2 nH	0.07 pF
Oz Tek	Pogo Pin	2.4 nH	0.4 nH	0.09 pF
Prime Yield	"Surface Mount Matrix"			
Synergetix	Pogo Pin	1.3 nH	0.1 nH	0.1 pF
Tecknit	"Fuzz Button"	2.7 nH	0.3 pF	0.3 pF

Note: Values supplied are typical values from manufacturer's catalog. Refer to manufacturer for specific information to support your specific needs.

Fine pitch packaging has increased the challenges associated with contactor manufacturing and lifetime. Packages such as TSSOP, SOT, SC70, and the new Micro Leadframe Flatpack (MLF) have pitches as small as 0.020 in. and may require a back-side ground connection. As contactor element size is reduced to support fine pitch packages, sacrifices are made in compliance and lifetime.

High frequency contactors are typically custom machined and assembled making them expensive. Suppliers are quoting $1000 to $4000 for a single contactor. If this expense is amortized over 500,000 parts, the cost per insertion is about one-half cent. This may be acceptable for some high value added part, but certainly not for all RF parts in general. Add to this the need to support your product mix and the need for spares and you will find that contactors can be more expensive than your capital test equipment. There is a true need for an industry solution to provide an affordable contactor with low parasitics, adequate compliance, tolerance to tin lead buildup.

The second half of the test interface is the test board, which interfaces the contactor to the test system. The test board can provide a simple circuit routing function or a matching circuit. It is common for RF circuits to utilize off-chip components for any non-active function. The production test board often requires tuning to compensate for contactor parasitics. This can result in a high Q matching circuit that increases measurement variability due to the interaction between the part, the contactor, and the test board. It is recommended to consider the contactor and test board during the product design cycle allowing the configuration to be optimized for robust performance.

20.3 High Volume Test Challenges

20.3.1 Required Infrastructure

The recommended facility for test of RF semiconductor components is a class 100,000 clean room with full ESD protection. RF circuits, especially Gallium Arsenide, are ESD sensitive to as little as 100 volts. Although silicon tends to be more robust than Gallium Arsenide, the same precautions should be taken. The temperature and humidity control aids test equipment stable operation and helps prolong the life of other automated equipment. Facility requirements include HVAC, lights, pressurized air and nitrogen, vacuum, various electrical resources, and network lines.

As volume increases the information system becomes a critical part of running the operation. The ideal system aids the decision process, communicates instructions, monitors inventory, tracks work in process, and measures equipment and product performance. The importance of information automation and integration cannot be overemphasized. It takes vision, skill, and corporate support to integrate all technical, manufacturing, and business systems.

The human resources are the backbone of any high volume operation. Almost any piece of equipment or software solution can be purchased, but it takes a talented core team to assemble a competitive operation and keep it running. Strengths are required in operations, software, and test systems, products, data analysis, and automation.

20.3.2 Accuracy and Repeatability Challenges

Measurement accuracy and repeatability are significant challenges for most high volume RF measurements. All elements of the setup may contribute to measurement inaccuracies and variability. The primary considerations are test system, the test board, the contactor, and the test environment.

For this discussion we will assume that all production setups are qualified for accuracy. This allows us to focus this discussion on variability.

20.3.2.1 Measuring Variability

Gauge Repeatability and Reproducibility (Gauge R&R) measurements can be used to measure variability. In this context the measurement system is referred to as the gauge. The gauge measurement is a structured approach that measures "x" products, "y" times, on "z" machines allowing the calculation of "machine" variability. Variability is reported in terms of repeatability and reproducibility. Repeatability describes variability within a given setup such as variability of contact resistance in one test lot. Reproducibility describes the variability between setups such as between different test systems or on different days. An overview of gauge measurement theory and calculations can be found in any statistical textbook.[2]

Figure 20.3 summarizes the sources of measurement variability within an automated test setup. The three locations are identified to allow easy gauge measurements.

Table 20.6 qualitatively rates the sources of measurement variability for repeatability and reproducibility. We can see that the system calibration and test board variations are large between setups while the contactor variations are large within a given setup. We will use these relationships in the case study to follow

Variability is expressed in terms of standard deviation. This allows normalized calculations to be made. For example, the variability of any measurement is a combination of the variability of the product and the gauge. This can be expressed as:

$$\sigma^2 \text{measured} = \sigma^2 \text{product} + \sigma^2 \text{gauge}$$

Based on Fig. 20.3 the total variability of an automated test can be described as:

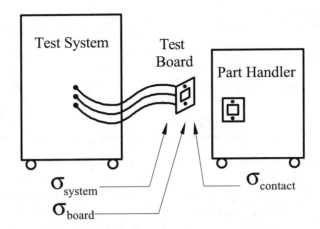

FIGURE 20.3 Sources of variability in an automated test setup.

TABLE 20.6 Repeatability and Reproducibility Comparison for the Complete Test Environment

Source	Description	Repeatability (within a setup)	Reproducibility (between setups)
Test System	Calibration	Low	High
Test Board	Matching Circuit	Low	High
Contactor	Contact Resistance	High	Low

TABLE 20.7 Gauge Test Design

"Machine"	# Machines	"Product"	# Products	# "Measurements"
Test System	4	Part soldered to test board	3	3
Test Board	4	Loose parts	3	3
Handler Contact	3	Loose parts	10	3

$$\sigma^2 total = \sigma^2 product + \sigma^2 system + \sigma^2 board + \sigma^2 contact$$

And for any expression of variability we can distinguish between repeatability and reproducibility as:

$$\sigma^2 gauge = \sigma^2 repeatability + \sigma^2 reproducibility$$

Table 20.7 recommends a gauge test design to characterize the components shown in Fig. 20.3. In this design we are measuring the "Machine" variation using "Products" and repetitive "Measurements." In all cases, stable product fixturing techniques are required for measurement accuracy. For the handler contact measurement, a single test setup is recommended.

20.3.2.2 Case Study

A low yielding product has been identified. Feedback from the test floor suggests the yield depends on the machine used and the day it was tested. These are the signs that yield is largely affected by variability. The following presents an analytical process that identifies the variability that must be addressed to improve yields.

Step 1: Identify the Failure Mode— For this product we found one gain measurement to be more sensitive than others. In fact this single parameter was driving the final yield result. This simplifies the analysis allowing us to focus on one parameter.

TABLE 20.8   Measurement and Product Variability

Data Source	Variability	Repeatability (one setup)	Reproducibility (across setups)	Total
Production Data	Total	1.6 dB	1.0 dB	1.89 dB
Gauge R&R	System	0.09 dB	0.23 dB	0.25 dB
	Board	0.13 dB	0.75 dB	0.76 dB
	Contact	0.54 dB	0.00 dB	0.54 dB
Calculation	Product	1.50 dB	0.62 dB	1.62 dB

Step 2: Quantify Measurement and Product Variability— A query of the test database showed 1086 production lots tested over a four-month span. For each production lot the average gain and standard deviation was reported. We define a typical gain value by taking the average of all production lot averages. Repeatability, or variability within any given test, was defined by finding the average of all production lot standard deviations. Reproducibility, or variability between tests, was found by taking the standard deviation of the average gain values for all production lots. Gauge R&R testing was conducted to determine the repeatability and reproducibility of the "system," "board," and "contact" as described previously. This allows calculation of product variability as shown in Table 20.8.

Step 3: Relate Variability to Yield— Relating variability to yield will define the product's sensitivity to the measurement. This will allow us to focus our efforts efficiently to maximize yield while supporting the customers' requirements. We can calculate yield to each spec limit using Microsoft Excel's NORMDIST function as follows:

Percent below upper spec limit = Y(USL) = NORMDIST(USL, μ, σ, 1)
Percent above lower spec limit = Y(LSL) = 1- NORMDIST(LSL, μ, σ, 1)

And we can calculate the final yield as follows:

Yield = Y(USL) − (1 − Y(LSL))

Prior to calculating yield we need to make some assumptions of how repeatability and reproducibility should be treated. For this analysis it is assumed that repeatability values will be applied to the standard deviation and reproducibility values will be used to shift the mean. Yield will be calculated assuming a worst case shift of the mean by one, two, and three standard deviations. The result will be plotted as Yield vs. Standard Deviation. The plot can be interpreted as the sensitivity of the parameter yield versus the measured variability of the test setup. This result is shown in Fig. 20.4 using the data in Table 20.8, the USL = 26.5 dB, the LSL = 21.5 dB, and the Average Gain = 23.1 dB.

Figure 20.4 quickly communicates the severity of the situation and identifies the test board as the most significant contributor. Looking at the product by itself we see that its yield can vary between 90% and 43%. Adding the test system variability makes matters worse. Adding the test board shows the entire process is not capable of supporting the specification. There are three solutions to this problem. Change the specifications, reduce the variability, or control the variability. Changing the specification requires significant customer involvement and communication. From the customer's point of view, if a product is released to production, specification changes are risky and avoided unless threat of line shutdown is evident. Reducing variability is where your effort needs to be focused. This may require new techniques and technology to achieve. In the process of reducing variability lessons learned can be applied across all products resulting in increased general expertise that can support existing and future products. The last method that can be applied immediately is to control variability. This is a process of tightly measuring and approving your measurement hardware from test systems to surface mount components. Everything gets qualified prior to use. This may take significant logistics efforts to put in place, but the yield improvements can be substantial.

This study is of an extreme case. To communicate this issue in a generic sense we can compare the same product case for various Cpk values. Figure 20.5 displays total variability vs. Cpk values of 0.5, 1.0,

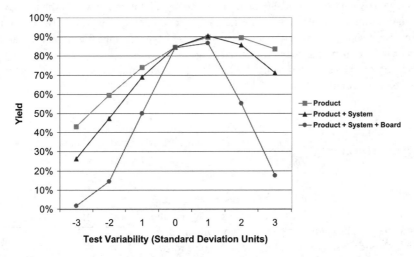

FIGURE 20.4 Yield vs. variability for test system elements

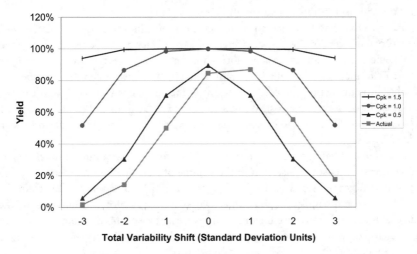

FIGURE 20.5 Yield vs. variability as function of Cpk.

and 1.5. We see that the case study shape is similar to the Cpk = 0.5 curve with a mean offset. It also shows that the process can be supported by a Cpk = 1.5 or greater. Anything less requires control of variability.

20.3.3 Volume and Cost Relationship

In general, cost of test reduces with increasing volume. Your ability to model available capacity will allow accurate estimation of cost. A generic capacity equation is:

$$Capacity = \frac{(Time\ Available)(Efficiency)}{Test\ Time + Handling\ Time} \tag{20.1}$$

Time available can be a day, month, or year as long as all time units are consistent. Efficiency is a measure of productive machine time. Efficiency accounts for all downtime within the time available due to equipment calibration, handler jams, material tracking operations, or anything else. For time intervals greater than a week you will find that efficiency converges. A typical range for initial estimates is 60% to 70%. Focus or lack of focus can swing the initial range by ±20%.

Cost of testing can be calculated using the estimated capacity and costs or with the actual cost and volume. The baseline result is shown in Eq. (20.2).

$$Unit\ Cost = \frac{Cost}{Volume} = \frac{Facility + Equipment + Labor + Materials}{(Capacity)(Yield)} \qquad (20.2)$$

Example Cost of Test: A complex part enters production. A $650,000 test system and a $350,000 handler are required and have been purchased. The estimated test and handling times are both one-half second. Based on Eq. (20.1) we can solve for the monthly capacity for varying efficiencies. This is shown in Table 20.9 for an average of 600 hours available per month.

We can see from Table 20.9 that there is a wide range of possible outcomes for capacity. In fact this is a very realistic result. If the objective was to install a monthly capacity of 1,600,000 parts, then the efficiency of operation defines if one or two systems are required. For this case an average of 74% efficiency will be required to support the job. Successful implementation requires consideration of machine design, vendor support, and operation skill sets to support 74% efficiency. If the efficiency cannot be met, then two systems need to be purchased.

TABLE 20.9   Efficiency vs. Capacity

Efficiency	Capacity
40%	864,000
60%	1,296,000
80%	1,728,000
100%	2,160,000

Efficiency has little impact on the cost of test unless the volume is increased. This can be shown by expanding our example to calculate cost. We will assume fixed facilities and capital costs; variable labor and material costs; and 100% yield to calculate the cost per test insertion. The assumptions are summarized in Table 20.10.

Cost per insertion calculations are shown in Table 20.11 for varying volume and efficiencies.

Columns compare the cost per insertion to the volume of test. The improvements in cost are due to amortizing facility and capital costs across more parts. The impact is significant due to the high capital cost of the test system and handler. Rows compare the cost per insertion as compared to efficiency.The difference in cost is relatively low since the only savings are labor. For this dedicated equipment example, improving efficiency only has value if the capacity is needed. Given efficiency or capacity, the cost of test can be reduced by increasing volume through product mix.

20.3.4   Product Mix Impact

Product mix adds several challenges such as tooling costs and manufacturing setup time. Tooling costs include test boards, mounting hardware, product standards, documentation, and training. These costs can run as high as $10,000 or as low as the documentation depending on product similarity and your approach to standardization. Tooling complexity will ultimately govern your product mix through resource limitations. Production output, on the other hand, will be governed by setup time. Setup time is the time to break down a setup and configure for another part number. This can involve test system calibration, test board change and/or handler change. Typical setup time can take from ten minutes to four hours. The following example explores product mix, setup time, and volume.

Example: Setup Time — Assume that setup can vary between ten minutes and four hours, equal volumes of four products are needed, test plus handing time is 1.0 second, and the efficiency is 60%. Calculate the optimum output assuring deliveries are required at monthly, weekly, or daily intervals. To

TABLE 20.10   Cost Assumptions

Cost	Assumption	Fixed or Volume Dependent
Facility	$ per square foot of floor space	Fixed
Capital	3 year linear depreciation	Fixed
Labor	Labor and fringe	Volume Dependent
Materials	General Consumables	Volume Dependent
Yield	Not used	

TABLE 20.11   Cost vs. Volume vs. Efficiency

Efficiency/Volume	100%	90%	80%	70%	60%
400,000	$0.096	$0.097	$0.098	$0.100	$0.101
800,000	$0.053	$0.054	$0.055	$0.056	$0.058
1,200,000	$0.038	$0.039	$0.040	$0.042	$0.043
1,600,000	$0.031	$0.032	$0.033	N/A	N/A
2,000,000	$0.027	N/A	N/A	N/A	N/A

TABLE 20.12   Monthly Capacity of Four Products with Varying Setup Time and Delivery Intervals

Setup/Delivery	10 min.	30 min.	1 hour	2 hours	4 hours
Monthly	1,294,531	1,291,680	1,287,360	1,278,720	1,261,440
Weekly	1,290,125	1,278,720	1,261,440	1,226,880	1,157,760
Daily	1,251,936	1,166,400	1,036,800	777,600	259,200

do this we subtract four setup periods from the delivery interval, calculate the test capacity of the remainder of the interval, and then normalize to one-month output. Table 20.12 summarizes the results.

As you may have expected, long setup times and regular delivery schedules can significantly reduce capacity. When faced with a high-mix environment everything needs to be standardized from fixturing to calibration files to equipment types and operating procedures.

20.4 Data Analysis Overview

20.4.1 Product Data Requirements and Database

Tested parameters for average RF devices can range from as little as 3 to as many as 30 depending on the functional complexity. In a high volume environment, where output can reach over 500,000 devices daily with a moderate product mix, methods to monitor and evaluate performance criteria have to provide efficient access to data sets with minimal user interaction. Questions such as "How high is the yield?" and "What RF parameters are failing most?" are important in any test facility, but can be very difficult to monitor and answer as volumes grow.

Many arguments have been made concerning the necessity of collecting parameter information on high yielding devices. To answer the two questions asked above, only limited information need be gathered. Most testers are capable of creating bin summary reports that can assign a bin number to a failure mechanism and output final counts to summarize the results.

The "binning" method may yield enough information for many circuits, but will not give insight into questions about tightness of parameter distributions, insufficient (or over-sufficient) amount of testing, test limits to change to optimize the yield, or possible change in part performance. These can only be answered with full data analysis packages either supplied by third parties or developed in-house. Standard histogram (Fig. 20.6) or wafer maps (Fig. 20.7) can answer the first question by providing distributions,

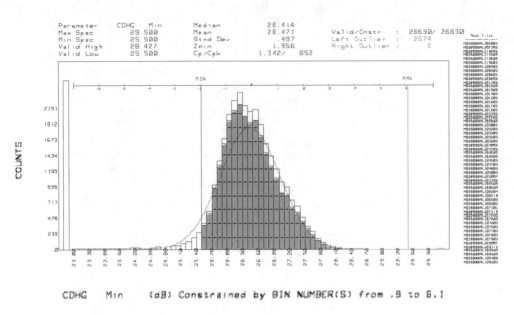

FIGURE 20.6   Histogram for distribution analysis.

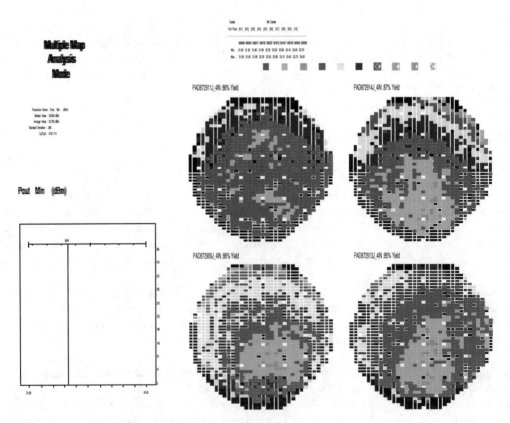

FIGURE 20.7   Wafer maps for yield pattern analysis.

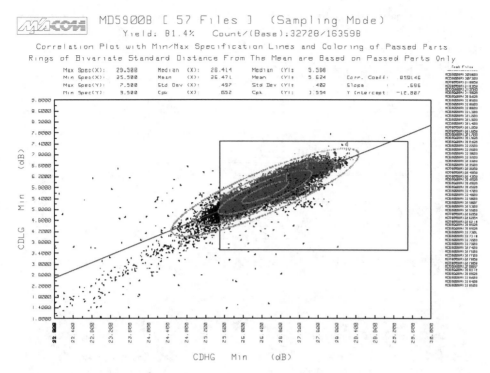

FIGURE 20.8   Scatter plot for parameter correlation analysis.

standard deviation, average values, and when supplied with limit specifications, CP and CPK values. XY or correlation plots (Fig. 20.8) can answer the second question, but when dealing with 20 or so parameters, this can be very time consuming to monitor.

The last questions require tools focusing on multivariable correlation and historical analysis. Changing of limit specifications to optimize yield is a tricky process and should not be performed on a small sample base. Nor should the interdependency of multiple parameters be ignored. Control charts such as Box Plots (Fig. 20.9) are ideal tools for monitoring performance variations over time.

These same tools when applied in real time can usually highlight problem parameters to help drill down to the problem at hand. Yield analysis tools displaying low yielding test parameters or single failure mechanisms are critical for efficient feedback analysis to the test floor as well as the product lines.

20.4.2   Database Tools

Analysis tools to quickly identify failure mechanisms are among the most important in high volume for quick feedback to the manufacturing floor. This requires that the database have full knowledge of not only the resulting data but also the high and low specifications placed on each individual parameter.

All databases, whether third party or custom, are depots for immense amounts of data with standard input and output utilities for organizing, feeding, and extracting information. The tools to display and report that information are usually independent of the database software.

Most third party database software packages can accommodate links to an exhaustive set of tools for extensive data analysis requirements. These external tools, again whether third party or custom, can be designed to provide fixed output reports for each device in question. But these databases usually require rigid data structures with fixed field assignments. Because of this, a high level of management for porting data, defining record structures, and organizing outputs is necessary when dealing with a continually changing product mix. Of course, if the application is needed for a few devices with compatible parameter tables, the management level will be minimal.

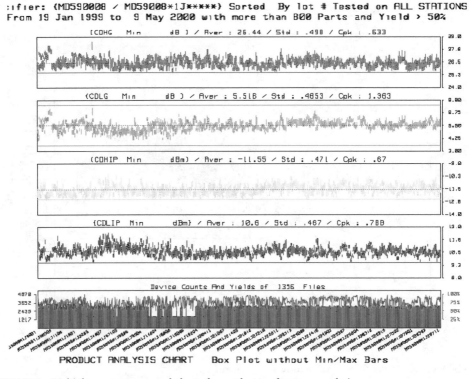

FIGURE 20.9   Multiple parameter control charts for product performance analysis.

The alternative is creating a custom database structure to handle the dynamics of a high product mix for your specific needs. This is neither easy or recommended when starting fresh in today's market since it requires in-house expertise in selecting the appropriate platforms and data structures. But if the capability already exists and can handle the increased demand, it may be a more cost-effective path considering the available resources.

An important note on the consideration of third party vs. in-house is the ability to implement software changes as the need arises. With third party platforms these changes may not be instituted until the next available revision or never if deemed highly custom. So be sure to select the appropriate mix to ensure this does not happen.

Regardless of the database option selected, data backups, network issues, and system integrity will still have to be maintained. Most systems today can use compression tools to maintain access to large amounts of data without the need to reload from externally archived tapes. Disc space is extremely cheap today. Even with high volume data collection requirements, information can be kept online for well over a year if necessary. More mature products can actually stop processing dense detailed information and only provide more condensed summary statistics used for tracking process uniformity.

20.4.3   Test Operation Data

To reduce the cost of testing and remain competitive in today's market, a constant monitoring of resource utilization is advantageous. A simple system utilization analysis can consist of a count test system, average cycle time of a device, and the quantity of parts in and parts out. This information is enough to get a rough idea of the average system utilization, but cannot give a complete picture when dealing with a large product base and package style mix. With detailed information of system throughput, pinpointing specific problem systems and focusing available resources to resolve the issues can be performed more efficiently. Output similar to the operational chart of Fig.20.10 can show information such as efficiency and system utilization within seconds to evaluate performance issues.

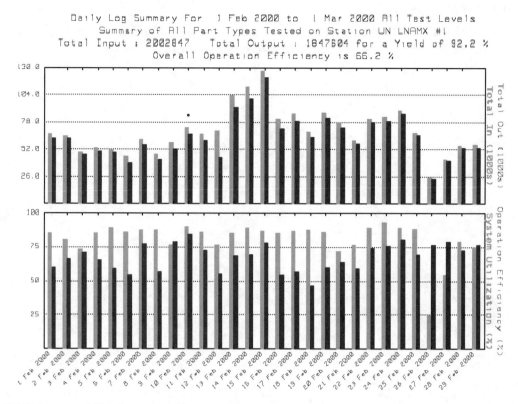

FIGURE 20.10   Yield and operation efficiency analysis tool.

Another important aspect of monitoring is the availability of resources to floor personnel to help them react to issues as fast as possible. During the course of a measurement sequence, potential problems could arise that require immediate response. A continuous yield display will react slowly to a degradation in contact or measurement performance, especially after thousands of devices have been tested. For this reason it is beneficial to have a sample or instantaneous yield reported during the test cycle to alert operators for quick reaction.

20.5 Conclusion

High volume microwave testing has become an everyday activity for all RFIC suppliers. Microwave test equipment vendors have developed equipment with acceptable accuracy and reproducibility, and satisfactory speed. Actual test software is robust and allows automatic revision tracking. Package handlers are improving although they are the throughput bottleneck for most standard RFICs, and do not accept module packages easily. Test contactors remain a technical difficulty, especially for high frequency or high power applications. In general, "hardware" solutions for microwave high volume testing exist today.

The remaining challenge is to reduce the customer's cost of quality and the supplier test cost with existing equipment. The ability to understand the customer specifications, the test system limitations, the test information available, and their interaction is key to test effectiveness improvement today. Analysis tools and methods to exploit the vast amount of data generated are essential to pinpoint the areas of possible improvement. These tools can highlight the fabrication process, the calibration process, the specification versus process limits, the package supplier, or the handler as the first area to focus upon for cost and quality improvement. This "software" side of people with the appropriate knowledge and tools to translate data into actionable information is where we expect the most effort and the most progress.

References

1. Crowley, R., Socket Developments for CSP and FBGA Packages, *Chip Scale Review,* May 1998.
2. Montgomery, D.C., Introduction to Statistical Quality Control, chap. 9.6, 455–460.

21

Computer-Aided Design of Passive Components

Daniel C. Swanson, Jr.
Bartley RF Systems

21.1 Introduction

Computer-aided design (CAD) of passive RF and microwave components has advanced slowly but steadily over the past four decades. The 1960s and 1970s were the decades of the mainframe computer. In the early years, CAD tools were proprietary, in-house efforts running on text-only terminals. The few graphics terminals available were large, expensive, and required a short, direct connection to the mainframe. Later in this period, commercial tools became available for use on in-house machines or through time sharing services. A simulation of a RF or microwave network was based on a combination of lumped and distributed elements. The elements were connected in cascade using ABCD parameters or in a nodal network using admittance- or Y-parameters. The connection between elements and the control parameters for the simulation were stored in a text file called a netlist. The netlist syntax was similar but unique for each software tool. The mathematical foundations for a more sophisticated analysis based on Maxwell's equations were being laid down in this same time period.[1-4] However, the computer technology of the day could not support effective commercial implementation of these more advanced codes.

The 1980s brought the development of the microprocessor and UNIX workstations. The UNIX workstation played a large role in the development of more sophisticated CAD tools. For the first time there was a common operating system and computer language (the C language) to support the development of cross-platform applications. UNIX workstations also featured large, bit mapped graphics displays for interaction with the user. The same microprocessor technology that launched the workstation also made the personal computer possible. Although the workstation architecture was initially more sophisticated, personal computer hardware and software has grown steadily more elaborate. Today, the choice between a workstation and a personal computer is largely a personal one. CAD tools in this time period were still based on lumped and distributed concepts. The innovations brought about by the cheaper, graphics-based hardware had largely to do with schematic capture and layout. Schematic capture replaced the netlist on the input side of the analysis and automatic or semi-automatic layout provided a quicker path to the finished circuit after analysis and optimization.

The greatest innovation in the 1990s was the emergence of CAD tools based on the direct solution of Maxwell's equations. Finally, there was enough computer horsepower to support commercial versions of the codes that had been in development since the late 1960s and early 1970s. These codes are in general labeled electromagnetic field-solvers although any one code may be based on one of several different numerical methods. Sonnet *em*,[5] based on the Method of Moments (MoM), was the first commercially viable tool designed for RF and microwave engineers. Only a few months later, Hewlett-Packard HFSS,[6] a Finite Element Method (FEM) code co-developed with Ansoft Corp., was released to the design community. All of these tools approximate the true fields or currents in the problem space by subdividing the problem into basic "cells" or "elements" that are roughly one tenth to one twentieth of a guide wavelength in size. For any guided electromagnetic wave, the guide wavelength is the distance spanned by one full cycle of the electric or magnetic field. The problem is to find the magnitude of the assumed current on each cell or the field at the junction of elements. The final solution is then just the sum of each small contribution from each basic unit. Most of these codes first appeared on UNIX workstations and then migrated to the personal computer, as that hardware became more powerful. In the later years of the 1990s, field-solver codes appeared that were developed on and for the personal computer. In the early years, the typical field-solver problem was a single discontinuity or some other structure that was small in terms of wavelengths. Today, groups of discontinuities, complete matching networks, or small parts of a multilayer printed circuit (PC) board are all suitable problems for a field-solver. Field-solver data in S-parameter form is typically imported into a circuit simulator and combined with lumped and distributed models to complete the analysis of the structure.

21.2 Circuit Theory Based CAD

CAD of low frequency circuits is at least 30 years old and microwave circuits have been analyzed by computer for at least 20 years. At very low frequencies, we can connect inductors, capacitors, resistors, and active devices in a very arbitrary way. The lumped lowpass filter shown in Fig. 21.1 is a simple example. This very simple circuit has only three nodes. Most network analysis programs will form an admittance matrix (Y-matrix) internally and invert the matrix to find a solution. The Y-matrix is filled using some fairly simple rules. A shunt element connected to node two generates an entry at Y_{22}. A series element connected between nodes two and three generates entries at Y_{22}, Y_{23}, Y_{32}, and Y_{33}. A large ladder network with sequential node numbering results in a large tri-diagonal matrix with many zeros off axis.

$$\mathbf{Y} = \begin{bmatrix} j\omega C_1 - j\dfrac{1}{\omega L_2} & j\dfrac{1}{\omega L_2} & 0 \\ j\dfrac{1}{\omega L_2} & j\omega C_3 - j\dfrac{1}{\omega L_2} - j\dfrac{1}{\omega L_4} & j\dfrac{1}{\omega L_4} \\ 0 & j\dfrac{1}{\omega L_4} & j\omega C_5 - j\dfrac{1}{\omega L_4} \end{bmatrix}$$

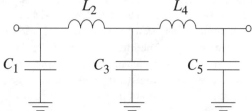

FIGURE 21.1   Lumped element lowpass filter or matching network.

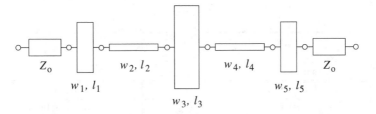

FIGURE 21.2   Distributed lowpass filter circuit. Step discontinuities are ignored.

The Y-matrix links the known source currents to the unknown node voltages. **I** is a vector of source currents. Typically the input node is excited with a one amp source and the rest of the nodes are set to zero. **V** is the vector of unknown node voltages. To find **V**, we invert the matrix **Y** and multiply by the known source currents.

$$\mathbf{I} = \mathbf{YV}$$

$$\mathbf{V} = \mathbf{Y}^{-1}\mathbf{I}$$

The time needed to invert an N × N matrix is roughly proportional to N^3. Filling and inverting the Y-matrix for each frequency of interest will be very fast, in this case, so fast it will be difficult to measure the computation time unless we specify a very large number of frequencies. This very simple approach might be good up to 1 MHz or so.

In our low-frequency model there is no concept of wavelength or even physical size. Any phase shift we compute is strictly due to the reactance of the component, not its physical size. There is also no concept of radiation; power can only be dissipated in resistive components. As we move into the HF frequency range (1 to 30 MHz) the real components we buy will have significant parasitics. Lead lengths and proximity to the ground plane become very important and our physical construction techniques will have a big impact on the results achieved.

By the time we reach VHF frequencies (50 to 150 MHz) we are forced to adopt distributed concepts in the physical construction and analysis of our circuits. The connections between components become transmission lines and many components themselves are based on transmission line models. Our simple lowpass circuit might become a cascade of low and high impedance transmission lines, as seen in Fig. 21.2.

If this was a microstrip circuit, we would typically specify the substrate parameters and the width and length of each transmission line. We have ignored the step discontinuities due to changes in line width in this simplified example. Internally, the software would use analytical equations to convert our physical dimensions to impedances and electrical lengths. The software might use a Y-matrix, a cascade of ABCD parameter blocks, or a cascade of scattering-parameter (S-parameter) blocks for the actual analysis. At the ports, we typically ask for S-parameters referenced to the system impedance.

Notice that we still have a small number of nodes to consider. Our circuit is clearly distributed but the solution time does not depend on its size in terms of wavelengths. Any phase shift we compute is directly related to the physical size of the network. Although we can include conductor and substrate losses, there is still no radiation loss mechanism. It is also difficult to include enclosure effects; there may be box resonances or waveguide modes in our physical implementation. There is also no mechanism for parasitic coupling between our various circuit models.

The boundary between a lumped circuit point of view and a distributed point of view can be somewhat fuzzy. A quick review of some rules of thumb and terminology might be helpful. One common rule of thumb says that the boundary between lumped and distributed behavior is somewhere between a tenth and an eighth of a guide wavelength. Remember that wavelength in inches is defined by

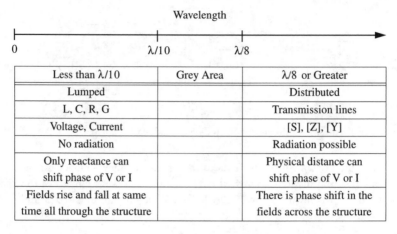

FIGURE 21.3   The transition between lumped and distributed behavior and some common terminology.

$$\lambda = \frac{11.803}{\sqrt{\varepsilon_{eff} \cdot f}}$$

where ε_{eff} is the effective dielectric constant of the medium and f is in GHz. At 1 GHz, $\lambda = 11.803$ inches in air and $\lambda = 6.465$ inches for a 50 ohm line on 0.014-inch thick FR4. FR4 is a common, low cost printed circuit board material for digital and RF circuits. In Fig. 21.3 we can relate the physical size of our structure to the concept of wavelength and to some common terminology. Again, the boundary between purely lumped and purely distributed behavior is not always distinct.

21.3 Field Theory-Based CAD

A field-solver based solution is an alternative to the previous distributed, circuit theory based approach. The field-solver takes a more microscopic view of any distributed geometry. Any field-solver we might employ must subdivide the geometry based on guide wavelength. Typically we need 10 to 30 elements or cells per guide wavelength to capture the fields or currents in our structure. Figure 21.4 shows a typical mesh generated by Agilent Momentum[7] for our microstrip lowpass filter example. Narrow cells are used on the edges of the strip to capture the spatial wavelength, or highly nonuniform current distribution across the width of the strips. This Method of Moments code has subdivided the microstrip metal and will solve for the current on each small rectangular or triangular patch. The default settings for mesh generation were used.

For this type of field-solver there is a strong analogy between the Y-matrix description we discussed for our lumped element circuit and what the field-solver must do internally. Imagine a lumped capacitor to ground at the center of each "cell" in our field-solver description. Series inductors connect these capacitors to each other. Coupling between non-adjacent cells can be represented by mutual inductances. So we have to fill and invert a matrix, but this matrix is now large and dense compared to our simple, lumped element circuit Y-matrix. For the mesh in Figure 21.4, N = 474 and we must fill and invert an N × N matrix.

One reason we turn to the field-solver is because it can potentially include all electromagnetic effects from first principles. We can include all loss mechanisms including surface waves and radiation. We can also include parasitic coupling between elements and the effects of compacting a circuit into a small space. The effects of the package or housing on our circuit performance can also be included in the field-solver analysis. However, the size of the numerical problem is now proportional to the structure size in

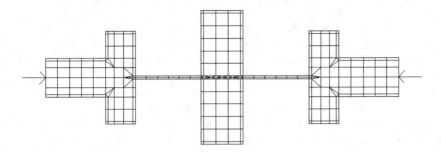

FIGURE 21.4   A typical model mesh for the distributed lowpass filter circuit. The number of unknowns, N is 474. *Agilent Momentum, ADS 1.3.*

wavelengths. The details of how enclosures are included in our analysis will vary from solver to solver. In some tools an enclosure is part of the basic formulation. In other tools, the analysis environment is "laterally open"; there are no sidewalls, although there may be a cover. One of the exciting aspects of field-solvers is the ability to observe fields and currents in the circuit, which sometimes leads to a deeper understanding of how the circuit actually operates. However, the size of the numerical problem will also be greater using a field-solver versus circuit theory, so we must carefully choose which pieces of global problem we will attack with the field-solver.

Although our discussion so far has focused on planar, distributed circuits there are actually three broad classes of field-solver codes. The 2D cross-section codes solve for the modal impedance and phase velocity of 1 to N strips with a uniform cross-section. This class of problem includes coupled microstrips, coupled slots, and conductors of arbitrary cross-section buried in a multilayer PC board. These tools use a variety of numerical methods including Method of Moments, the Finite Element Method, and the Spectral Domain Method. Field-solver engines that solve for multiple strips in a layered environment are built into several linear and nonlinear simulators. A multistrip model of this type is a building block for more complicated geometries like Lange couplers, spiral inductors, baluns, and many distributed filters. The advantage of this approach is speed; only the 2D cross-section must be discretized and solved.

The second general class of codes mesh or subdivide the surfaces of planar metals. The assumed environment for these surface meshing codes is a set of homogeneous dielectric layers with patterned metal conductors at the layer interfaces. Vertical vias are available to form connections between metal layers. There are two fundamental formulations for these codes, closed box and laterally open. In the closed box formulation the boundaries of the problem space are perfectly conducting walls. In the laterally open formulation, the dielectric layers extend to infinity. The numerical method for this class of tool is generally Method of Moments (MoM). Surface meshing codes can solve a broad range of strip- and slot-based planar circuits and antennas. Compared to the 2D cross-section solvers, the numerical effort is considerably higher.

The third general class of codes meshes or subdivides a 3D volume. These volume meshing codes can handle virtually any three-dimensional object, with some restrictions on where ports can be located. Typical problems are waveguide discontinuities, various coaxial junctions, and transitions between different guiding systems, such as transitions from coax to waveguide. These codes can also be quite efficient for computing transitions between layers in multilayer PC boards and connector transitions between boards or off the board. The more popular volume meshing codes employ the Finite Element Method, the Finite Difference Time Domain (FDTD) method, and the Transmission Line Matrix (TLM) method. Although the volume meshing codes can solve a very broad range of problems, the penalty for this generality is total solution time. It typically takes longer to set up and run a 3D problem compared to a surface meshing or cross-section problem. Sadiku[8] has compiled a very thorough introduction to many of these numerical methods.

21.4 Solution Time for Circuit Theory and Field Theory

When we use circuit theory to analyze a RF or microwave network, we are building a Y-matrix of dimension N, where N is the number of nodes. A typical amplifier or oscillator design may have only a couple of dozen nodes. Depending on the solution method, the solution time is proportional to a factor between N^2 and N^3. When we talk about a "solution" we really mean matrix inversion. In Fig. 21.5 we have plotted solution time as a function of matrix size N. The vertical time scale is somewhat arbitrary but should be typical of workstations and personal computers today.

When we use a MoM field-solver, a "small" problem has a matrix dimension of N = 300–600. Medium size problems may be around N = 1500 and large problems quickly get into the N = 2000–3000 range. Because of the N^2/N^3 effect, the solution time is impacted dramatically as the problem size grows. In this case we can identify two processes, filling the matrix with all the couplings between cells and inverting or solving that matrix. So we are motivated to keep our problem size as small as possible. The FEM codes also must fill and invert a matrix. Compared to MoM, the matrix tends to be larger but more sparse.

The time domain solvers using FDTD or TLM are exceptions to the N^2/N^3 rule. The solution process for these codes is iterative; there is no matrix to fill or invert with these solvers. Thus the memory required and the solution time grow more linearly with problem size in terms of wavelengths. This is one reason these tools have been very popular for radar cross-section (RCS) analysis of ships and airplanes. However, because these are time stepping codes, we must perform a Fast Fourier Transform (FFT) on the time domain solution to get the frequency domain solution. Closely spaced resonances in the frequency domain require a large number of time samples in the time domain. Therefore, time stepping codes may not be the most efficient choice for structures like filters, although there are techniques available to speed up convergence. Veidt[9] presents a good summary of how solution time scales for various numerical methods.

21.5 A Hybrid Approach to Circuit Analysis

If long solution times prevent us from analyzing complete circuits with a field-solver, what is the best strategy for integrating these tools into the design process? I believe the best approach is to identify the key pieces of the problem that need the field-solver, and to do the rest with circuit theory. Thus the final result is a "hybrid solution" using different techniques, and even different tools from different vendors. As computer power grows and software techniques improve, we can do larger and larger pieces of the problem with a field-solver. A simple example will help to demonstrate this approach. The circuit in

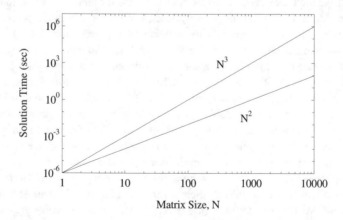

FIGURE 21.5   Solution time as a function of matrix size, N. Solution time for circuit simulators, MoM field-solvers, and FEM field-solvers is roughly proportional to N^3.

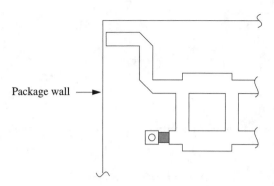

FIGURE 21.6   Part of an RF printed circuit board which includes a branchline coupler, a resistive termination to ground, and several mitered bends.

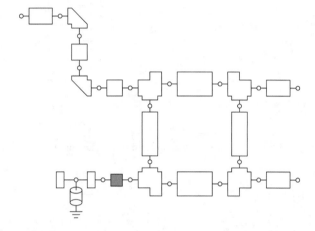

FIGURE 21.7   The layout in Fig. 21.6 has been subdivided for analysis using the standard library elements found in many circuit-theory-based simulations.

Fig. 21.6 is part of a larger RF printed circuit board. In one corner of the board we have a branchline coupler, a resistive termination, and several mitered bends.

Using the library of elements in our favorite linear simulator, there are several possible ways to subdivide this network for analysis (see Fig. 21.7). In this case we get about 21 nodes in our circuit. Solution time is roughly proportional to N^3, so if we ignore the overhead of computing any of the individual models, we would expect the solution to come back very quickly. But we have clearly neglected several things in our analysis. Parasitic coupling between the arms of the coupler, interaction between the discontinuities, and any potential interaction with the package have all been ignored. Some of our analytical models may not be as accurate as we would like, and in some cases a combination of models may not accurately describe our actual circuit. If this circuit were compacted into a much denser layout, all of the effects mentioned above would become more pronounced.

Each one of the circuit elements in our schematic has some kind of analytic model inside the software. For a transmission line, the model would relate physical width and length to impedance and electrical length through a set of closed form equations. For a discontinuity like the mitered bend, the physical parameters might be mapped to an equivalent lumped element circuit (Fig. 21.8), again through a set of closed form equations. The field-solver will take a more microscopic view of the same mitered bend discontinuity. Any tool we use will subdivide the metal pattern using 10 to 30 elements per guide wavelength. The sharp inside corner where current changes direction rapidly will force an even finer subdivision. If we want to solve the bend discontinuity individually, we must also connect a short length of series line to each port. Agilent

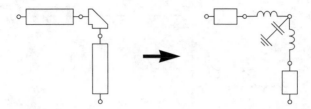

FIGURE 21.8   The equivalent circuit of a microstrip mitered bend. The physical dimensions are mapped to an equivalent lumped element circuit.

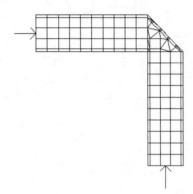

FIGURE 21.9   A typical MoM mesh for the microstrip mitered bend. The number of unknowns, N, is 221. *Agilent Momentum, ADS 1.3.*

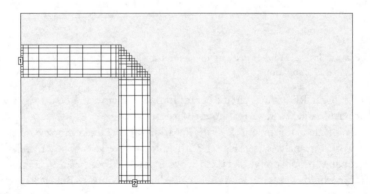

FIGURE 21.10   An analysis of the input line and the mitered bend in the presence of the package walls. The number of unknowns, N, is 360. *Sonnet **em** 6.0.*

Momentum generated the mesh in Fig. 21.9. The number of unknowns is 221. If the line widths are not variable in our design, we could compute this bend once, and use it over and over again in our circuit design.

Another potential field-solver problem is in the corner of the package near the input trace. You might be able to include the box wall effect on the series line, but wall effects are generally not included in discontinuity models. However, it is quite easy to set up a field-solver problem that would include the microstrip line, the mitered bend, and the influence of the walls. The project in Fig. 21.10 was drawn using Sonnet *em*. The box walls to the left and top in the electromagnetic simulation mimic the true location of the package walls in the real hardware. There are 360 unknowns in this simulation.

One of the more interesting ways to use a field-solver is to analyze groups of discontinuities rather than single discontinuities. A good example of this is the termination resistor and via[10,11] in our example circuit. A field-solver analysis of this group may be much more accurate than a combination of individual analytical models. We could also optimize the termination, then use the analysis data and the optimized geometry over and over again in this project or other projects. The mesh for the resistor via combination (Fig. 21.11) was generated using Sonnet *em* and represents a problem with 452 unknowns.

Our original analysis scheme based on circuit theory models alone is shown in Fig. 21.7. Although this will give us the fastest analysis, there may be room for improvement. We can substitute results in our field-solver for the elements near the package walls and for the resistor/via combination (Fig. 21.12). The data from the field-solver would typically be S-parameter files. This "hybrid" solution mixes field theory and circuit theory in a cost-effective way.[12] The challenge for the design engineer is to identify the critical components that should be addressed using the field-solver.

The hybrid solution philosophy is not limited to planar components; three-dimensional problems can be solved and cascaded as well. The right angle coax bend shown in Fig. 21.13 is one example of a 3D component that was analyzed and optimized using Ansoft HFSS.[13] In this case we have taken advantage of a symmetry plane down the center of the problem in order to reduce solution time. This component includes a large step in inner conductor diameter and a Teflon sleeve to support the larger inner conductor. After optimizing two

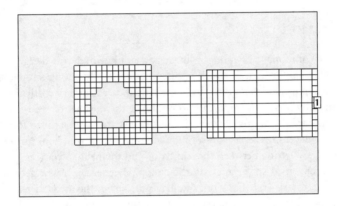

FIGURE 21.11 A MoM analysis of a group of discontinuities including a thin-film resistor, two steps in width, and a via hole to ground. The number of unknowns, N, is 452. *Sonnet em 6.0.*

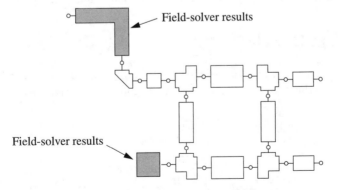

FIGURE 21.12 Substituting field-solver results into the original solution scheme mixes field-theory and circuit theory in a cost effective way.

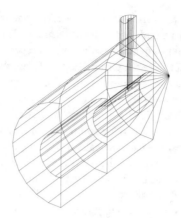

FIGURE 21.13   A right angle coax-to-coax transition that was optimized for return loss. The number of unknowns, N, is 8172. *An soft HFSS 7.0.*

dimensions, the computed return loss is greater than –30 dB. The coax bend is only one of several problems taken from a larger assembly that included a lowpass filter, coupler, amplifier, and bandpass filter.

21.6 Optimization

Optimization is a key component of modern linear and nonlinear circuit design. Many optimization schemes require gradient information, which is often computed by taking simple forward or central differences. The extra computations required to find gradients become very costly if there is afield-solver inside the optimization loop. So it is important to minimize the number of field-solver analysis runs. It is also necessary to capture the desired changes in the geometry and pass this information to the field-solver. Bandler et al.[14,15] developed an elegant solution to both of these problems in 1993. The key concept was a "data pipe" program sitting between the simulator and the field-solver (see Fig. 21.14). When the linear simulator calls for a field-solver analysis, the data pipe generates a new geometry file and passes it to the field-solver. In the reverse direction, the data pipe stores the analysis results and interpolates between data sets if possible. The final iterations of the optimization operate entirely on interpolated data without requiring any new field-solver runs. This concept was applied quite successfully to both surface meshing[16] and volume meshing solvers. The same basic rules that lead to successful circuit theory based optimization apply when a field-solver is in the loop as well. First, a good starting point leads to more rapid and consistent convergence. Second, it is important to limit the number of variables.

21.7 The Next Decade

The need for inexpensive wireless systems has forced the RF community to rapidly adopt low cost, multilayer PC board technology. In the simpler examples, most circuitry and components are mounted

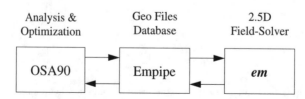

FIGURE 21.14   The first commercially successful optimization scheme which included a field-solver inside the optimization loop.

on the top layer while inner layers are used for routing of RF signals and DC bias. However, more complex examples can be found where printed passive components and discontinuities are located in one or more buried layers. Given the large number of variables in PC board construction it will be difficult for vendors of linear and nonlinear circuit simulators to support large libraries of passive models that cover all possible scenarios. However, a field-solver can be used to generate new models as needed for any novel layer stack up. Of course the user is also free to use the field-solver data to develop custom, proprietary models for his or her particular technology.

The traditional hierarchy of construction for RF systems has been a chip device, mounted to leaded package, mounted to printed circuit board located in system cabinet or housing. Today however, the "package" may be a multilayer Low Temperature Co-fired Ceramic (LTCC) substrate or a multilayer PC board using Ball Grid Array (BGA) interconnects. Thus the boundary between package and PC board has blurred somewhat. No matter what the technology details, the problem remains to transfer a signal from the outside world into the system, onto the main system board, through the package, and into the chip. And of course there is an analogous connection from the chip back to the outside world. From this point of view, the problem becomes a complex, multilevel passive interconnect that must support not only the signal currents but also the ground currents in the return path. It is often the ground return path that limits package isolation or causes unexpected oscillations in active circuits.[17] The high-speed digital community is faced with very similar passive interconnect challenges at similar, if not higher frequencies and typically much higher signal densities. Again, there is ample opportunity to apply field-solver technology to these problems, although practical problem size is still somewhat limited. The challenge to the practitioner is to identify and correct problems at multiple points in the signal path.

21.8 Conclusion

At very low frequencies we can use lumped element models to describe our circuits. Connection lengths and device parasitics are not issues. At higher frequencies we use distributed models to capture the effects of guide wavelength, but spurious couplings between elements and other effects due to circuit compaction are typically not captured. A field-solver can potentially capture all the macro and micro aspects of our circuit. It should capture spatial wavelength effects, guide wavelength, spurious couplings among elements, and interference among elements due to dense packing. Although the size of a practical field-solver problem is still somewhat small, there are many useful and cost effective problems that can be identified and solved using a combination of circuit theory based and field theory based CAD.

References

1. K. S. Yee, Numerical solution of initial boundary-value problems involving Maxwell's equations in isotropic media, *IEEE Trans. Ant. Prop.*, AP-14, 302–207, May 1966.
2. R. F. Harrington, *Field Computation by Moment Methods*, Macmillan, New York, 1968.
3. P. B. Johns and R. L. Beurle, Numerical solution of 2-dimensional scattering problems using a transmission-line matrix, *Proc. Inst. Electr. Eng.*, 118, 1203–1208, Sept. 1971.
4. P. Silvester, Finite element analysis of planar microwave networks, *IEEE Trans. Microwave Theory Tech.*, MTT-21, 104–108, Feb. 1973.
5. *em*™, Sonnet Software, Liverpool, NY.
6. HFSS, Hewlett-Packard, Santa Rosa, CA and Ansoft, Pittsburgh, PA.
7. Momentum, Agilent EEsof EDA, Santa Rosa, CA.
8. M. Sadiku, *Numerical Techniques in Electromagnetics*, CRC Press, Boca Raton, 1992.
9. B. Veidt, Selecting 3D electromagnetic software, *Microwave Journal*, 126–137, Sept. 1998.
10. M. Goldfarb and R. Pucel, Modeling via hole grounds in microstrip, *IEEE Microwave and Guided Wave Letters*, 1, 135–137, June 1991.

11. D. Swanson, Grounding microstrip lines with via holes, *IEEE Trans. Microwave Theory Tech.*, MTT-40, 1719–1721, Aug. 1992.

12. D. Swanson, Using a microstrip bandpass filter to compare different circuit analysis techniques, *Int. J. MIMICAE*, 5, 4–12, Jan. 1995.

13. HFSS, Ansoft Corp., Pittsburgh, PA.

14. J. W. Bandler, S. Ye, R. M. Biernacki, S. H. Chen, and D. G. Swanson, Jr., Minimax microstrip filter design using direct em field simulation, *IEEE MTT-S Int. Microwave Symposium Digest*, 889–892, 1993.

15. J. W. Bandler, R. M. Biernacki, S. H. Chen, D. G. Swanson, Jr., and S. Ye, Microstrip filter design using direct em field simulation, *IEEE Trans. Microwave Theory Tech.*, MTT-42, 1353–1359, July 1994.

16. D. Swanson, Optimizing a microstrip bandpass filter using electromagnetics, *Int. J. MIMICAE*, 5, 344–351, Sept. 1995.

17. D. Swanson, D. Baker, and M. O'Mahoney, Connecting MMIC chips to ground in a microstrip environment, *Microwave Journal*, 58–64, Dec. 1993.

22

Nonlinear RF and Microwave Circuit Analysis

Michael B. Steer
North Carolina State University

John F. Sevic
Ultra RF, Inc.

22.1 Introduction

The two most popular circuit-level simulation technologies are embodied in SPICE-like simulators, operating entirely in the time domain, and in Harmonic Balance (HB) simulators, which are hybrid time and frequency domain simulators. Neither is ideal for modeling RF and microwave circuits and in this chapter their concepts and bases of operation will be explored with the aim of illuminating the limitations and advantages of each. All of the technologies considered here have been implemented in commercial microwave simulators. An effort is made to provide sufficient background for these to be used to full advantage.

Simulation of digital and low frequency analog circuits at the component level is performed using SPICE, or commercial equivalents, and this has proved to be very robust. The operation of SPICE will be considered in detail later, but in essence SPICE solves for the state of the circuit at a time point and then uses this state to estimate the state of the circuit at the next time point (and so is referred to as a time-marching technique). The state of the circuit at the new time point is iterated to minimize error. This process captures the transient response of a circuit and the algorithm obtains the best waveform estimate. That is, the best estimate of the current and voltages in the circuit at each time point are obtained. The accurate calculation of the waveform in a circuit is what we want in low pass circuits such as digital and low frequency analog circuits. However with RF and microwave circuits, especially in communications, it is more critical to accurately determine the spectrum of a signal (i.e., the frequency components and their amplitudes) than the precise waveform. In part this is because regulations require strict control of spurious spectral emissions so as not to interfere with other wireless systems, and also because the generation of extraneous emissions compromises the demodulation and detection of communication signals by other radios in the same system. The primary distortion concern in radio is spectrum spreading or more specifically, adjacent channel interference. In-band distortion is also important especially with base station amplifiers where filtering can be used to eliminate spectral components outside the main channel. Distortion is largely the result of the nonlinear behavior of transmitters and so characterization of this phenomenon is important in RF design. In addition, provided that the designer has confidence in the stability and well-behaved transient response of a circuit, it is only necessary to determine its steady-state response. In order to determine the steady-state response using a time-marching approach, it is necessary to determine the RF waveform for perhaps millions of RF cycles, including the full transient interval, so as to extract the superimposed modulated signal. The essential feature of HB is that a solution form is assumed, in particular, a sum of sinusoids and the unknowns to be solved for are the amplitudes and phases of these sinusoids. The form of the solution then allows simplification of the equations and determination of the unknown coefficients. HB procedures work well when the signal can be described by a simple spectrum. However, it does not enable the transient response to be determined exactly.

In the following sections we will first look at the types of signals that must be characterized and identify the information that must be extracted from a circuit simulation. We will then look at transient SPICE-like simulation and HB simulation. Both types of analyses have restrictions and neither provides a complete characterization of an RF or microwave circuit. However, there are extensions to each that improve their basic capabilities and increase applicability. We will also review frequency domain analysis techniques as this is also an important technique and forms the basis of behavioral modeling approaches.

22.2 Modeling RF and Microwave Signals

The way nonlinear effects are modeled and characterized depends on the properties of the input signal. Signals having frequency components above a few hundred megahertz are generally regarded as RF or microwave signals. However, the distinguishing features that identify RF and microwave circuits are the design methodologies used with them. Communication systems generally have a a small operating fractional bandwidth — rarely is it much higher than 10%. Generated or monitored signals in sensing systems (including radar and imaging systems) generally have small bandwidths. Even broadband systems including instrumentation circuits and octave (and more) bandwidth amplifiers have passband characteristics. Thus RF and microwave design and modeling technology has developed specifically for narrowband systems.

The signals to be characterized in RF and microwave circuits are either correlated, in the case of communication and radar systems, or uncorrelated noise in the case of many imaging systems. We are principally interested in handling correlated signals as uncorrelated noise is nearly always very small and can be handled using relatively straightforward linear circuit analysis techniques. There are two families of correlated signals, one being discrete tone and the other being digitally modulated. In the following, three types of signals will be examined and their response to nonlinearities described.

22.2.1 Discrete Tone Signals

Single tone signals, i.e., a single sinewave, are found in frequency sources but such tones do not transmit information and must be modulated. Until recently, communication and radar systems used amplitude, phase, or frequency modulation (AM, PM, and FM, respectively) to put information on a carrier and transmission of the carrier was usually suppressed. These modulation formats are called analog modulation and the resulting frequency components can be considered as being sums of sinusoids. The signal and its response are then deterministic and a well-defined design methodology has been developed to characterize nonlinear effects. With multifrequency sinusoidal excitation consisting possibly of nonharmonically related (or non-commensurable) frequency components, the waveforms in the circuit are not periodic yet the nonlinear circuit does have a steady-state response. Even considering a single-tone signal (a single sinewave) yields directly usable design information. However, being able to model the response of a circuit to a multitone stimulus increases the likelihood that the fabricated circuit will have the desired performance.

In an FM modulated scheme the transmitted signal can be represented as

$$x(t) = \cos\left\{\left[\omega_c + \omega_i(t)\right]t\right\} + \sin\left\{\left[\omega_c + \omega_q(t)\right]t\right\} \tag{22.1}$$

where the signal information is contained in $\omega_i(t)$ and can be adequately represented as a sum of sinewaves. The term $\omega_q(t)$ is the quadrature of $\omega_i(t)$, meaning that it is 90° out of phase. The net result is that $x(t)$ can also be represented as a sum of sinusoids. Other forms of analog modulation can be represented in a similar way. The consequence of this is that all signals in a circuit with analog modulation can be adequately represented as comprising discrete tones.

With discrete tones input to a nonlinear circuit, the output will also consist of discrete tones but will have components at frequencies that were not part of the input signal. Power series expansion analysis of a nonlinear subsystem illustrates the nonlinear process involved. When a single frequency sinusoidal signal excites a nonlinear circuit, the response "usually" includes the original signal and harmonics of the input sinewave. We say "usually," because if the circuit contains nonlinear reactive elements, subharmonics and autonomous oscillation could also be present. The process is more complicated when the excitation includes more than one sinusoid, as the circuit response may then include all sum and difference frequencies of the original signals. The term *intermodulation* is generally used to describe this process, in which power at one frequency, or group of frequencies, is transferred to power at other frequencies. The term intermodulation is also used to describe the production of sum and difference frequency components, or intermodulation frequencies, in the output of a system with multiple input sinewaves. This is a macroscopic definition of intermodulation as the generation of each intermodulation frequency component derives from many separate intermodulation processes. Here a treatment of intermodulation is developed at the microscopic level.

To begin with, consider a nonlinear system with output $y(t)$ described by the power series

$$y(t) = \sum_{l=1}^{\infty} a_l x(t)^l \tag{22.2}$$

where $x(t)$ is the input and is the sum of three sinusoids:

$$x(t) = c_1 \cos(\omega_1 t) + c_2 \cos(\omega_2 t) + c_3 \cos(\omega_3 t). \tag{22.3}$$

Thus

$$x(t)^l = \left[c_1 \cos(\omega_1 t) + c_2 \cos(\omega_2 t) + c_3 \cos(\omega_3 t)\right]^l. \tag{22.4}$$

This equation includes a large number of components the radian frequencies of which are the sum and differences of ω_1, ω_2, and ω_3. These result from multiplying out the term $[\cos(\omega_1 t)]^k [\cos(\omega_3 t)]^{l-p}$. For example

$$\cos(\omega_1 t)\cos(\omega_2 t)\cos(\omega_3 t) = \Big[\cos(\omega_1 + \omega_2 + \omega_3)t + \cos(\omega_1 + \omega_2 - \omega_3)t + \cos(\omega_1 - \omega_2 + \omega_3)t$$

$$+ \cos(\omega_1 - \omega_2 - \omega_3)t\Big]/4 \tag{22.5}$$

where the (radian) frequencies of the components are, in order, $(\omega_1 + \omega_2 + \omega_3)$, $(\omega_1 + \omega_2 - \omega_3)$, $(\omega_1 - \omega_2 + \omega_3)$, and $(\omega_1 - \omega_2 - \omega_3)$. This mixing process is called intermodulation and the additional tones are called intermodulation frequencies with each separate component of the intermodulation process called an intermodulation product or IP. Thus when a sum of sinusoids is input to a nonlinear element additional frequency components are generated. In order to make the analysis tractable, the number of frequency components considered must be limited. With a two-tone input, the frequencies generated are integer combinations of the two inputs, e.g., $f = mf_1 + nf_2$. One way of limiting the number of frequencies is to consider only the combinations of m and p such that

$$|m| + |n| \le p_{MAX} \tag{22.6}$$

assuming that all products of order greater than p_{MAX} are negligible. This is called a triangular truncation scheme and is depicted as shown in Fig.22.1. The alternative rectangular truncation scheme is shown in Fig. 22.2 and is defined by

$$|m| \le m_{MAX} \quad \text{and} \quad |n| \le n_{MAX}. \tag{22.7}$$

With one-tone excitation, the spectra of the input and output of a nonlinear circuit consists of a single tone at the input and the original, fundamental tone, and its harmonics. Here intermodulation converts power at f_1 to power at DC (this intermodulation is commonly referred to as rectification), and to power at the harmonics ($2f_1, 3f_1, \ldots$), as well as to power at f_1. Simply squaring a sinusoidal signal will give rise to a second harmonic component. The measured and simulated responses of a class A amplifier operating at 2 GHz are shown in Fig. 22.3. This exhibits classic responses. At low signal levels the fundamental response has a slope of 1:1 with respect to the input signal level — corresponding to the linear response. Initially the second harmonic varies as the square of the input fundamental level and so has a 2:1 slope on the log-log plot. This is because the dominant IP contributing to the second harmonic level at low input powers is second order. Similarly the third harmonic response has a 3:1 slope because the dominant IP here is third order. As the input power increases, the second harmonic exhibits classic nonlinear behavior which is observed with many intermodulation tones and results from the production of a second,

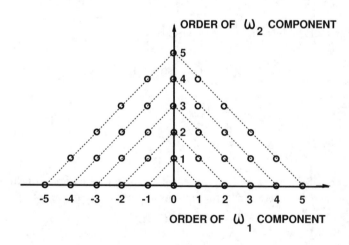

FIGURE 22.1   A triangular scheme for truncating higher order tones.

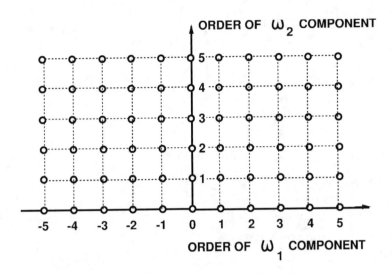

FIGURE 22.2 A rectangular scheme for truncating higher order tones. Here $m_{max} = 5 = n_{max}$.

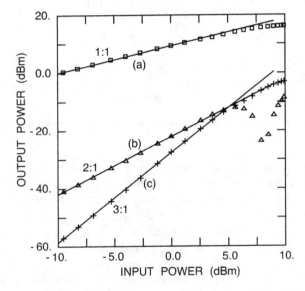

FIGURE 22.3 Measured (markers) and simulated (lines) response of a class A MESFET amplifier to a single tone input: (a) is the fundamental output; (b) is the second harmonic response; and (c) is the third harmonic response.

or more significant IP tone, which is due to higher order intermodulation than the dominant IP. In this situation, the dominant and additional IPs vectorially combine, with the result that the tone almost cancels out.

It is much more complicated to describe the nonlinear response to multifrequency sinusoidal excitation. If the excitation of an analog circuit is sinusoidal, then specifications of circuit performance are generally in terms of frequency domain phenomena, e.g., intermodulation levels, gain, and the 1 dB gain compression point. However, with multi-frequency excitation by signals that are not harmonically related, the waveforms in the circuit are not periodic, although there is a steady-state response often called quasi-periodic.

Consider the nonlinear response of a system to the two-tone excitation shown in Fig. 22.4. The frequencies f_1 and f_2 are, in general, nonharmonically related and components at all sum and difference

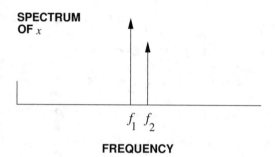

FIGURE 22.4   The spectrum of a two-tone signal.

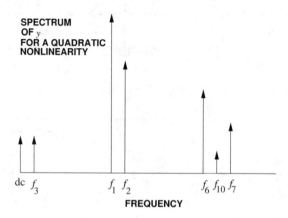

FIGURE 22.5   The spectrum at the output of a quadratic nonlinear system with a two-tone input.

frequencies $mf_1 + nf_2$, ($m, n = -\infty,\ldots,-1,0,1,\ldots,\infty$) of f_1 and f_2 will appear at the output of the system. If the nonlinear system has a quadratic nonlinearity, the spectrum of the output of the system is that of Fig. 22.5. With a general nonlinearity, the spectrum of the output will contain a very large number of components. An approximate output spectrum is given in Fig. 22.6. Also shown is a truncated spectrum that will be used in the following discussion. Most of the frequency components in the truncated spectrum of Fig. 22.6 have names: DC (f_6) results from rectification; $f_3, f_4, f_5, f_8, f_9, f_{10}$, and f_{11} are called intermodulation frequency components; f_4, f_5 are commonly called image frequencies, or "third order" intermods, as well; f_1, f_2 are the input frequencies; and f_7, f_8 are harmonics.

All of the frequencies in the steady-state output of the nonlinear system result from intermodulation — the process of frequency mixing. A classification of nonlinear behavior that closely parallels the way in which nonlinear responses are observed and specified is given below.

Gain Compression/Enhancement: Gain compression can be conveniently described in the time domain or in the frequency domain. Time domain descriptions refer to limited power availability or to limitations on voltage or current swings. At low signal levels, moderately nonlinear devices such as class A amplifiers behave linearly so that there is one dominant IP with a zero saturation term. As signal levels increase, other IPs become important as harmonic levels increase. Depending on the harmonic loading condition, these IPs could be in phase with the original IP contributing to gain enhancement or out of phase contributing to gainl compression.

Desensitization: Desensitization is the variation of the amplitude of one of the desired components due to the presence of another noncommensurable signal. This is an over-riding saturation effect affecting all output tones and comes out of the power series expansion.

Harmonic Generation: Harmonic generation is the most obvious result of nonlinear distortion and is identical to the process with a single-tone input.

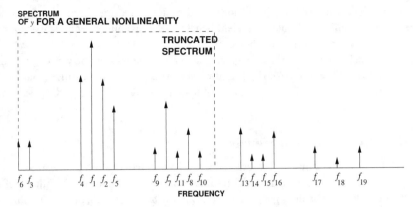

FIGURE 22.6 The approximate spectrum for a general nonlinearity with a two-tone input.

Intermodulation: Intermodulation is the generation of spurious frequency components at the sum and difference frequencies of the input frequencies. In the truncated spectrum $f_3, f_4, f_5, f_9, f_{10}$, and f_{11} are intermodulation frequencies. Numerically $f_4 = 2f_1 - f_2$ and so this intermodulation tone is commonly called the lower third order intermod. There are other IPs that can contribute to the "third order" intermod that are not due to third order intermodulation. A particularly important intermodulation process begins with the generation of the difference frequency component $f_3 = f_2 - f_1$ as a second order IP. This is also referred to as the baseband component, envelope frequency, intermediate frequency, or difference frequency. This component then mixes with one of the original tones to contribute to the level of the "third order" intermod, e.g., $f_4 = f_1 - f_3$, again a second order process. The corresponding contribution to the upper third order intermod f_5, i.e., $f_5 = f_2 + f_3$, can (depending on the baseband impedance) have a phase that differs from the phase of the f_4 contribution and, in general, the result is that there can be asymmetry in the lower and upper third order intermod levels as the various IPs, at their respective frequencies, add vectorially.

Cross-modulation: Cross-modulation is modulation of one component by another noncommensurable component. Here it would be modulation of f_1 by f_2 or modulation of f_2 by f_1. However, with cross-modulation, information contained in the sidebands of one non-commensurable tone can be transferred to the other non-commensurable tone.

Detuning: Detuning is the generation of DC charge or DC current resulting in change of an active device's operating point. The generation of DC current with a large signal is commonly referred to as rectification. The effect of rectification can often be reduced by biasing using voltage and current sources. However, DC charge generation in nonlinear reactances is more troublesome as it can neither be detected nor effectively reduced.

AM-PM Conversion: The conversion of amplitude modulation to phase modulation (AM-PM conversion) is a troublesome nonlinear phenomenon in high frequency analog circuits and results from the amplitude of a signal affecting the delay through a system. Alternatively, the process can be understood by considering that at higher input levels, additional IPs are generated at the fundamental frequency and when these vectorially contribute to the fundamental response, phase rotation occurs.

Subharmonic Generation and Chaos: In systems with memory effects, i.e., with reactive elements, subharmonic generation is possible. The intermodulation products for subharmonics cannot be expressed in terms of the input non-commensurable components. (Components are non-commensurable if they cannot be expressed as integer multiples of each other.) Subharmonics are initiated by noise, possibly a turn-on transient, and so in a steady-state simulation must be explicitly incorporated into the assumed set of steady-state frequency components. The lowest common denominator of the subharmonic frequencies then becomes the basis non-commensurable component. Chaotic behavior can only be simulated in the time domain. The nonlinear frequency domain methods as well as the conventional harmonic balance methods simplify a nonlinear problem by imposing an assumed steady-state on the nonlinear circuit solution problem. Chaotic behavior is not periodic and so the simplification is not valid in this case. Together with the ability to simulate transient behavior, the capability to simulate chaotic behavior is the unrivaled realm of time domain methods.

Except for chaotic behavior, all nonlinear behavior with discrete tones can be viewed as an intermodulation process with IPs (the number of significant ones increasing with increasing signal level) adding vectorially. Understanding this process provides valuable design insight and is also the basis of frequency domain nonlinear analysis.

22.2.2   Digitally Modulated Signals

A digitally modulated signal cannot be represented by discrete tones and so nonlinear behavior cannot be adequately characterized by considering the response to a sum of sinusoids. Nonlinear effects with digital are difficult to describe as the signals themselves appear to be random, but there is an underlying correlation. It is more appropriate to characterize a digitally modulated signal by its statistics, such as power spectral density, than by its component tones. Most current (and future) wireless communication systems use digital modulation, in contrast to first-generation radio systems, which were based on analog modulation. Digital modulation offers increased channel capacity, improved transmission quality, secure channels, and the ability to provide other value-added services. These systems present significant challenges to the RF and microwave engineer with respect to representation and characterization of digitally modulated signals, and also with respect to nonlinear analysis of digital wireless communication systems.

Amplifier linearity in the context of digital modulation is therefore most suitably characterized by measuring the degree of spectrum regeneration. This is done by comparing the power in the upper and lower adjacent channels to the power in the main channel: the adjacent-channel power ratio (ACPR). The spectrum of a digitally modulated signal is shown in Fig. 22.7. This is the spectrum of a finite bit length digitally modulated signal and not the smooth spectrum of an infinitely long sequence often depicted.

22.3 Basics of Circuit Modeling

The solution, or simulation, of a circuit is obtained by solving a number of network equations developed by applying Kirchoff's current law (KCL) and Kirchoff's voltage law (KVL). There are two basic methods for developing the network equations for DC analysis, or steady-state analysis of linear circuits with sinusoidal excitation, based on Kirchoff's laws. These are the nodal formulation and mesh formulation of the network equations. The nodal formulation is best for electronic circuits as there are many fewer nodes than there are elements connecting the nodes. The nodal formulation, specifically node-voltage analysis, requires that the current in an element be expressed as a function of voltage. Some elements cannot be so described and so there is not a node-voltage description for them. Then the modified nodal approach is most commonly used wherein every element that can be described by an equation for current in terms of voltages is described in this way, and only for the exceptional elements are other constitutive relations considered. However, the general formulation approach can be illustrated by considering node-voltage analysis.

The nodal formulation of the network equations is based on the application of KCL, which in its general form states that if a circuit is partitioned, then the total instantaneous current flowing into a partition is zero. This is an instantaneous requirement — physically it is only necessary that the net current flow be zero on average to ensure charge conservation. So this is an artificial constraint imposed by circuit analysis technology. The approach used in overcoming this restriction is to cast this issue as a modeling problem: it is the responsibility of the device modeler to ensure that a model satisfies KCL instantaneously. This results in many of the modeling limitations that are encountered. A general network is shown in Fig. 22.8. The concept here is that every node of the circuit is pulled to the outside of the main body of the network. The main body contains only the constitutive relations and the required external nodes have the connectivity information to implement Kirchoff's laws. The result is that the

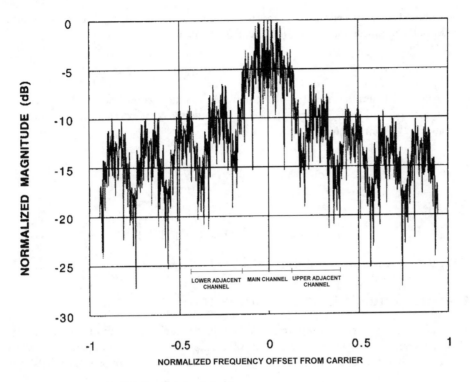

FIGURE 22.7 Spectrum of a digitally modulated signal.

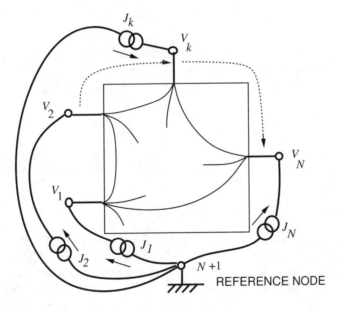

FIGURE 22.8 General network.

constitutive relations are contained in the main body, but the variables, the node-voltages, and the external currents are clearly separated. This representation of a network enables as uniform a treatment as possible. It makes it very easy to add one element at a time to the network as variables are already defined. Indeed this is how all general purpose simulators work and the network equations are built up by inspection. Initially the network is defined with nothing in the main body and only the variables defined. Then each element is considered in turn and the describing relations added to the evolving network equation matrix.

This representation serves us well when it comes to harmonic balance. Applying KCL to each of the nodes of the network the following matrix network equation is obtained:

$$\mathbf{YV} = \mathbf{J}. \tag{22.8}$$

Here $\mathbf{Y}$ is the nodal admittance matrix of the network, $\mathbf{V}$ is the vector of node voltages (i.e., voltages at the nodes each referred to the reference node), and $\mathbf{J}$ is the vector of external current sources at each node. Expanding the matrix equation:

$$
\begin{bmatrix}
y_{11} & y_{12} & \cdots & y_{1N} \\
y_{21} & y_{22} & \cdots & y_{2N} \\
\vdots & \vdots & \ddots & \vdots \\
y_{N1} & y_{N2} & \cdots & y_{NN}
\end{bmatrix}
\begin{bmatrix}
V_1 \\ V_2 \\ \vdots \\ V_N
\end{bmatrix}
=
\begin{bmatrix}
J_1 \\ J_2 \\ \vdots \\ J_N
\end{bmatrix}. \tag{22.9}
$$

We will see this utilized in the formulation of SPICE and HB analyses.

22.4 Time-Domain Circuit Simulation

The principal advantage of simulating circuits in the time domain is that it most closely resembles the real world. Phenomena such as chaos, instability, subharmonic generation, and parametric effects can be accurately simulated without the *a priori* knowledge of the spectral components of the signals in a circuit.

22.4.1   Direct Integration of the State Equations

The most direct method for analyzing nonlinear circuits is numerical integration of the differential equations describing the network. By applying Kirchoff's voltage and current laws and using the characteristic equations for the circuit elements (generally using the modified nodal formulation), the state equations can be written as a set of coupled first-order differential equations:

$$\dot{\mathbf{X}} = f(\mathbf{X}, t) \tag{22.10}$$

where, for example, the time derivative of a quantity such as voltage or current is a function of time and of the voltages and currents in the circuit. More generally the state equations are rearranged and written in the implicit form

$$g(\dot{\mathbf{X}}, \mathbf{X}, t) = 0 \tag{22.11}$$

where $\mathbf{X} = [X_1, X_2, \ldots, X_N]^T$ is a set of voltages and currents, typically at different nodes and different time instants. The general formulation of Eq. (22.11) is discretized in time and solved using a numerical integration procedure. This modeling approach can be used with many systems as well as circuits and was the only approach considered in the early days of circuit simulation (in the 1960s). Unfortunately, it was not robust except for the simplest of circuits. SPICE-like analysis, considered next, solves the same problem but in a much more robust way.

22.4.2   SPICE: Associated Discrete Circuit Modeling

SPICE is the most common of the time domain methods used for nonlinear circuit analysis. This method is fundamentally the same as that just described in that the state equations are integrated numerically, however the order of operations is changed. The time discretization step is applied directly to the equations describing the circuit element characteristics. The nonlinear differential equations are thereby converted to nonlinear algebraic equations. Kirchoff's voltage and current laws are then applied to form a set of algebraic equations solved iteratively at each time point.

Converting the differential equations describing the element characteristics into algebraic equations changes the network from a nonlinear dynamic circuit to a nonlinear resistive circuit. In effect, the differential equations describing the capacitors and inductors, for example, are approximated by resistive circuits associated with the numerical integration algorithm. This modeling approach is called associated discrete modeling or just companion modeling. The term "associated" refers to the model's dependence upon the integration method while "discrete" refers to the model's dependence on the discrete time value.

The numerical integration algorithm is the means by which the element characteristics are turned into difference equations. Three low order numerical integration formulas are commonly used: the Forward Euler formula, the Backward Euler formula, and the Trapezoidal Rule. A generalization of these to higher order is called the weighted integration formula from which the Gear Two method, available in some SPICE simulators, is derived. In all methods the aim is to estimate the state of a circuit at the next time instant from the current state of the circuit and derivative information. In one dimension and denoting the current state by x_0 and the next state by x_1, the basic integration step is

$$x_1 = x_0 + hx'. \tag{22.12}$$

The formulas differ by the method used to estimate x'.

In the Forward Euler Formula, $x' = x'_\phi$ is used and the basic numerical integration step [Eq. (8.58)] becomes

$$x_1 = x_0 + hx'_0. \tag{22.13}$$

Numerical integration using the forward Euler formula is called a predictor method as information about the behavior of the waveform at time t_0, x'_0, is used to predict the waveform at t_1.

In the Backward Euler Formula, $x' = x'_1$ is used and the discretized numerical integration equation becomes

$$x_1 = x_0 + hx'_1. \tag{22.14}$$

The obvious problem here is how to determine x'_1 when x_1 is not known. The solution is to iterate as follows: (1) assume some initial value for x_1 (e.g., using the Forward Euler formula); and (2) iterate to satisfy the requirement $x'_1 = f(x_1, t)$. Discretization using the Backward Euler formula is therefore called a predictor-corrector method.

In the Trapezoidal Rule, $x' = (x'_0 + x'_1)/2$ is used and the discretized numerical integration equation becomes

$$x_1 = x_0 + h\left(x'_0 + x'_1\right)/2. \tag{22.15}$$

So the essence of the trapezoidal rule is that the slope of the waveform is taken as the average of the slope at the beginning of the time step and the slope at the end of the time step determined using the Backward Euler formula.

There is a significant difference in the numerical stability, accuracy, and run times of these methods, although all will be stable with a small enough step size. Note that stability is a different issue than whether or not the correct answer is obtained. The Backward Euler and Trapezoidal Rules will always be stable and these are the preferred integration methods. The Forward Euler method of discretization does not always result in a numerically stable method. This can be understood by considering that the Forward Euler method always predicts the response into the future and does not improve on the guess using other information that can be obtained. The Backward Euler and Trapezoidal Rule approaches use a prediction of the future state of a waveform, but then require iteration to correct any error and use derivative information as well as instantaneous information to achieve this. Generally, when any simulation strategy is first developed, predictor methods are used. However, in the long run, predictor-corrector methods are always adopted as these have much better overall performance in terms of stability and accuracy but do require much more development effort. Except for the Forward Euler method, none of the other methods are clearly the best choice in all circumstances, and experimentation should occur. Generally, we can say that for RF and microwave circuits that have resonant bandpass-pass characteristics, the Trapezoidal Rule tends to result in an over-damped response and the Backward Euler method results in an under-damped response. The effect of this on accuracy, the prime requirement, is not consistent and must be investigated for a specific circuit.

22.4.3   Associated Discrete Model of a Linear Element

The development of the associated discrete model (ADM) of an element begins with a time discretization of the constitutive relation of the element. The development for a linear capacitor is presented here as an example. The simplest algorithm to use in developing this discretization is the Backward Euler integration formula. The Backward Euler algorithm for solving the differential equation

$$\dot{x} = f\left(x\right) \tag{22.16}$$

with step size $h = t_{n+1} - t_n$ is

$$x_{n+1} = x_n + hf\left(x_{n+1}\right) = x_n + h\dot{x}_{n+1} \tag{22.17}$$

where the subscript n refers to the nth time sample. The discretization is performed for each and every element independently by replacing the differential equation of Eq. (22.16) by the constitutive relation of the particular elements.

For a linear capacitor, the charge on the capacitor is linearly proportional to the voltage across it so that $q = Cv$. Thus

$$i\left(t\right) = \frac{dq}{dt} = C\frac{dv}{dt} = C\dot{v}$$

or

$$v_{n+1} = \frac{1}{C}i_{n+1} \tag{22.18}$$

where the reference convention for the circuit quantities are defined in Fig. 22.9.

Substituting Eq. (22.18) into Eq. (22.19) and rearranging leads to the discretized Backward Euler model of the linear capacitor:

$$i_{n+1} = \frac{C}{h}v_{n+1} - \frac{C}{h}v_n. \tag{22.19}$$

This equation has the form

$$i_{n+1} = g_{eq}v_{n+1} - i_{eq} \qquad (22.20)$$

and so is modeled by a constant conductance $g_{eq} = C/h$ in parallel with a current source $i_{eq} = -C/hv_n$ that depends on the previous time step, as shown in Fig. 22.10. The associated discrete circuit models for all other elements are developed in the same way, but of course the development is usually much more complicated, especially for nonlinear and multiterminal elements, but the approach is the same. The final circuit combining the ADM of all of the elements is linear with resistors and current sources, as well as a few special elements such as voltage sources. This circuit is especially compatible with the nodal-formulation described by Eq. (22.9). The linear circuit is then solved repeatedly with the circuit elements updated at each step and, if the circuit voltage and current quantities change by less than a specified tolerance, the time step advanced.

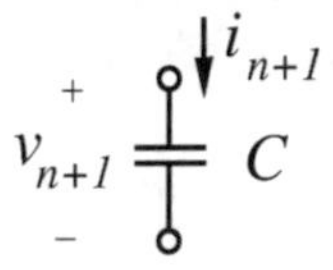

FIGURE 22.9   Reference direction for the circuit quantities of a capacitor.

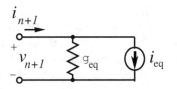

FIGURE 22.10   The associated discrete model of a two-terminal element.

The feature that distinguishes the associated discrete modeling approach from integration of the state equations for the system is that the discretization and particularly the Newton iteration is performed at the individual element level rather than at the top system level. The most important aspect of this is that special convergence treatments can be applied locally. For example, a diode has an exponential relationship between current and voltage and is the most difficult characteristic to handle. With the top-level systems-of-equations approach, any convergence scheme developed would need to be applied to all elements in a circuit, not just to the problem elements. In the associated discrete modeling, many local steps can be taken to improve convergence properties. This can include limiting the voltage and current changes from one iteration step to another. The scheme adopted depends on the characteristics of a particular element and heuristics developed in using it. It is this focus on local convergence control and embedding specific element knowledge in the element code that makes the SPICE approach so successful.

22.4.4   The Shooting Method

As has been mentioned, time-marching simulation has problems in determining the steady-state response because of the long simulation times that are involved. There is one elegant solution when the excitation is a sinusoid so that the response is known to be periodic. For strictly periodic excitation, shooting methods are often used to bypass the transient response altogether. This is advantageous in situations that would require many iterations for the transient components to die out. It is assumed that the nonlinear circuit has a periodic solution and that the solution can be determined by finding an initial state such that transients are not excited. If $\mathbf{x}(t)$ is the set of state variables obtained by a time-domain analysis, the boundary value constraint for periodicity is that $\mathbf{x}(t) = \mathbf{x}(t + T)$, where T is the known period. A series of iterations at time points between t and $t + T$ can be performed for a given set of initial conditions, and the condition for periodicity checked. Thus, in the shooting method, the problem of solving the state equations is converted into the two-point boundary value problem

$$\mathbf{x}(0) = \mathbf{x}(T)$$

$$\mathbf{x}(T) = \int_0^T \mathbf{f}(\mathbf{x}, \tau)d\tau + \mathbf{x}(0). \qquad (22.21)$$

If $\mathbf{x}(t) \neq \mathbf{x}(t + T)$ then a new set of initial conditions can then be determined using a gradient method based upon the error in achieving a periodic solution. Once the sensitivity of the circuit to the choice of initial conditions is established in this way, a set of initial conditions that establishes steady-state operation can be determined; this set is, of course, the desired solution. This iterative procedure can be implemented using the Newton's method iteration

$$\mathbf{x}^{k+1} = \mathbf{x}^k - \left[\mathbf{I} - \frac{\partial \mathbf{x}^k(T)}{\partial \mathbf{x}^k(0)}\right]^{-1} \left[\mathbf{x}^k(0) - \mathbf{x}^k(T)\right] \qquad (22.22)$$

where the superscripts refer to iteration numbers and $\mathbf{x}^k(T)$ is found by integrating the circuit equations over one period from the initial state $\mathbf{x}^k(0)$.

To begin the analysis, the period (T) is determined and the initial state $(\mathbf{x}^k(0))$ is estimated. Using these values, the circuit equations are numerically integrated from $t = 0$ to $t = T$ and the necessary derivatives calculated. Then, the estimate of the initial state is updated using the Newton iteration [Eq. (22.22)]. This process is repeated until $\mathbf{x}(0) = \mathbf{x}(T)$ is satisfied within a reasonable tolerance.

Shooting methods are attractive for problems that have small periods. Unlike the direct integration methods, the circuit equations are only integrated over one period (per iteration). They are therefore more efficient, provided that the initial state can be found in a number of iterations that is smaller than the number of periods that must be simulated before steady-state is reached in the direct methods. Unfortunately, shooting methods can only be applied to find periodic solutions. Also, shooting methods become less attractive for cases where the circuit has a large approximate period, for example, when several nonharmonic signals are present. The computation becomes further complicated when transmission lines are present, because functional initial conditions are then required to establish the initial conditions at every point along the line (corresponding to the delayed instants in time seen at the ports of the line).

In multitone situations when only one signal is large and when operating frequencies are not so high that distributed effects are important, the large tone response can be captured using the shooting method and then the frequency conversion method described in the next section can be used to determine the response with the additional small signals present.

22.4.5 Frequency Conversion Matrix Methods

In many multitone situations, one of two or more impressed non-commensurate tones is large while the others are much smaller. In a mixer, a large local oscillator, LO, (which is generally 20 dB or more larger than the other signals) pumps a nonlinearity, while the effect of the other signals on the waveforms at the nonlinearities is negligible. The pumped time-invariant nonlinearity can be replaced by a linear time-varying circuit without an LO signal. The electrical properties of the time-varying circuit are described by a frequency domain conversion matrix. This conversion matrix relates the current and voltage phasors of the first order sidebands with each other. In other words, by performing a fast, single-tone shooting method or harmonic balance analysis with only the LO impressed upon it, the AC operating point of the mixer may be determined and linearized with respect to small-signal perturbations about this point. This information is already available in the Jacobian, which is essentially a gradient matrix relating the sensitivity of one dependent variable to another independent variable. A two-tone signal can be rewritten to group the LO waveform, $x_{LO}(t)$ terms and the first order sidebands as

$$x(t,j) = x_{LO}(t) + \mathrm{Re}\left\{\sum_{p=0}^{N_A} \mathbf{X}_{p,1} e^{j\left(p\omega_{LO} + \omega_{RF}\right)} + \sum_{p=0}^{N_A} \mathbf{X}_{p,-1} e^{j\left(p\omega_{LO} - \omega_{RF}\right)}\right\} \qquad (22.23)$$

where $\mathbf{X}_{p,1}$ and $\mathbf{X}_{p,-1}$ are vectors of the spectral components at the first order sidebands of the pth harmonic of the LO. For voltage controlled nonlinearities, the output quantities (the $\mathbf{X}$'s) are current phasors so that the expression relating the IF current to the RF voltage is

$$\left[\mathbf{I}_{p,1},\mathbf{I}_{p,-1}\right]^{\mathrm{T}} = \mathbf{Y}_C\left[\mathbf{V}_{0,1},\mathbf{V}_{0,-1}\right]^{\mathrm{T}}. \tag{22.24}$$

Here $\mathbf{Y}_C$ is the admittance conversion matrix and can be used in much the same manner as a nodal admittance matrix. Alternatively, for current-controlled nonlinearities the following could be used:

$$\left[\mathbf{V}_{0,1},\mathbf{V}_{0,-1}\right]^{\mathrm{T}} = \mathbf{Z}_C\left[\mathbf{I}_{0,1},\mathbf{I}_{0,-1}\right]^{\mathrm{T}} \tag{22.25}$$

where $\mathbf{Z}_C$ is the impedance conversion matrix. Nonlinearities with state variable descriptions or mixed voltage-controlled and current-controlled descriptions require a combination of Eqs. (22.24) and (22.25) to derive a modified nodal admittance formulation.

22.4.6 Convolution Techniques

The fundamental difficulty encountered in integrating RF and microwave circuits in a transient circuit simulator arises because circuits containing nonlinear devices or time dependent characteristics must be characterized in the time domain while distributed elements such as transmission lines with loss, dispersion, and interconnect discontinuities are best simulated in the frequency domain. Convolution techniques are directed at the simulation of these circuits.

The procedure begins by partitioning the network into linear and nonlinear subcircuits as shown in Fig.22.11. In a typical approach the frequency domain admittance (y) parameter description of the distributed network is converted to a time domain description using a Fourier transform. This time domain description is then the Dirac delta impulse response of the distributed system. Using the method of Green's function, the system response is found by convolving the impulse response with the transient response of the terminating nonlinear load. Normally this requires that the impulse response be extended in time to include many reflections. While this technique can handle arbitrary distributed networks, a difficulty arises as the y parameters of a typical multiconductor array have a wide dynamic range. For a low loss, closely matched, strongly coupled system, the y parameters describing the coupling mechanism approach zero at low frequencies and become very large at high frequencies. Conversely, the transmission and self-admittance y parameters approach infinity at DC and zero at resonance frequencies. Both numerical extremes are important in describing the physical process of reflections and crosstalk. The dynamic range of the time domain solution is similarly large and values close to zero are significant in determining reflections and crosstalk. Consequently, aliasing in the frequency domain to time domain transformation can cause appreciable errors in the simulated transient response. The problem is considerably reduced by using resistive padding at the linear-nonlinear circuit interface to reduce the dynamic range of the variables being transformed. The effect of the padding can then be removed in subsequent iteration.

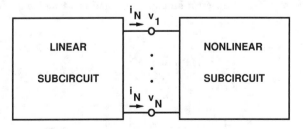

FIGURE 22.11 Circuit partitioned into linear and nonlinear sub-circuits.

22.5 Harmonic Balance: Mixed Frequency and Time Domain Simulation

The Harmonic Balance (HB) procedure has emerged as a practical and efficient tool for the design and analysis of nonlinear circuits in steady-state with sinusoidal excitation. The harmonic balance method is a technique that allows efficient solution of the steady-state response of a nonlinear circuit. For example, the steady-state response of a circuit driven by one or more sinusoidal signals is also a sum of sinusoids and includes tones at frequencies other than those of the input sinusoids (e.g., harmonics and difference frequencies). The response does not need to be periodic to be steady-state and with narrowband systems it is common to call the response to a complicated narrowband input as being quasi-periodic. Usually we are not interested in the transient response of the circuit such as when the power supply is turned on or when a signal is first applied. Thus much of the behavior of the circuit is not of interest. The harmonic balance procedure is a technique to extract just the information that is required to describe the steady-state response. The method may also be compared to the solution of a homogeneous, ordinary differential equation. A solution that is the sum of sinusoids of unknown amplitudes is substituted into the differential equation. Using the orthogonality of the sinusoids, the resulting problem simplifies to solving a set of nonlinear algebraic equations for the amplitudes of the sinusoids. There are several methods of solving for (complex) amplitudes, which will be discussed later in this section.

The HB method formulates the system of nonlinear equations in either domain (although more typically the time domain), with the linear contributions calculated in the frequency domain and the nonlinear contributions in the time domain. This is a distinct advantage for microwave circuits, in that distributed and dispersive elements are then much more readily modeled analytically or using alternative electromagnetic techniques based in the frequency domain.

While it is common to refer to the nonlinear calculations as being in the time domain, the most usual HB implementations require that the nonlinear elements be described algebraically, that is without derivatives or other memory effects. Thus a nonlinear resistor is described, for example, as a current as a nonlinear function of instantaneous voltage. So given the voltage across the nonlinear resistor at a particular time, the current that flows at the same instant can be calculated. A nonlinear capacitive element must be expressed as a charge which is a nonlinear function of instantaneous voltage. Then a sequence of charge values in time is Fourier transformed so that phasors of charge are obtained. Each phasor of charge is then multiplied by the appropriate $j\omega$ to yield current phasors.

22.5.1 Problem Formulation

The harmonic balance method seeks to match the frequency components (harmonics) of current at the interface of two sub-circuits — one linear and one nonlinear. The sub-circuits are chosen in such a way that nonlinear elements are partitioned into one sub-circuit, linear elements into another, and (in some approaches) sources into a third (see Fig. 22.12). The edges at the linear/nonlinear interface connect the two circuits and define corresponding nodes; current flowing out of one circuit must equal that flowing into the other. Every node in the nonlinear circuit is "pulled out" of the nonlinear sub-circuit so that it is at the interface and becomes part of the error function formulation. Matching the frequency components

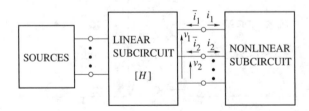

FIGURE 22.12   Circuit partitioned into linear, nonlinear, and source sub-circuits.

in each edge satisfies the continuity equation for current. The current at each edge is obtained by a process of iteration so that dependencies are satisfied for both the linear and nonlinear sides of the circuit.

The unknowns are found by forming an error function — typically the Kirchoff's Current Law (KCL) error at the linear/nonlinear interface. This error function is minimized by adjusting the voltages at the interface. Every node in the nonlinear sub-circuit is therefore considered to be connected to the linear sub-circuit. If the total circuit has N nodes, and if $\mathbf{v}$ is the vector of node voltage waveforms, then applying KCL to each node yields a system of equations

$$f\left(\mathbf{v},t\right)=\mathbf{i}\left[\mathbf{v}\left(t\right)\right]+\frac{\mathrm{d}}{\mathrm{d}t}\mathbf{q}\left[\mathbf{v}\left(t\right)\right]+\int_{-\infty}^{t} y\left(t-\tau\right)\mathbf{v}\left(\tau\right)\mathrm{d}\tau+\mathbf{i}_{s}\left(t\right)=0 \qquad (22.26)$$

where the nonlinear circuit is chosen to contain only voltage-controlled resistors and capacitors for representational ease. The quantities $\mathbf{i}$ and $\mathbf{q}$ are the sum of the currents and charges entering the nodes from the nonlinearities, y is the matrix impulse response of the linear circuit with all the nonlinear devices removed, and $\mathbf{i}_s$ are the external source currents.

In the frequency domain, the convolution integral maps into $\mathbf{YV}$, where $\mathbf{V}$ contains the Fourier coefficients of the voltages at each node and at each harmonic, and $\mathbf{Y}$ is a block node admittance matrix for the linear portion of the circuit. The system of Eq. (22.26) then becomes, on transforming into the frequency domain

$$\mathbf{F}\left(\mathbf{V}\right)=\mathbf{IV}+\Omega\mathbf{QV}+\mathbf{YV}+\mathbf{I}_{S}=0 \qquad (22.27)$$

where Ω is a matrix with frequency coefficients (terms such as $j\Omega_k$) representing the differentiation step. The notation here uses small letters to represent the time domain waveforms and capital letters for the frequency domain spectra. This equation is, then, just KCL in the frequency domain for a nonlinear circuit. HB seeks a solution to Eq. (22.27) by matching harmonic quantities at the linear-nonlinear interface. The first two terms are spectra of waveforms calculated in the time domain via the nonlinear model, i.e.,

$$\mathbf{F}\left(\mathbf{V}\right)=\Im i\left(\Im^{-1}\mathbf{V}\right)+\Omega\Im q\left(\Im^{-1}\mathbf{V}\right)+\mathbf{YV}+\mathbf{I}_{s}=0 \qquad (22.28)$$

where $\Im$ is the Fourier transform and $\Im^{-1}$ is the inverse Fourier transform.

The solution of Eq. (22.28) can be obtained by several methods. One method, known as relaxation, uses no derivative information and is relatively simple and fast, but is not robust. In a relaxation method the error function is taken to zero by adjusting current phasors or voltage phasors on successive iterations using what is in effect very limited derivative information. Alternatively, gradient methods can be used to solve either a system of equations (e.g., using a quasi-Newton method) or to minimize an objective function using a quasi-Newton or search method.

The Newton and quasi-Newton methods require derivative information to guide the error minimization process. Calculation of these derivatives is computationally intensive and generally the equations for these require considerable development. As with all the harmonic balance methods, the number of nodes used can be reduced by "burying" internal nodes within the linear network, which then becomes a single multiterminal sub-circuit as far as harmonic balance is concerned. The system of equations is then reduced accordingly. Once the "interfacial" node voltages are known, any internal node voltage can be found by using simple linear analysis and the full y matrix for the linear circuit.

22.5.2 Multitone Analysis

The problem with multitone analysis reduces to implementing a method to perform the multifrequency Fourier transform operations required in solving Eq. (22.28). This also requires developing the multifrequency Jacobian required in a Newton-like procedure. Time-frequency conversion for multitone signals can be achieved using nested Fourier transform operations. This is implemented using the multidimensional Fast Fourier Transform, or MFFT. Application of the multidimensional Fourier transform (MFFT), in which the Fourier coefficients are themselves periodic in the other dimensions, requires that the multiple tones (in each dimension) be truly orthogonal, i.e., not integer multiples of each other. If the two tones are frequency degenerate, then the method fails because orthogonality of the bases is a requirement for determining the Fourier coefficients in that basis. In such a case, one of the tones is slightly shifted to ensure that the technique can be applied.

The most general and easily programmed of the Fourier transform techniques applied to the HB method is the Almost-Periodic Discrete Fourier Transform (APDFT) algorithm. After truncation, consider the K arbitrarily spaced frequencies 0, ω_1, ω_2, ..., ω_{K-1} generated by the nonlinearity. Then

$$\sum_{k=0}^{K-1} X_k^C \cos\omega_1 t_1 + X_k^2 \sin\omega_1 t_1 = x(t)$$ may be sampled at S time points, resulting in a set of S equations and $2K-1$ unknowns:

$$
\begin{bmatrix}
1 & \cos\omega_1 t_1 & \sin\omega_1 t_1 & \cdots & \cos\omega_{K-1} t_1 & \sin\omega_{K-1} t_1 \\
1 & \cos\omega_2 t_2 & \sin\omega_2 t_2 & \cdots & \cos\omega_{K-1} t_2 & \sin\omega_{K-1} t_2 \\
\vdots & \vdots & \vdots & \ddots & \vdots & \vdots \\
1 & \cos\omega_1 t_S & \sin\omega_2 t_S & \cdots & \cos\omega_{K-1} t_S & \sin\omega_{K-1} t_S
\end{bmatrix}
\begin{bmatrix}
X_0 \\
X_1^C \\
X_1^S \\
\vdots \\
X_{K-1}^C \\
X_{K-1}^S
\end{bmatrix}
=
\begin{bmatrix}
x(t_1) \\
x(t_1) \\
\vdots \\
x(t_S)
\end{bmatrix}.
\tag{22.29}
$$

The number of samples S must be at least $2K-1$ to uniquely determine the coefficients. This equation may compactly be written as

$$\Gamma^{-1}X = x \text{ or } \Gamma x = X \tag{22.30}$$

where Γ and Γ^{-1} are known as an almost-periodic Fourier transform pair. Thus the multifrequency transform can be performed as a matrix operation but spectrum mapping and Fast Fourier transformation is much faster.

Combining the above procedures yields the time invariant form of Harmonic Balance. This is referred to as just Harmonic Balance. This technique is very efficient for simulating circuits with just a few active devices and a few tones, as then there are only a few unknowns. Problems arise as the number of active devices increases or the number of tones becomes large as the size of the problem increases significantly. Still, digitally modulated signals can be reasonably modeled by considering a very large number of tones.

22.5.3 Method of Time-Varying Phasors

Harmonic balance using time-variant phasors is ideally suited to the representation and characterization of circuits with digitally modulated signals. In contrast to time-variant harmonic balance, where the assumed phasor solution was time invariant, we instead assume a solution of the form

$$V_k(j\omega) = \text{real}\left\{\sum_{m=0}^{n} V_m(t)\exp\left[jm\omega(t) + \phi_m(t)\right]\right\} \tag{22.31}$$

where in general the amplitude, frequency, and phase of each term are allowed to vary with respect to time. If $V_m(t)$ varies slowly with respect to the carrier frequency, we are in essence solving for the envelope of the signal at each node without the requisite memory requirements of time-invariant harmonic balance, or the frequency domain dynamic range and resolution problems of time domain methods. Taking the Fourier transform of each summation term in Eq. (22.31) results in a highly resolved power spectral density distribution approximation of the digitally modulated signal, not an ill-conditioned approximation, as with time-invariant harmonic balance.

22.6 Frequency Domain Analysis of Nonlinear Circuits

Frequency domain characterization of RF and microwave circuits directly provides the types of performance parameters required in communication systems as well as many other applications of RF and microwave circuits. The Harmonic Balance (HB) method uses Fourier transformation to relate sequence of instantaneous current, voltage, and charge to their (frequency domain) phasor forms. In frequency domain nonlinear analysis techniques alternative mappings are used. There are many types of mappings for arriving at a set of (say, current) phasors as a nonlinear function of another set of (say, voltage) phasors.

The common underlying principle of frequency domain nonlinear analysis techniques is that the spectrum of the output of a broad class of nonlinear circuits and systems can be calculated directly given the input spectrum input to the nonlinear system. The mapping operation is depicted in Fig. 22.13 and is the concept behind most RF and microwave behavioral modeling approaches. Some techniques determine an output frequency component by summing calculations of individual intermodulation products. For example, the product of two tones is, in the time domain, the product of two sinusoids. The trigonometric expansion of this yields two intermodulation products with frequencies that are the sum and difference, respectively, of the frequencies of the tones. Power series techniques use trigonometric identities to expand the power series and calculate each intermodulation product individually. Algorithms sum these by frequency to yield the output spectrum. At the coarse end of the scale are Volterra series-based techniques that evaluate groups of intermodulation products at a single frequency. Some frequency-domain nonlinear analysis techniques are noniterative, although these are restricted to unilateral systems. Others, known as frequency domain spectral balance techniques, are iterative, being the frequency domain equivalent of the harmonic balance techniques. The term spectral balance is used to distinguish the frequency domain techniques from the harmonic balance techniques as the latter term has come to be solely applied to mixed time and frequency domain methods in which Fourier transformation is used. Intermediate between these extremes are techniques that operate by converting a nonlinear element into a linear element shunted by a number of controlled current sources. This process is iterative and at each iteration a residual nonlinear element is left that reduces from one iteration to another. This is the basis of one of the Volterra series analysis techniques called the method of nonlinear currents, which is also discussed in the next section.

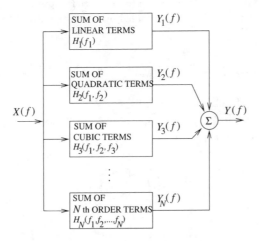

FIGURE 22.13   Mapping concept of frequency domain analysis.

22.6.1   Volterra Analysis

Expanding on Volterra analysis illustrates the concepts behind functional analysis of circuits in the frequency domain. Volterra series have the form

$$G(x) = \sum_{n=0}^{\infty} F_n(x) \tag{22.32}$$

where $F_n(x)$ is a regular homogeneous functional such that

$$F_n(x) = \int_a^b \cdots \int_a^b h_n(\chi_1, \chi_2, \ldots, \chi_n) x(\chi_1) x(\chi_2) \cdots x(\chi_n) d\chi_1 d\chi_2 \cdots d\chi_n \tag{22.33}$$

and the functions $h_n(\chi_1, \chi_2, \ldots, \chi_n)$ are known as the nth order Volterra kernels. It can be used as a time domain description (with the χ's replaced by t) of many nonlinear systems including nonlinear microwave circuits that do not exhibit hysteresis. In this case, the nth order kernel, h_n, is called the nonlinear impulse response of the circuit of order n. Equation (22.33) is then interpreted as an n dimensional convolution of an nth order impulse response (h_n) and the input signal (x). The total response $G(x)$ is the summation of the different order responses $F_n(x)$. Note that for a linear system there is only a first order response so that the total response of the system is the conventional convolution integral

$$G(x) = F_0 + \int_a^b h_1(\tau) d\tau \tag{22.34}$$

where F_0 is just a DC offset.

The important concept here is that the total response of a signal is the summation of a number of responses of different order. This scheme only works if, as the order n increases, the contribution to the response gets smaller and eventually insignificant. The reason this works for many RF and microwave circuits is that the response is close to linear and nonlinear behavior is a departure from linearity. A weak nonlinearity could be represented with just the first few terms of such a series.

In analyzing nonlinear circuits it is not necessary to deal with the Volterra series which, here, is in the time domain. Mostly the frequency domain form is used, which is expressed in terms of Volterra nonlinear transfer functions. Mathematically these are obtained by taking the n-fold Fourier transform of h_n:

$$H_n(f_1, f_2, \ldots, f_n) = \int_{-\infty}^{\infty} \cdots \int_{-\infty}^{\infty} h_n(\tau_1, \tau_2, \ldots, \tau_n) e^{-j2(f_1\tau_{1+} \ldots \tau_n)} d\tau_1 d\tau_2 \cdots d\tau_n \tag{22.35}$$

where H_n is called the nonlinear transfer function of order n. The time-domain input-output relation $y(t) = f[x(t)]$ can be put in the form

$$y(t) = \sum_{n=1}^{\infty} y_n(t) \tag{22.36}$$

where

$$y_n(t) = \int_{-\infty}^{\infty} \cdots \int_{-\infty}^{\infty} h_n(\tau_1,\ldots,\tau_n) x(t-\tau_1) \cdots x(t-\tau_n) d\tau_1 \cdots d\tau_n \qquad (22.37)$$

and $x(t)$ is the input. Taking the n-fold Fourier transform of both sides we have an expression for the spectrum of the nth order component of the output

$$\mathbf{Y}_n = \int_{-\infty}^{\infty} \cdots \int_{-\infty}^{\infty} H_n(f_1,\ldots,f_n) \delta(f - f_1 - \cdots - f_n) \prod_{i=1}^{n} X(f_i) df_i \qquad (22.38)$$

where $\mathbf{X}_n(f)$ is the Fourier transform of $x(t)$, $\mathbf{Y}_n(f)$ is the Fourier transform of $y_n(t)$, and $\delta(\cdot)$ is the delta function. This expresses the nth order terms of the output as a function of the input spectrum. The order of the terms refers to the fact that multiplication of the input by a constant A results in multiplication of the nth order terms by A^n. Then a frequency domain series for the output can be written as

$$Y(f) = \sum_{n=1}^{\infty} Y_n(f) \qquad (22.39)$$

in terms of the input spectrum and the nonlinear transfer functions. $\mathbf{Y}_n(f)$ is the nth order response and corresponds to the response of the nth order term in the power series description of the nonlinearity.

The method of nonlinear currents enables the direct calculation of the response of a circuit with nonlinear elements that are described by a power series. Here a circuit is first solved for its linearized response described by zero and first order Volterra nonlinear transfer functions. Considering only the linearized response allows standard linear circuit nodal admittance matrix techniques to be used. The second order response, described by the second order Volterra nonlinear transfer functions can then be represented by controlled current sources. Thus the second order sources are used as excitations again enabling linear nodal admittance techniques to be used. The process is repeated for the third- and higher-order node voltages and is easily automated in a general purpose microwave simulator. The process is terminated at some specified order of the Volterra nonlinear transfer functions. This is a noniterative technique, but relies on rapid convergence of the Volterra series restricting its use to moderately strong nonlinear circuits.

22.7 Summary

SPICE is at its best when simulating large circuits as memory and computation time increase a little more than linearly after a circuit reaches a certain size. However, to determine the response to sinusoidal excitation requires simulation over a great many cycles until the transient response has died down. A major problem in itself is determining when the steady state has been achieved. A similar problem occurs with narrowband modulated signals, which can have many millions of RF cycles before the response appears to be steady state. For example, a typical sequence length for the (digitally modulated) DAMPS format is 10 ms, although the time step would be on the order of 100 ps to capture the fundamental and harmonics of the 850 MHz carrier. This results in 10^8 time-points and hence the same order discrete Fourier transforms. Fourier transformation, e.g., using a fast Fourier transform, of the simulated waveform is required to determine its spectral content. This is not too complex a task if the exciting signal is a single frequency, but if the signal driving the nonlinear circuit has non-commensurable frequency components or is digitally modulated, then the procedure is more difficult and the effect of numerical noise is exaggerated. Even low-level numerical noise may make it impossible to extract a low-level tone in the presence of a large tone. The ability to detect a small tone defines the dynamic range of a simulator in RF and microwave applications and SPICE-analysis has poor performance in this case.

There is also a fundamental approximation error present in the SPICE algorithm due to what amounts to a z-domain approximation to the frequency domain characteristics of the circuit. The consequence is that time steps must be short for reasonable dynamic range. This also makes it particularly difficult to represent circuits with strongly varying narrowband frequency response. Recent extensions to SPICE — the shooting method with the frequency conversion method and convolution techniques — increase the applicability of SPICE to RF and microwave circuits. In spite of the difficulties, SPICE remains the only method of determining the transient response of a circuit.

Harmonic Balance analysis of circuits achieves significant computation savings by assuming that the signals in a circuit are steady state, described by a sum of sinusoids. The coefficients and phases of these sinusoids are solved for and not the transient response. Harmonic balance has a significant computation time advantage over SPICE for small to medium RF and microwave circuits. However, the time increases rapidly as circuit size increases. HB lends itself well to optimization and to analysis of multifunction circuits including amplifiers, oscillators, mixers, frequency converters, and numerous types of control circuits such as limiters and switches, if transient effects are not of concern. Another major advantage of the harmonic balance method is that linear circuits can be of practically any size, with no significant decrease in speed if additional internal nodes are added, or if elements of widely varying time constants are used (such is not the case with time domain simulators). Two extensions, separately implemented, also increase the usefulness of Harmonic Balance. The method of time-variant phasors enables digitally modulated signals to be handled. The second extension using matrix-free methods enables Harmonic Balance to handle very rich spectra and thus also approximately treat digitally modulated signals.

All of the techniques discussed here have been implemented in circuit simulators developed for RF and microwave circuit modeling. Many other simulator technologies exist, but these are within the overall framework of the discussion here. The reader is directed to the Further Information list for exploration of other technologies and for greater detail on those treated here.

Further Information

The bases of circuit simulation are described in J. Vlach and K. Singhal, *Computer Methods for Circuit Analysis and Design*, Van Nostrand Reinhold, 1983, ISBN 0442281080; and L. T. Pillage, R. A. Rohrer, and C. Visweswariah, *Electronic Circuit and System Simulation Methods*, McGraw-Hill, 1995, ISBN 0070501696. These two books are oriented toward SPICE-like analysis. Details on the algorithms used in SPICE are given in A. Vladimirescu, *The SPICE Book*, J. Wiley, 1994, ISBN 0471609269, and the techniques used in developing the associated discrete models used in SPICE in P. Antognetti and G. Massobrio, *Semiconductor Device Modeling with SPICE*, McGraw-Hill, 1988, ISBN 0070021538. In addition to the above, a short discussion of SPICE errors relevant to modeling RF and microwave circuits is contained in A. Brambilla and D. D'Amore, The simulation errors introduced by the SPICE transient analysis, *IEEE Trans. on Circuits and Systems-I: Fundamental Theory and Application*, 40, 57–60, January 1993. Circuit simulations oriented toward microwave circuit simulation are described in J. Dobrowolski, *Computer-Aided Analysis, Modeling, and Design of Microwave Networks: the Wave Approach*, Artech House, 1996, ISBN 0890066698; P. J. C. Rodrigues, *Computer-Aided Analysis of Nonlinear Microwave Circuits*, Artech House, 1998, ISBN 0890066906; and G. D. Vendelin, A. M. Pavio, and U L. Rohde, *Microwave Circuit Design Using Linear and Nonlinear Techniques*, Wiley, 1990, ISBN 0471602760. As well as providing a treatment of microwave circuit simulation, the following book provides a good treatment of Volterra analysis: S. A. Maas, *Nonlinear Microwave Circuits*, IEEE Press, 1997, ISBN 0780334035. Simulation of microwave circuits with digitally modulated signals is given in J. F. Sevic, M. B. Steer and A. M. Pavio, Nonlinear analysis methods for the simulation of digital wireless communication systems, *Int. J. on Microwave and Millimeter Wave Computer Aided Engineering*, 197–216, May 1996. A review of frequency domain techniques for microwave circuit simulation is given in M. B. Steer, C. R. Chang and G. W. Rhyne, Computer aided analysis of nonlinear microwave circuits using frequency domain spectral balance techniques: the state of the art, *Int. J. on Microwave and Millimeter Wave Computer Aided Engineering*, 1, 181–200, April 1991.

23

Computer-Aided Design of Microwave Circuitry

Ron Kielmeyer

Motorola, Inc.

23.1 Introduction

The growth of personal communication and Internet industries along with the need for portability has resulted in an ever-increasing demand for low cost, high volume microwave circuitry. The commercialization of GaAs wafer processing and the simultaneous reduction in the physical size of silicon devices has enable the development of complex microwave circuitry which can no longer be designed without the aid of sophisticated CAD circuit simulators. This article will discuss the typical steps involved in a design cycle, some basic requirements for a CAD program, a look at the theory behind the most popular CAD programs in use today, and some emerging CAD technologies.

23.2 Initial Design

The design cycle shown in Fig. 23.1 starts with a circuit and/or system function such as an amplifier, a mixer, or a whole receiver along with appropriate specifications for that function. Then active devices such as transistors or diodes, if required to achieve the function, are chosen. Circuit topologies may be explored simultaneously with device selection. With these active devices will come a computer representation for the device, usually from the device vendor. This computer representation is in the form of a mathematical model or measured S-parameters.

Synthesis programs, if available, are used to determine the best possible performance. Many different topologies can be explored or identified as possible candidates for realizing the function. The ideal topologies generated by the synthesis program must exceed the design specifications since performance will only deteriorate from the idealizedcase.

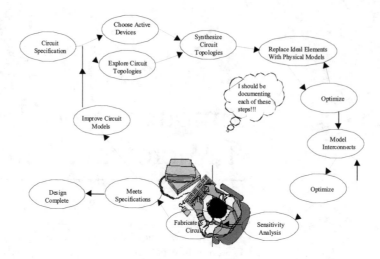

FIGURE 23.1   The design cycle.

23.3 Physical Element Models

Ideal elements must be replaced by models of the physical devices. These models are typically sub-circuits made up of ideal elements. For example, an ideal capacitor, physically realized by a chip capacitor, must have a model that can account for the finite inductance of the terminals and the finite resistive loss inherent in the physical device. For example, Fig. 23.2 shows a model for a physical capacitor mounted on a printed circuit board. In this model, C is the desired or ideal capacitance while Cpad represents the shunt capacitance of the metal pads on the circuit board to which the capacitor is soldered. Rs and Ls take into account the inductance and conductivity of the metal plates that form the capacitance. Rpar and Cpar account for the parallel resonant

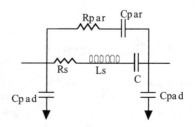

FIGURE 23.2   Physical model of a chip capacitor.

characteristics of the capacitor. It is common to refer to the extra elements Rs, Rpar, Cpar, Cpad, and Ls as "parasitic" elements. By replacing the ideal elements with nonideal models one at a time, the designer can accomplish two important practicalities. First, the sensitivity of the circuit to the nonideal element can be evaluated. Second, an optimization step can be performed on the remaining ideal elements in order to bring the circuit back within the design specifications. The actual value of the nonideal elements can also be optimized as long as the resulting optimized values do not change significantly from the pre-optimized values.

23.4 Layout Effects

As the process of replacing the idealized elements continues, the physical layout of the circuit must be introduced into the analysis. Two physical elements cannot be connected with zero length metal patterns. How close the elements can be placed is usually determined by how close the manufacturing process can place them. The metal patterns used to interconnect the physical devices must be introduced in the form of transmission lines and/or transmission line junctions. Figure 23.3 illustrates a simple PI resistive attenuator as might be realized using MMIC or MIC technology. The metal interconnects are modeled as a microstrip transmission line, followed by a Tee junction model. The resistors in this model are modeled as ideal resistors. The ground vias are represented as inductors. As each of these physical effects are introduced, an optimization step is performed in an attempt to meet or exceed the design specifications.

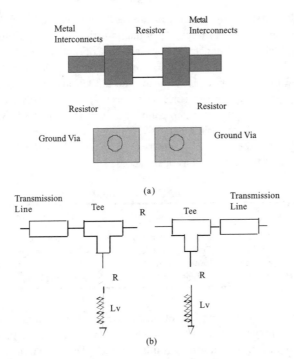

FIGURE 23.3 Simple PI pad resistive attenuator; (a) physical layout, (b) CAD model.

23.5 Sensitivity to Process Variation

After transforming many possible design topologies, sensitivity of the circuit to the nonideal or parasitic elements of the circuits as well as to the actual element values must be determined. One measure of the sensitivity of a circuit is the percent yield. To determine the percent yield of a circuit, the element values, parasitic elements, and/or known physical tolerances are treated as independent random variables. A range for each random variable is given and the computer analysis program iterates through random samples of the variables with their given ranges. This process is called a Monte Carlo analysis. The ideal outcome of the Monte Carlo analysis would be that the circuit passes all specifications under any combination of the random variable values. The percent yield can be improved by performing an optimization on the circuit. The optimization includes the ranges assigned to the random variables in the analysis. At each step in the optimization process, the mean values of the random variables are varied. The end result is an overall increase in the percent yield.

Design of Experiments (DOE) is another way to measure the sensitivity of the circuit to the random variables. In performing the DOE, each of the random variable values are independently incremented from their mean values. The computer keeps track of the analyzed responses and does the tedious bookkeeping task of incrementing all of the variables. Once this process is complete, statistical techniques are used to determine the sensitivity of each of the random variables to each of the specifications.

23.5.1 Design Tool Requirements

There are certain features that a microwave CAD tool must have in order to improve the productivity of the microwave designer. The most important is that it must be accurate. Once a circuit has been simulated, it must be fabricated and tested against the specifications for that circuit. If the computer simulation is inaccurate, the reasons for this must be determined. The fabrication of the circuit and the debugging of the circuit are by far the most time consuming part of the design cycle. Therefore, the goal of any CAD design is to model the circuit so that only one fabrication cycle needs to be implemented.

Inaccurate simulations that are tool dependent can result from numerical difficulties, model inaccuracies, or a lack of acceptable models. Inaccurate simulations that are directly caused by the user usually result from failure to model the circuit correctly or the misuse of models by using them outside the range for which they are valid. The designer has little control over the numerical problems. In some cases, the designer can overcome model accuracy by knowing the limitations of the models being used and if possible compensating for them. However, since model accuracy and availability usually are the features that distinguish one program from the other, companies tend to treat the models as proprietary and therefore do not want to release information about the models and how they are implemented. This can be a great disadvantage since the designer has no way of knowing how the model is implemented and when the range of validity for the model has been exceeded.

TABLE 23.1   Summary of Desirable CAD Features

Analysis Features	Synthesis Features	User Interface Features
Accurate models	Filter/impedance matching	Schematic input
Optimization	Constant noise circles	Text and graphical output
Import/export data	Constant gain circles	Easy documentation
Robust model library	Stability circles	Drawing complex shapes

Programs usually provide a way to implement custom models. For example, many programs will allow the user to import S-parameter data files to represent a part of the circuit that cannot be accurately represented by standard transmission line discontinuity models. These S-parameter files can be generated from a measurement of the actual discontinuity. Likewise, active device S-parameters can be measured at a specific bias point and imported for use in the design process. Measured data is usually limited to two-port networks since multiport network analyzers are not typically available to the designer. For multiport passive networks, electromagnetic simulators are often used to create data file representations of the passive network. In some cases, programs will allow the designer access to the code or hooks into the code that can be used to implement custom models.

Ease of use is a program feature that can greatly increase the productivity of a designer. Early implementations of circuit analysis programs used a descriptive language called a netlist to describe the circuit to be analyzed. Most modern programs however, use a graphical interface to create a schematic representation of the circuit. Schematics are much easier to create and are less prone to error than netlists. For electromagnetic simulators, a drawing package is used to draw the geometric shapes. These drawing packages, at a minimum, must provide a means to implement complex, difficult-to-draw structures from basic shapes that are easy for a designer to draw. In addition to this, the ability to change dimensions without the need to redraw the shape can greatly decrease the time it takes to perform tolerance studies or to modify the structure in order to meet the desired performance criteria.

Because microwave circuit performance specifications can be both time and frequency domain quantities, the circuit analysis program must provide an easy means to display the resulting data in both the time and frequency domains. The program must also be able to export the results into standard text or graphic formats for importation into word processors or view cell generating programs for documentation and presentation purposes.

Electromagnetic simulators need to be able to display field quantities such as current density on the conductors, E and H field intensity and direction, as well as outputting S, Z, or Y parameters for importation into circuit simulators. Data must be displayed in either graphical or text format.

Synthesis programs are available that provide impedance matching network topologies for amplifier/mixer design and filter response functions. These programs could be simple spreadsheet implementations of well-known design equations, an electronic Smith chart, or sophisticated implementations of impedance matching theory or filter design. The most sophisticated programs can provide networks based on both ideal and nonideal elements. The ability to display constant gain, noise circles, and stability circles can also be considered part of the synthesis capability since these are often used to determine the matching network impedances [1].

23.6 Time Domain vs. Frequency Domain Simulation

A circuit can be analyzed in either the frequency domain or the time domain. SPICE is a program developed at the University of California at Berkeley that analyzes a circuit in the time domain. Harmonic balance programs solve the circuit equations partially in the frequency domain and partially in the time domain. The choice of which program is used depends on the parameters that are specified. Since SPICE does the computations in the time domain, it is quite naturally used when the design parameters involve time-dependent quantities. SPICE is used for transient parameters such as the turn-on time of an oscillator or amplifier, the switching time of a switch, or perhaps the impulse response of a circuit. SPICE can also solve for the DC bias point of the circuit and then perform a small-signal, frequency domain analysis of the circuit about this bias point. By adding some special circuit elements to the file description, S-parameters can be computed from the results of this small-signal analysis. However, many SPICE based programs have added direct S-parameter output capability in order to accommodate the needs of microwave designers.

SPICE can be used to predict the effects of noise and distortion within the circuit. Using small-signal, frequency domain analysis, the linear noise parameters of a circuit can be predicted. However, the noise due to mixing effects of nonlinearities within the circuit can not be predicted. Likewise the up- or downconversion noise and phase noise of an oscillator are not analyzed. Distortion analysis using SPICE is usually performed through the use of a transient analysis followed by conversion of a part of the time waveform into the frequency domain using a Discrete Fourier Transform (DFT). The microwave designer is typically interested in two types of distortions — the harmonic content of the time waveform and the intermodulation products caused by the excitation of the circuit by two signals typically close to each other in frequency.

Harmonic balance is used when the circuit is driven by periodic sources and when the design parameters, input and output, are specified in the frequency domain. The assumed periodicity of the circuit response avoids the need to compute the circuit response from time zero until the steady-state response is obtained. Therefore, much less computer time is required to predict the circuit response. Since the harmonic balance techniques were developed specifically to aid the microwave designer in the design of nonlinear circuits, the available programs are custom tailored to provide the results in a format familiar to the designer. For example, the input source for an amplifier can be swept in both frequency and power, and nonlinear parameters such as gain, 1 db compression point, saturated power output, power-added efficiency, and harmonic levels can all be displayed in a graphical or text form. Indeed these parameters are all natural artifacts of the computations. Other parameters that can easily be computed are intermodulation products, third order intercept point, noise side bands, and mixer conversion.

The harmonic balance method is a hybrid of the small-signal, frequency domain analysis and a nonlinear time domain analysis. The circuit is divided into two subcircuits. One sub-circuit contains only those circuit elements that can be modeled in the small signal frequency domain. This sub-circuit results in a Y matrix representation that relates the frequency domain currents to the frequency domain voltages. At this point, the matrix can be reduced in size by eliminating voltages and currents for the nodes that are not connected to the nonlinear devices or the input and output nodes. These matrix computations need only be performed once for each frequency and harmonic frequency of interest. The second subcircuit includes all of the active or nonlinear elements that are modeled in the time domain. These circuit models relate the instantaneous branch currents to the instantaneous voltages across the device nodes. These models are the same models used in SPICE programs.

The two sub-circuits make up two systems of equations having equal node voltages whose branch currents must obey Kirchhoff's current law. The system of equations corresponding to the linear sub-circuit is now solved by making an initial guess at the frequency spectrum of the node voltages. The node voltage frequency spectrum is then used to solve for the frequency spectrum of the branch currents of the linear sub-circuit. In addition to this, the node voltage frequency spectrum is transformed into the time domain using a FFT algorithm. The result of this operation is a sampling of the periodic time voltage waveform. The sample voltages are applied to the time domain sub-circuit resulting in time domain current waveforms, which are then transformed into the frequency domain again using a FFT algorithm.

The two frequency domain current spectrums are compared, and based on the error between them, the voltage spectrums are updated. This process is repeated until the error is sufficiently small.

Early implementations of harmonic balance programs used either an optimization routine to solve for the node voltage spectrums [2] or Newton's method [3]. The main advantage of Newton's method is that it uses the derivatives of the nonlinear device currents with respect to the node voltages to predict the next increment in the node voltages. By taking advantage of these derivatives, convergence can be achieved for a relatively large number of nonlinear devices. This method appears to work well as long as the nonlinearity of the system is not too severe.

23.6.1 Emerging Simulation Developments

Current research directed toward improving the implementation of harmonic balance programs is concentrating on techniques that can handle the large number of nonlinear devices typically found in integrated circuits. Currently, Krylov-subspace solutions have been implemented [4]. When Krylov subspace techniques are used, the harmonic balance method can be used to solve circuit problems containing hundreds of transistors.

When the excitation of a circuit consists of multiple sinusoids, closed spaced in frequency, both SPICE and conventional harmonic balance methods tax computer hardware resources as they require large amounts of memory and computer time. A program must be able to efficiently handle this type of excitation in order to be able to predict the effects of spectral reqrowth in digitally modulated circuits, as well as noise-power ratio simulations for these circuits. For these types of circuit analysis, the excitation consists of multiple sinusoids, closely spaced in frequency. Borich [5] has proposed a means of overcoming these problems for harmonic balance programs by adjusting the sampling rate and the spacing between excitation carriers in order to reduce the computations of the multitone distorted spectra to an efficient one-dimensional FFT operation.

Envelope following methods [6] have been implemented to solve for circuits in which the excitation consists of a high frequency carrier modulated by a much slower information signal. The method performs a transient analysis consistent with the time scales of the information signal. At each time step, a harmonic balance analysis is performed at the harmonic frequencies of the carrier. This method can be used to study PPL phase noise, oscillator turn-on time, and mixer spectral regrowth due to digital modulation on the RF carrier [7].

References

1. George D. Vendelin, *Design of Amplifiers & Oscillators by the S-Parameter Method.* John Wiley & Sons, New York, 1982.
2. M. Nakhla and J. Vlach, A Piecewise Harmonic Balance Technique for Determination of Periodic Response of Nonlinear Systems, *IEEE Transactions on Circuits and Systems,* CAS-23, 2, February 1976.
3. K. Kundert and A. Sangiovanni-Vincentelli, Simulation of Nonlinear Circuits in the Frequency Domain, *IEEE Transactions Computer-Aided Design,* CAD-5, 4, October 1986.
4. R. Telichevesky, K. Kundert, I. Elfadel, and J. White, Fast Simulation Algorithms for RF Circuits, IEEE 1996 Custom Integrated Circuits Conference.
5. V. Borich, J. East, and G. Haddad. An Efficient Fourier Transform Algorithm for Multitone Harmonic Balance, *IEEE Transactions Microwave Theory and Techniques,* 47, 2, February 1999.
6. P. Feldmann and J. Roychowdhury, Computation of Circuit Waveform Envelopes Using an Efficient, Matrix-Decomposed Harmonic Balance Algorithm, in *Proc. IC-CAD,* November 1996.
7. K. Mayaram, D.C. Lee, S. Moinian, D. Rich, and J. Roychowdhury, Overview of Computer-Aided Analysis Tools for RFIC Simulation: Algorithms, Features, and Limitations, IEEE 1997 Custom Integrated Circuits Conference.

24

Nonlinear Transistor Modeling for Circuit Simulation

Walter R. Curtice
W.R. Curtice Consulting

24.1 Modeling in General

By definition, a transistor model is a simplified representation of the physical entity, constructed to enable analysis to be made in a relatively simple manner. It follows that models, although useful, may be wrong or inaccurate for some application. Designers must learn the useful range of application for each model.

It is interesting to note that all transistors are fundamentally nonlinear. That is, under any bias condition, one can always measure harmonic output power or intermodulation products at any input RF power level, as long as the power is above the noise threshold of the measurement equipment. In that sense, the nonlinear model is more physical than the linear model.

The purpose of this work is give a tutorial presentation of nonlinear transistor modeling. After reviewing the types of models, we will concentrate on equivalent circuit models, of the type used in SPICE.[1] Recent improvements in models will be described and the modeling of temperature effects and the effects of traps will be discussed. Finally, parameter extraction and model verification is described.

24.1.1   Two-Dimensional Models

The models constructed for describing the nonlinear behavior of transistors fall into several
distinctly different categories, as depicted in
Table 24.1. The most complex is the "physics-
based" model. Here electron and hole transport
is described by fundamental transport and current continuity relationships and the physical
geometry may be described in one-, two-, or
even three-dimensional space. The electric field

TABLE 24.1   Types of Large-Signal Transistor Models

I. Physical or "Physics-Based" Device Models
II. Measurement-Based Models
1 - Anaytical Models, such as SPICE Models
2 - Black Box Models
Table-Based Models
Artificial Neural Network (ANN) Models

is found by solution of Poisson's equation consistent with the distribution of charge and boundary
conditions. Such a model may use macro-physics, such as drift-diffusion equations,[2-4] or more detailed
descriptions, such as a particle-mesh model with scattering implemented using Monte-Carlo methods.[5,6]
Two- and three-dimensional models must be used if geometrical effects are to be included. The matter
of how much detail to put into the model is often decided by the time it takes for the available computer
to run a useful simulation using the model. In fact, as computers have increased their speed, modelers
have increased the complexity of the model simulated.

The lengthy execution time required for Monte-Carlo analysis can be reduced by using electron
temperature[7] as a measure of electron energy. Electron temperature is determined by the standard
deviation of the energy distribution function and is well defined in the case of the displaced Maxwellian
distribution function. Electron and hole transport coefficients are developed as a function of electron
temperature, and nonequilibrium effects, such as velocity overshoot in GaAs, may be simulated in a more
efficient manner. However, the solution of Poisson's equation still require appreciable computational time.

BLAZE[8] is a good example of a commercial, physically-based device simulator that uses electron
temperature models. BLAZE is efficient enough to model interaction of a device with simple circuits.

The physics-based model would be constructed with all known parameters and simulations of current
control for DC and transient or RF operation, then compared with measured data. Using the data, some
transport coefficients or physical parameters would be fine-tuned for best agreement between the model
and the data. This is the process of calibration of the model.[9] After calibration, simulations can be trusted
to be of good accuracy as long as the model is not asked to produce effects that are not part of its
construction. That is, if trapping effects[10] have not been incorporated into the model, the model will
disagree with data when such effects are important. With the physics-based model, as with all others, a
range of validity must be established.

Present physics-based models still require too much computational time to be used to any extent in
circuit design work. Optimization of a circuit design will involve invoking the device model frequently
enough to be impractical with physics-based models. These models can be used if only one or two
nonlinear transistors are used in a specific circuit, but typically, the circuit designer has a larger number
of nonlinear devices.

Several quasi two-dimensional models[11,12] have been developed that execute more efficiently. Initial
results look good, but accuracy may depend upon the simplifications made in the development of the
code and will vary with the application.

24.1.2   Measurement-Based Models

The next general category is that of "measurements-based" models. These are empirical models either
constructed using analytical equations and called "analytical models" or based upon a lookup table
developed from the measured data. The latter are called "table-based" models. Multidimensional spline
functions are used to fit the data in some of these models[13] and only the coefficients need be stored.

In the case of analytical models, the coefficients of the equations serve as fitting parameters to permit
the equations to approximate the measured data. Functions are usually chosen with functional behavior
similar to measured data so that the number of fitting parameters is reduced.

The advantages of analytical models are: computational efficiency, automatic data smoothing, accommodation of device statistics, physical insight, and the ability to be modified in a systematic manner. Disadvantages are: restriction of behavior often due to use of over-simplified expressions, difficulty in parameter extraction, and guaranteed nonphysical behavior in some operating condition. The nonphysical behavior is often associated with the use of a function, such as a polynomial, to fit data over a specific range of voltages and subsequent application of the model to voltages outside this range. The function may not behave well outside the fitting range. The best example of analytical models is the set of transistor models used in the various forms of the SPICE program. A major advantage of analytical models is that all the microwave nonlinear simulators provide some sort of user-defined model interface for analytical model insertion.

Table-based models have some properties of black-box models. The equations used result from fitting to the data, using splines, or other such functions. These models can therefore "learn" the behavior of the nonlinear device and are ideal for applications where the functional form of the behavior is unknown. Table-based models are efficient but do not provide the user with any insight, since there is a minimal "circuit model." They have difficulty incorporating dispersive effects, such as "parasitic gating" due to traps (see the section on Modeling the Effects Due to Traps) and do not accommodate self-heating effects.

The model cannot be accurately extrapolated into regions where data was not taken, and the models are often limited in their application due to the particular coding used by their author. This means that the model cannot be tailored by users other than the author. Customization of models is important to improve the "performance" of a model. The first table-based model that has been widely used is the Root model.[13]

24.1.3 Physical Parameter Models

One may argue that there is a class of models between physics-based and analytic models, namely, physical-parameter models. A good example would be the Gummel-Poon model.[14] Here, analytical equations are used but the fitting parameters or equation coefficients have physical significance. For example, NF, the ideality factor of the emitter-base junction, is one model parameter. This model is an analytical model but more useful device information may be gleaned from the value of the coefficients. This is often the case and the model is widely used for various forms of bipolar devices.

A useful physical parameter model for the AlGaAs/InGaAs/GaAs PHEMT has been published and verified by Daniel and Tayrani.[15] No information has been given on the range of validity of the analytical model, and it is expected that such a simple model will have inaccuracy when two-dimensional effects or nonequilibrium effects are important. It is interesting that the HEMT structure has less two-dimensional effects than the MESFET because of the sheet current layer produced in the HEMT.

There are many analytical models for which the coefficient has the name and dimensions of a physical parameter but the coefficient is only a fitting parameter and is not strongly related to the physical parameter. Khatibzadeh and Trew[16] have presented one commonly called the Trew model and Ladbrooke[17] has presented a second model commonly called the Ladbrooke model. Ladbrooke's model is an extension of the much earlier Lehovec and Zuleeg[18] model and it is more empirical than physical, as Bandler has shown.[19] Such models must be tested to see how strong the relationship is between the model parameter and the physical parameter. It is a matter of the degree of correlation between the two quantities. It is dangerous to attach too much physical significance to these coefficients. One should verify the relationship before doing so.

There is a unique case where physical parameters have been installed in a previously developed analytical model. The Statz-Pucel (analytical) GaAs MESFET model[20] has been converted to a physical parameter model by D'Agostino et al.[21]

24.1.4 Neural Network Modeling

Rather recently, a new approach has been developed for the modeling of nonlinear devices and networks. It utilizes artificial neural networks, or ANN. ANN models are similar to table-based or black-box models in that there is no assumption of particular analytical functions. As with table-based models, the ANN model "learns" the relationship between current and voltage from the data, and model currents are efficiently calculated after application of voltages. ANN analysis can treat linear or nonlinear operation of devices or complex circuits. ANN models have many of the advantages and deficiencies of table-based models. An excellent special issue of *RF and Microwave Computer-Aided Engineering*[22] has been devoted to this modeling method. Unfortunately, discussion of this approach is beyond the scope of this work.

24.2 Scope of This Work

The purpose of this work is to present a tutorial on the modeling of the nonlinear behavior of transistors. The scope is limited to nonlinear models useful for the development of circuit designs. It would not be possible to cover all the important material on other types of models, such as physical models, in this article. This paper will deal primarily with nonlinear analytical models for MESFETs, (P)HEMTs and HBTs. The emphasis is on GaAs device models, although many of these models are also used for transistors fabricated in InP, silicon, and other materials. The RF LDMOS power device is also discussed because it can be treated as a three-terminal device, much like a MESFET.

We address the concerns of analog and digital circuit designers who must choose between a wide variety of nonlinear models for transistors. Of particular concern here is the MMIC designer who must select the proper model to use for his GaAs microwave transistor. Even with very complex models presently supplied in circuit simulators, some specific behaviors are not modeled and new model features are required. We will spend much time on SPICE models and SPICE-type models. We will inspect some of the recent models that incorporate important device effects, omitted in previous models. We will discuss the modeling of gate charge as a function of local and remote voltages, the modeling of self-heating effects, the modeling of trapping effects, and model verification.

Unfortunately, the references presented will only be representative of prior work because there is a wealth of papers in each area of transistor modeling. I apologize to any author not included, as there are now and have been many people working in this area. I do recommend some modeling tutorial articles, previously published. Trew[23] and Snowden[24] and Dortu et al.[25] have presented excellent reviews of SPICE type transistor models and are recommended reading.

24.3 Equivalent Circuit Models

We concern ourselves here with equivalent circuit models because they are formulated to be efficiently exercised in a circuit simulator, and thus are efficient for circuit design and optimization. This is because the simulator is accustomed to dealing with resistors, capacitors, inductors, and voltage or current-controlled sources.

One problem is that "de facto" standard models have evolved in the industry, and often these models are inadequate to describe the device behavior. This is even true for small-signal equivalent-circuit models. Still, all circuit simulators utilize standard model topologies for small-signal and large-signal MESFET and PHEMT models. These models represent a minimum number of elements and are efficient for evaluation of transistor characteristics. Figures 24.1 and 24.2 show the conventional topologies for small-signal and large-signal simulation, respectively. It is conventional to separate the extrinsic parameters from the intrinsic device parameters, as shown in Fig. 24.1. The intrinsic parameters are assumed to contain all the bias-dependent behavior and the extrinsic parameters are assumed to be of constant values.

Curtice and Camisa[26] and Viakus[27] have discussed the trade-offs that exist between simple models with a small number of parameters, and complex models with a large number of parameters. A primary concern is the significant increase in the uncertainty for each model parameter in a complex model. Viakus showed that a small increase in the number of elements in a small-signal model could easily increase the uncertainty of critical elements beyond the standard deviation of the element value resulting from the fabrication process.

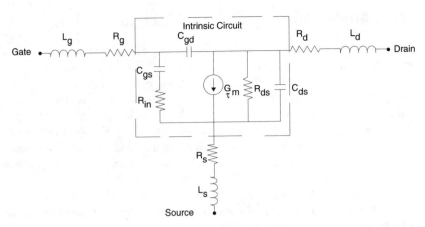

FIGURE 24.1 The conventional small-signal model for a MESFET, showing intrinsic and extrinsic elements.

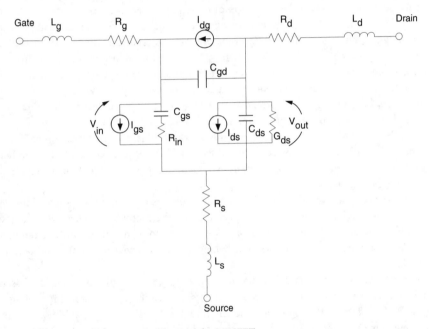

FIGURE 24.2 The conventional large-signal model for MESFET.

Byun[28] and others assert that source resistance should be taken as bias dependent. This decision is actually a choice made by the modeler. If source and drain resistance are taken as constant, then the reference planes defining these resistances are taken as being close to the metal ohmic contacts and not too close to the Schottky contact. That is, no region that may become depleted of change is included. All bias-dependent behavior is then lumped into the intrinsic elements. This is the convention followed by most modelers. It results in a simpler model, with fewer parameters.

Unfortunately, many of the published nonlinear device models have inconsistencies with the conventional small-signal model. Some of the inconsistencies may go unnoticed but can cause design errors. A good example that will be described later is the modeling of capacitance as a function of two independent voltages. The small-signal model must contain "transcapacitance" elements to be consistent with the large-signal model.

The large-signal model should agree with the small-signal model, but is not expected to be as efficient. Because transconductance and some resistances and capacitances must be evaluated from functions of voltages, numerical evaluation will take more computational time. However, the behavior of the large-signal model will gracefully go from small-signal to large-signal in a good model.

24.4 SPICE Models and Application-Specific Models

SPICE was developed to help in the design of switching circuits. Thus, the SPICE transistor models were developed to model the time domain behavior of devices in such circuits. However, transistors operating in RF analog circuits have a different locus of operation. For example, a simple class A amplifier will have locus of operation around its quiescent bias point. This should be compared with a logic circuit where the transistors go from a biased-off condition (high voltage, low current) to a strongly turned on condition (low voltage, high current). If the same transistor is used in these two applications, one would expect the SPICE model would approximate both behaviors, but not be optimum for either. In fact, if the model is fine-tuned to be more accurate for one of these applications, it will by default, be less accurate for the other.

For these reasons, accurate nonlinear transistor models will be application specific. In order to make the model more general, the model can be made more complex and more model parameters will be added. This may result in poorer execution efficiency.

Clearly one goal of transistor modeling should be to keep the model simple and to keep the number of model parameters small so that the extraction of these parameters is more efficient. My experience is that designers like simple models for initial work and are willing to work with more complicated models for difficult design specifications. The SPICE transistor models serve the function of the initial, simple models. These models are also universally known by name and have history and familiarity associated with them. A designer attempting to use a new GaAs foundry would not be intimidated by obscure nonlinear models if SPICE transistor models are used in that foundry.

Many modelers have attempted to extend or enhance the SPICE models so that their accuracy is improved, particularly in microwave analog applications. Usually, the default model is the original SPICE model and the designer will feel comfortable with this approach. There are many examples. The Gummel-Poon BJT model has been extended by Samelis and Pavlidis[29] and others[30] for application to heterojunction bipolar devices. The JFET SPICE model was extended by Curtice[31] in 1980 for better application to GaAs MESFET logic circuits.

Because of the increasing use of harmonic-balance simulators for microwave applications, many new equivalent circuit models have been developed specifically for these simulators. Nevertheless, these models are "SPICE-type" models, and can also be executed in a time domain simulation. The requirements of a model for SPICE are the same as for harmonic balance since the device is operated in the time domain in both simulators.

The producers of commercial harmonic balance (HB) software recognized the need of users to customize their transistor models. All commercial packages now contain user-defined modeling interfaces that permit the installation of customized models into the transistor model library. The process of installing or customizing a model in SPICE is much more difficult and not available to the average user. However, the ease of installation of models into HB software has produced a rash of new models for many transistor types.

Table 24.2 shows the typical array of SPICE equivalent circuit models available as part of a commercial simulator software package. The models are categorized, in general, as diode models, GaAs MESFET or (P)HEMT models, MOS models, and bipolar device models. The list is not complete for any specific product but representative of the models available. The models listed in Table 24.2 are in most commercial simulator products whether a version of SPICE or a harmonic balance simulator.

24.5 Improved Transistor Models for Circuit Simulation

Early SPICE models have shown a number of deficiencies. One problem is that the models developed before 1980 were developed for silicon devices and they do not reflect the behavior of GaAs devices. Most SPICE models need to be customized to be accurate enough for present design requirements. With regard to GaAs MESFET and PHEMT modeling, the strong dependency of gate-source capacitance upon drain-source voltage as well as gate-source voltage is not modeled in the early SPICE

TABLE 24.2   SPICE Models

GaAs MESFET/HEMT	MOS Models
Curtice (Cubic and quadratic)	BSIM 1,2,3 3v3
STATZ (Raytheon)	UC Berkeley 2 and 3
JFET (N & P)	HSPICE
TOM (TriQuint's Own Model)	MOSFET (various levels)
Materka	
Diode Models	BJT Models
P/N diode	Gummel-Poon
PIN diode	METRAM
	VBIC

models. None of the standard SPICE models accommodate self-heating effects. These effects are more important in GaAs applications due to the poorer thermal conductivity of GaAs compared to silicon. Some GaAs transistors exhibit important dispersion effects in transconductance as well as in drain admittance. All the GaAs models in Table 8.8 were added during the 1980s. These models represented major improvements; however, deficiencies remain. These deficiencies are summarized below for SPICE large-signal models:

- Insufficient accuracy for GaAs applications.
- Poor modeling of nonlinear capacitance.
- Poor modeling of self-heating effects.
- No modeling of dispersion of transconductance.
- Model parameter extraction not defined.
- Poor modeling of nonlinear effects dependent upon higher order derivatives.

The improved large-signal models of the 1990s exhibit some common features. It is quite popular to utilize analytical functions that have an infinite number of derivatives. For example, in the modeling of PHEMTs, Angelov et al.[32] have relied heavily on the hyperbolic tangent function for current because its derivative with respect to gate-source voltage is a bell-shaped curve, much like transconductance in PHEMTs. All further derivatives also exit. The Parker[33] model also utilizes higher order continuity in the drain current description and its derivatives.

Some SPICE models do have continuous derivatives but may not be accurate. The differences between the COBRA[34] model and the previous Materka[35] model are more evident when the derivatives of current (first through third) are compared. Since the COBRA model has derivatives closer to the data, Cojocaru and Brazil[34] show that the model predicts intermodulation products more accurately.

Many SPICE models use a simple expression for junction capacitance in MESFETs and PHEMTs. However, capacitance values extracted from data show that the gate-source and gate-drain capacitance depends strongly on the remote voltage as well as the local voltage (or capacitance terminal voltage). The Statz (Raytheon), the TOM, and the EEFET3 SPICE models[36] all have detailed equations for the gate, drain, and source charge as a function of local and remote voltages. Extraction of coefficients for these expressions is not simple and this will be discussed in the next section.

24.6 Modeling Gate Charge as a Function of Local and Remote Voltages in MESFETS and PHEMTS

Small-signal modeling of GaAs and InP MESFETs and PHEMT show that both C_{gd}, the gate-drain capacitance, and C_{gs}, the gate-source capacitance, vary with change of both V_{gs}, the gate-source voltage, and V_{ds}, the drain-source voltage. Thus, these capacitances are dependent upon the local, or terminal voltage, and a remote voltage. The dependency upon the local voltage is expected for capacitances, but the dependency upon the remote voltage leads to a term called "transcapacitance." The modeling of these capacitances can lead to nonphysical effects if not handled properly, as Calvo et al.[37] have shown.

Harmonic balance simulators work with charge functions whose derivatives are capacitive terms. The conventional approach is to find some charge function for total gate charge, such as $Q_g (V_{gs}, V_{ds})$. Then,

$$C_{11} = \text{Partial Derivative of } Q_g \text{ with respect to } V_{gs}$$

$$C_{12} = \text{Partial Derivative of } Q_g \text{ with respect to } V_{ds}$$

For consistency with small-signal models:

$$C_{11} = C_{gs} + C_{gd}$$

$$C_{12} = -C_{gd}$$

The problem remaining is to partition the total gate charge Q_g into charge associated with the gate-source region, Q_{gs}, and charge associated with the gate-drain region, Q_{gd}. Then, for large signal modeling, the gate node has charge Q_g, the source node has charge $-Q_{gs}$, and the drain node has charge $-Q_{gd}$, and

$$Q_g = Q_{gs} + Q_{gd}.$$

One scheme for partitioning the charge is used in the EEFET model and described in the HP-EEsof Manual.[36] One advantage of this approach is that the model becomes symmetrical, meaning that drain and source may be interchanged and the expressions are still valid.

A simpler approach is presented by Jansen et al.[38] Jansen assumes that all of the gate charge is associated with the gate source region and:

$$Q_{gd} = 0$$

$$Q_g = Q_{gs}.$$

The topology for this approach is different than the conventional model. There is no drain-gate capacitance element. Instead, the transcapacitance term accounts for the conventional drain-to-gate capacitive effects. That is, a change in V_{ds} produces current in the gate source region through the transcapacitance term. Figure 24.3 shows the new topology for the intrinsic circuit.

The procedure for modeling the capacitive effects is the same for both approaches and is the following:

1. Measure small-signal values of C_{gs} and C_{gd} as a function of V_{gs} and V_{ds}
2. Choose a Q_g function and optimize the coefficients of the Q_g expression for best fit of

$$C_{11} = C_{gs} + C_{gd}$$

$$C_{12} = -C_{gd}$$

A good example of the fitting functions and the type of fit obtained is given by Mallavarpu, Teeter, and Snow[39] for a PHEMT.

In summary, large-signal capacitive effects are modeled by constructing a total gate charge function, $Q_g(V_{gs}, V_{ds})$, whose partial derivative approximates the measured capacitance functions. If device symmetry is important, the EEFET charge partitioning scheme may be used. For amplifier applications, the Jansen model is simplest to code and implement because there is no charge partitioning expression. However, the Jansen model uses a topology that is not conventional.

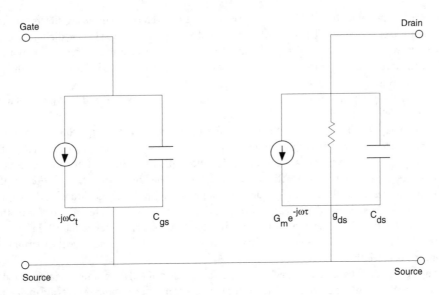

FIGURE 24.3 Jansen's topology for the intrinsic circuit.

24.7 Modeling the Effects Due to Traps

Electron and hole traps exist in GaAs materials and cause numerous effects during operation of a GaAs MESFET or PHEMT transistor. The following is a brief listing of these effects:

- Dispersion in transconductance and output admittance
- Backgating
- Parasitic bipolar effects
- "Kinks" in the I/V relationship
- Surface gating
- Gate and drain lag effects during switching
- Light sensitivity
- Substrate current or lack of current pinchoff

These effects have been studied and circuit-level models developed to simulate the effects. In many cases, the details of the behavior have been made clear using two-dimensional simulation modeling. For example, Li and Dutton[10] used PISCES-IIB to show that the common EL2 trap causes dispersion in the output conductance of a GaAs MESFET up to several hundred Hz.

The circuit-level modeling of dispersion is described by Cojocaru and Brazil.[34] They extend the previous conventional modeling of dispersion of the output conductance to include dispersion of the transconductance. The circuit is very simple. A second voltage-control current source in parallel with a resistance is capacitively coupled to the internal drain-source terminals. This enables the model's transconductance and drain-source conductance to be tailed for high frequencies using these new elements.

Horio and Usarni[40] use two-dimensional simulation to show that a small amount of avalanche breakdown in the presence of traps causes excess hole charge in the substrate that produces "kinks" in the low-frequency I/V data. Since the traps cannot be easily eliminated, the kinks may be removed by removing the conditions initiating avalanche breakdown.

Upon switching the gate voltage, the drain current of a MESFET will have lag effects in the microsecond and millisecond regions that are produced by traps. Curtice et al.[41] have shown the circuit-level modeling of such gate lag effects as well as drain lag effects. Others, such as Kunihiro and Ohno[42] have also presented circuits for the modeling of drain lag effects.

The transistor model with such circuits may then be used to determine if the lag effects interfere with proper operation of the circuit. The work of Curtice et al. was directed toward GaAs digital circuits where the switching waveform must be of high quality. Using the new transistor model, one may determine not only if the circuit will perform, but also circuit changes that will permit operation in the presence of strong lag effects.

Light sensitivity has been described and modeled by Chakrubarti et al.[43] and by Madjar et al.[44] Some circuit-level models are presented in their discussions.

In the microwave application arena, the principal difficulty with traps is that they cause difficulty in determining an accurate microwave model for the transistor. An excellent experimental study of surface gating effects is given by Teyssier et al.[45] Surface gating means that the charge stored in surface states and traps influences the I/V behavior by acting as a second gate. The amount of charge stored in the traps will vary, depending upon the applied voltages, the ambient light, the temperature, and trapping time constants. Teyssier et al. show that measured trap capture time constants are quite different than trap emission time constants. They show how they are able to accurately characterize the I/V behavior for RF operation by using 150 ns bias pulse width. They also describe how they characterize the thermal behavior using longer pulse width.

Many others have published data showing surface gating effects (see, for example, Platzker et al.[46]). The approach a modeler should use is to first determine how important such trapping effects are in the transistor operation. In many transistor designs, the active region is shielded sufficiently from surface charge so that negligible surface gating occurs. In that case, low-frequency I/V data and the resulting transconductance may be very predictive of microwave-frequency behavior. If the trapping effects are found to be of importance, then short-pulsed characterization is required and low-frequency I/V data will not suffice.

In any case, device characterization must include the behavior changes due to self-heating and ambient temperature effects. The modeling of heating effects will be discussed next.

24.8 Modeling Temperature Effects and Self-Heating

Anholt and Swirhun[47] and others have documented the changes of GaAs MESFETs and HEMTs at elevated temperatures. However, the modeling of a device over a temperature range is often accomplished using temperature coefficients. With regard to drain current and DC transconductance, the effects of elevated temperature is quite different at large channel current as compared to operation near pinch off. At large channel current, the electron mobility decrease with temperature increase is most important, whereas at low current, the decrease in the pinch-off voltage with temperature is most important. This produces the interesting effect that transconductance decreases with temperature at large current but increases with temperature at very low currents. This behavior is best modeled using a temperature analog circuit that will be described later.

Figure 24.4 shows the behavior of DC drain current for a 0.25-μm PHEMT at four different ambient temperatures. Heating effects are obvious in Fig. 24.4 where current (and transconductance) near positive V_{gs} is reduced and current (and transconductance) near pinch off is increased.

The effects of temperature upon drain current in MESFETs and PHEMTs may be considered second order but they are of first order in bipolar devices. The reason is due to the current exponential dependence upon temperature in a bipolar device. The gate current effects due to temperature in MESFETs and PHEMTs are of first order for the same reason. First order and even some second order changes with temperature may be modeled using linear temperature coefficients if the changes are reasonably linear.

Many devices are operated such that their self-heating effects are quite important. This can occur, for example, in a power amplifier design where the standby biasing current is low, but the advent of input RF power turns on the drain current and output RF power. The ambient temperature may not change but the device will operate with more elevated temperature with the application of the RF power. The use of temperature coefficients may not be accurate for such simulations.

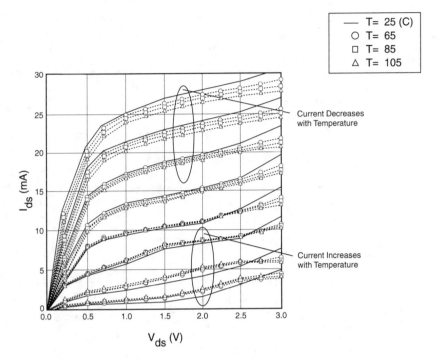

FIGURE 24.4   The behavior of DC drain current for a 0.25-μm PHEMT at four different temperatures.

More accurate modeling of self-heating effects in transistors circuits simulators has been done for about ten years using a thermal analog circuit, first used in SPICE applications. It has been found to work well for bipolar simulation as well as for MESFET and PHEMT simulations. The early studies were reported by Grossman and Oki,[48] F. Q. Ye,[49] and others.[50-52] The main differences in these early studies relate to the description of temperature effects in the transistor and not to the CAD model used for simulation in SPICE.

Figure 24.5 shows an HBT transistor model with the additional thermal analog circuit. The device can be of any type but must have well-defined descriptions of the model coefficients as a function of temperature. The analog circuit consists of a current source, a resistor, and a capacitor. R_{th} is the value of thermal resistance and the $R_{th}*C_{th}$ time constant is the thermal time constant of the device. The current source to the thermal circuit, I_{th}, is equal in magnitude to the instantaneous internal dissipated power to the device. For DC biasing, I_{th} to the thermal circuit would be equal to the total biasing power to the device and the temperature rise would be numerically equal to I_{th} times R_{th}. For RF or transient conditions, the average temperature rise would be that evaluated over some period of time including the effects of the thermal time constant. Thus, for an RF amplifier application, I_{th} would be equal to the DC biasing power plus the RF heating effects less the net RF power leaving the device. Convergence in a simulator is not assured since the value of the model coefficients must be consistent for the temperature of the device.

Harmonic balance simulators usually find the steady-state RF condition efficiently. Using this method, the simulator must find the solution with temperature rise consistent with the device parameters producing the temperature rise. Surprisingly, the harmonic balance simulators do not seem to be much less efficient when the thermal circuit is used, unless a thermal runaway condition exists. In that case, no solution will be found.

Figure 24.6 shows the collector I-V relationship for an HBT exhibiting self-heating effects. The current curves without heating effects are flat in the saturation region. Heating of the lattice reduces the electron mobility, and thus reduces the collector current.

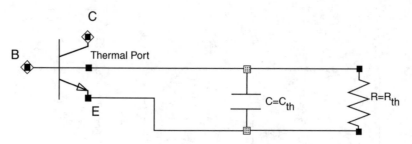

FIGURE 24.5 The HBT transistor model with the thermal analog circuit.

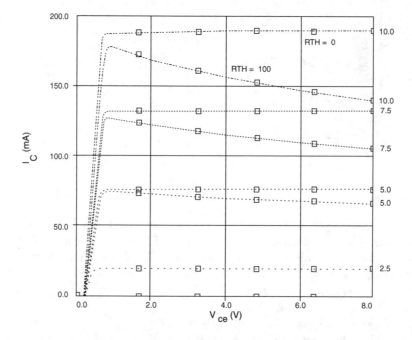

FIGURE 24.6 The collector I-V relationship for an HBT with and without self-heating effects.

24.9 Enhancing the Gummel-Poon Model for Use with GaAs and InP HBTs

The Gummel-Poon model, or the GP model,[14] is a complex, physical parameter model with 55 parameters, and is widely used. It was developed early for SPICE, and all colleges and universities teach their electrical engineering students to use this model. Although the GP model has many parameters, the current expressions are relatively simple. In addition, the current parameters are more closely tied to material parameters rather than manufacturing tolerances, so that there is less variation in current control characteristics than with MESFETs and PHEMTs. The standard bipolar device has less two-dimensional effects than do MESFETs and PHEMTs.

Much effort has been expended to improve the accuracy of compact BJT circuit models for silicon devices. Fossum[53] has reviewed the effort to 1989 and it continues to this day.

Whether the bipolar device is all silicon or a heterojunction device made with SiGe on Si, AlGaAs on GaAs, GaInP on GaAs, or InP on InGaAs, the bipolar action with current gain is the same physical process. So one expects some similarities in the analysis and modeling of the device. However, the heterojunction with the wide bandgap emitter causes the details of the analysis and model to have significant differences from a homojunction device.

The standard SPICE GP model has a number of major deficiencies that must be addressed before it can be used to accurately model a GaAs-based HBT in a large-signal microwave application. First of all, the SPICE code has silicon bandgap parameters hard coded into it and this must be changed to produce the correct temperature effects upon the bandgap. Next, collector-to-base avalanche breakdown must be added because it is important to most GaAs applications. The GP model uses PTF, a phase function, to accommodate the time delay associated with transconductance. It is more convenient for microwave engineers to use the time delay term TAU, as used in MESFET models.

The parameter "Early Voltage" is not as important in GaAs modeling, as it is usually very large. This is because the base doping can be made an order of magnitude larger than for silicon devices, because of the wide bandgap emitter. The large base doping reduces the importance of collector biasing upon the base region, and thus upon the collector current.

There may be dispersion in the collector admittance in HBTs, so some RF conductive element may be needed between collector and emitter.

The manner in which F_t, the frequency for unity current gain, changes with voltage and current is quite different in GaAs devices than with silicon devices. Therefore, a new functional form is need here. Most of the behavior of F_t with respect to collector voltage is related to the electron velocity-electric field (v-E) curve for the material. In the case of silicon devices, F_t generally increases with collector voltage and saturates until heating effects cause a decrease. Figure 24.7 shows such behavior for a SiGe HBT. The I/V characteristic of the device is given in Fig. 24.8.

In the case of GaAs HBTs, F_t peaks at a low voltage and monotonically decrease with further increase of collector voltage. Figure 24.9 shows such data and the devices I/V characteristics are given in Fig. 24.10. This difference in behavior reflects the striking differences between silicon and GaAs v-E curves in the high field region.

There are a multitude of new CAD models formulated to model the behavior of GaAs and InP HBTs. Two examples are the VBIC model[54] and the MEXTRAM model[55] and these are installed on a number of circuit simulators. Most of the new models include self-heating effects because of their importance to device operation and accurate modeling. Because of the significantly poorer thermal conductivity in GaAs compared to silicon and because of the higher power density for best operation in GaAs, self-heating effects are usually important to the operation of the GaAs-based HBTs.

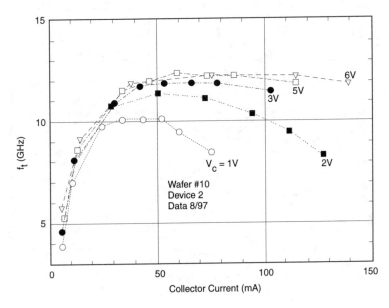

FIGURE 24.7 The Behavior of F_t with biasing for a SiGe HBT.

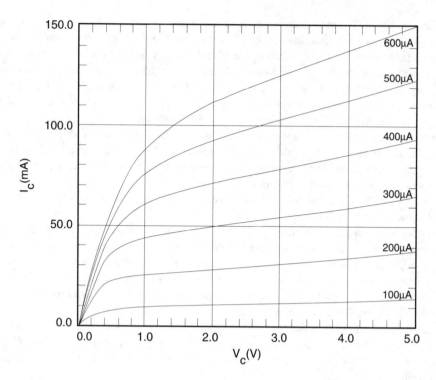

FIGURE 24.8   The I-V relationship for the device of Fig. 8.66.

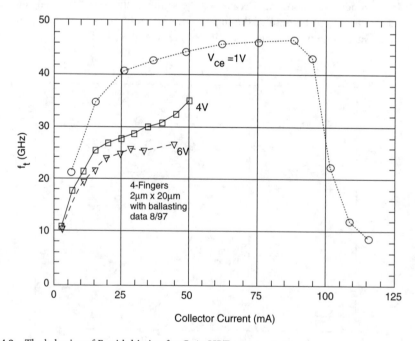

FIGURE 24.9   The behavior of F_t with biasing for GaAs HBT.

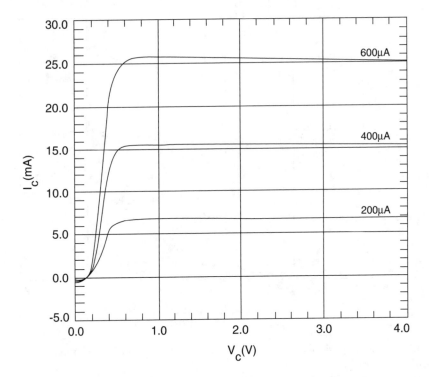

FIGURE 24.10   The I-V relationship for the device of Fig. 24.9.

24.10 Modeling the RF LDMOS Power Transistor

There are numerous silicon MOS models available for DC and RF modeling of silicon transistors. Because the silicon LDMOS transistor has become important for cost-effective consumer applications, many companies have developed nonlinear models specifically for this device. The device incorporates a p-type sinker diffusion to ground the source to the substrate, and thus can be treated as a three-terminal device. This makes it possible to construct a much simpler model, one very similar to the SPICE models developed for the GaAs MESFET.

Perugupalli et al.[56] have used a SPICE circuit network incorporating the standard NMOS SPICE element. Motorola uses the Root model developed for GaAs MESFETs for characterizing the device. However, self-heating effects, important for power applications, cannot be incorporated into this model. The Ho, Green, and Culbertson model[57] based upon the SPICE BSIM3v3 model suffers from the same problem.

Miller, Dinh, and Shumate[58] developed analytical current equations for the device, which led to the development of a new, simpler SPICE model by Curtice, Pla, Bridges, Liang, and Shumate.[59] The model includes self-heating effects, is accurate for both small and large-signal simulations, and operates in transient or harmonic balance simulators. Figure 24.11 shows the model predicts power-added efficiency in excellent agreement with the data for an RF power sweep. A similar model based upon the same equations has been developed and verified by Heo et al.[60]

24.11 Parameter Extraction for Analytical Models

The extraction of parameters for a device model has become less laborious since the advent of new extraction and equipment control programs, such as IC-CAP by Hewlett Packard and UTMOST by Sylvaco International. These programs provide data acquisition and parameter extraction. Such software first enables the engineer to collect I/V and RF data in a systematic fashion on each device tested. This

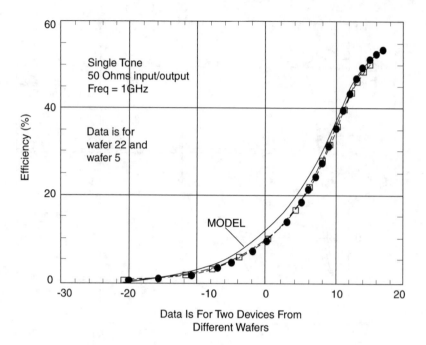

FIGURE 24.11 Efficiency prediction by the model and data for two RF LDMOS devices.

provides consistency between I/V and RF data. The parameter extraction routines then permit the extraction to specific, standard (SPICE) models, such as the Gunnel-Poon, or to new models for which equations may be user defined. Optimizers are used to provide the best fit between the data and the model. Testing can be with pulsed biasing or DC. Heating effects can be separately studied using thermal chucks during testing. Teyssier et al.[45] have discussed the merits of long- and short-pulse testing.

It is usually necessary to use devices of small sizes for characterization and then scale the model to devices actually used in the circuit design. Most SPICE models provide scaling with device area. However, the scaling laws for devices should be verified. It is usually possible to scale MESFETs and HEMTs accurately for a larger number of fingers of the same size. Golio[61] shows the scaling rules if the finger width is different. Because the biasing power may not be uniform on large devices and because the interelectrode capacitance does not scale simply, more complicated scale rules may be found for relatively large devices at high frequencies, i.e., above 5 GHz.

24.12 The Vector Nonlinear Network Analyzer

A large-signal, waveform measurement system has been used by many researchers to measure device characteristics dynamically. The equipment provides time domain voltage and current waveforms during RF excitation of the transistor. The equipment is often called the Vectorial Nonlinear Network Analyzer, or VNNA.

Demmler and Tasker[62] have shown that it is possible to accurately determine the drain current relationship to gate voltage for RF excitation at 2 GHz of a MODFET. The characteristic time delay is found by adding delay until "looping" is minimized. Furthermore, the drain-source I-V relationship can also be evaluated at 2 GHz. There is some difference from that obtained from the DC data. Thus the VNNA provides large-signal transfer characteristics from which more accurate model extraction can be done. Wei et al.[63] have also utilized this techniques to provide the data used for device model parameter extraction for a GaAs HBT.

24.13 Model Verification

The usual approach to verification of a large-signal model is to compare measured device performance with simulations under the same conditions. Initial verification should be comparison of a power sweep of the transistor at the application frequency and with no matching at the input and the output. For this test, we know that all harmonics at the input and output see 50-ohm impedance.

One should measure not only the output power at the fundamental, but also the power at second and third harmonics. If the model is not fully optimized, it will usually agree well with fundamental output power, gain, and efficiency, but not agree with the harmonic power production. The usual cause is due to poor modeling of the I/V relationship; however, in some cases, the nonlinear capacitive modeling may be the problem. After this problem is fixed, testing of third- and fifth-order IMD (intermodulation distortion) should be made, again in a 50-ohm system. If the harmonics now agree, the third-order IMD should agree and some further work may be required to get agreement for the fifth-order IMD.

In the previous test, it is important that the large-signal model be reasonably accurate at small-signal levels. It need not be as accurate as the best small-signal model for most applications.

Further verification work would involve power sweeps under tuned conditions. That is, the transistor may be tuned for best efficiency and the tuner impedance measured. It is important to measure the tuner impedance for fundamental, second, and third harmonics and to use these values in the simulation. The effects of the second harmonic voltage at either input[64] or output can be extremely important.

Testing of the load-pull characteristics should be made and compared with the model's behavior. Here, again, one has to be careful about the effects of harmonics. There are load-pull systems that operate separately on fundamentals and harmonics.

Further tests that may be important to the application may be testing with various ambient temperatures, noise testing, switch-on testing, and others. The specific application of the transistor will dictate the importance of the agreement for each test as well as the RF frequencies, power, and modulation to use for testing.

24.14 Foundry Models and Statistics

GaAs chip foundries provide design manuals that utilize small-signal as well as large-signal models. These are developed from measurements and statistical analysis of the data. However, these are guidelines for the designer, and often the best procedure is to obtain foundry test devices and develop more accurate models based upon new data. Device uniformity has improved greatly and yield prediction is becoming more accurate.

In addition, software programs such as IC-CAP and others provide statistical analysis of models extracted from test wafers. Both corner models and standard statistical patterns are available.

There is the fundamental problem that most foundries continue to tune their processes. It is often the case that the process has been changed and previous statistics are no longer valid. However, the design engineer is better off with approximate guidelines as to statistical patterns or corner models, than none at all.

24.15 Future Nonlinear Transistor Models

One can expect that with the ever-increasing speed of computers, circuit simulators will be able to utilize more physics-based models. This will aid in determining the effect of device design parameters upon chip yield and performance.

Improvements will be made in nonlinear model extraction software. The extraction parameters will be much less dependent upon the expertise of the tester. There will be improved collection schemes for transistor model statistics.

Finally, one expects that the nonlinear models will be made to be more easily tailored for adaptation to specific device behaviors. One often would like to start with a template for one of the standard nonlinear models and then tailor its behavior. Future simulators should make this procedure simpler than present procedures.

References

1. Nagle, L. W., SPICE 2: A computer program to simulate semiconductor circuits, Electronics Research Laboratory, College of Engineering, University of California, Berkeley, Memo, ERL-M520, 1975.
2. Pinto, M. R., Conor, R. S., and Dutton, R. W., PIECES2 - Poisson and continuity equation solver, Stanford Electronics Laboratory, Technical Report, Stanford University, 1984.
3. Wada, T. and Frey, J., Physical basis of short-channel MESFET operation, *IEEE Trans. on Electron Devices*, ED-26, 476, 1979.
4. Curtice, W. R., Analysis of the properties of three-terminal transferred electron logic gates, *IEEE Trans. on Electron. Devices*, ED-24, 1553, 1977.
5. Warriner, R. A., Computer simulation of gallium arsenide field-effect transistors using Monte-Carlo methods, *Solid-State Electron Devices*, 1, 105, 1977.
6. Moglestue, C., A self-consistent Monte Carlo particle model to analyze semiconductor microcomponents of any geometry, *IEEE Trans. on CAD*, CAD-5, 326, 1986.
7. Curtice, W. R. and Yun, Y-H, A temperature model for the GaAs MESFET, *IEEE Trans. on Electron Devices*, ED-28, 954, 1981.
8. *ATLAS User's Manual*, Silvaco International, Santa Clara, CA, Version 4.0, 1995.
9. Curtice, W. R., Direct comparison of the electron-temperature model with the particle-mesh (Monte-Carlo) model for the GaAs MESFET, *IEEE Trans. on Electron Devices*, ED-29, 1942, Dec. 1982.
10. Li, Q. and Dutton, R. W., Numerical small-signal AC modeling of deep-level-trap related frequency-dependent output conductance and capacitance for GaAs MESFETs on semi-insulating substrates, *IEEE Trans. on Electron Devices*, 38, 1285, 1991.
11. Snowden, D. M. and Pantoia, R. R., Quasi-two-dimensional MESFET simulation for CAD, *IEEE Trans. on Electron Devices*, 36, 1989.
12. Morton, C. G., Atherton, J. S., Snowden, C. M., Pollard, R. D., and Howes. M. J., A large-signal physical HEMT model, 1996 *International Micrwave Symposium Digest*, 1759, 1996.
13. Root, D. E. et al., Technology independent large-signal non quasi-static FET models by direct construction from automatically characterized device data, 21st *European Microwave Conference Proceedings*, 927, 1991.
14. Gummel and Poon, An integral charge-control relationship for bipolar transistors, *Bell System Tech. Journal*, 49, 115, 1970.
15. Daniel, T. T. and Tayrani, R., Fast bias dependent device models for CAD of MMICs, *Microwave Journal*, 74, 1995.
16. Khatibzadeh, M. A. and Trew, R. J., A large-signal analytical model for the GaAs MESFET, *IEEE Trans. on Microwave Theory and Tech.*, 36, 231, 1988.
17. Ladbrooke, P. H., *MMIC Design: GaAs FETs and HEMTs*, Artech House, Inc., Boston, 1989, chap. 6.
18. Lehovec, K. and Zuleeg, R., Voltage-current characteristics of GaAs JFETs in the hot electron range, *Solid State Electron.*, 13, 1415, 1970.
19. Bandler, J. W. et al., Statistical modeling of GaAs MESFETs, 1991 *IEEE MTT-S International Microwave Symposium Digest*, 1, 87, 1991.
20. Statz, H, Newman, P., Smith, I. W., Pucel, R. A., and Haus, H. A., GaAs FET device and circuit simulation in SPICE, *IEEE Trans. Electron Devices*, 34, 160, 1987.
21. D'Agostino, S. et al., Analytic physics-based expressions for the empirical parameters of the Statz-Pucel MESFET model, *IEEE Trans. on MTT*, MTT-40, 1576, 1992.
22. *International Journal of Microwave and Millimeter-Wave CAE*, 9, No. 3, 1999.
23. Trew, R. J., MESFET models for microwave CAD applications, *Internation Journal of Microwave and Millimeter-Wave CAE*, 1, 143, 1991.
24. Snowden, C. M., Nonlinear modeling of power FETs and HBTs, *International Journal of Microwave and Millimeter-Wave CAE*, 6, 219, 1996.

25. Dortu, J-M, Muller, J-E, Pirola, M., and Ghione, G. Accurate large-signal GaAs MESFET and HEMT modeling for power MMIC amplifier design, *International Journal of Microwave and Millimeter-Wave CAE*, 5, 195, 1995.

26. W. R. Curtice and R. L. Camisa, Self-consistent GaAs FET models for amplifier design and device diagnostics, *IEEE Trans. on Microwave Theory and Tech.*, MTT-32, 1573, 1984.

27. R. L. Vaitkus, Uncertainty in the Values of GaAS MESFET Equivalent Circuit Elements Extracted from Measured Two-Port Scattering Parameters, Presented at 1983 IEEE Cornell Conference on High Speed Semiconductor Devices and Circuits, Cornell University, Ithaca, NY, 1983.

28. Byun, Y. H., Shur, M. S., Peczalski, A., and Schuermeyer, F. L., Gate voltage dependence of source and drain resistances, *IEEE Trans. on Electron Devices*, 35, 1241, 1998.

29. Samelis, A. and Pavlidis, D., Modeling HBT self-heating, *Applied Microwave & Wireless*, Summer Issue, 56, 1995.

30. Teeter, D. A. and Curtice, W. R., Comparison of hybrid pi and tee HBT circuit topologies and their relationship to large-signal modeling, *1997 IEEE MTT-S International Microwave Symposium Digest*, 2, 375, 1997.

31. Curtice, W. R., A MESFET model for use in the design of GaAs integrated circuits, *IEEE Trans. on Microwave Theory and Techniques*, 23, 448, 1980.

32. Angelov, I., Zirath, H., and Rorsman, N., New empirical nonlinear model for HEMT and MESFET and devices, *IEEE Trans. on Microwave Theory and Techniques*, 40, 2258, 1992.

33. Qu, G. and Parker, A. E., Continuous HEMT model for SPICE, *IEE Electronic Letters*, 32, 1321, 1996.

34. Cojocaru, V. I. and Brazil, T. J., A scalable general-purpose model for microwave FETs including the DC/AC dispersion effects, *IEEE Trans. on Microwave Theory and Techniques*, 12, 2248, 1997.

35. Materka, A. and Kacprzak, T., Computer calculation of large-signal GaAs FET amplifier characteristics, *IEEE Trans. on Microwave Theory and Tech.*, 33, 129, 1985.

36. Circuit Network Items, Series IV, Hewlett Packard, HP Part. No. E4605-90038, 1161, 1995.

37. Calvo, M. V., Snider, A. D., and Dunleavy, L. P., Resolving Capacitor Discrepancies Between Large and Small Signal FET Models, 1995 IEEE MTT-S International Microwave Symposium, 1251, 1995.

38. Jansen, P. et al., Consistent small-signal and large-signal extraction techniques for heterojunction FET's, *IEEE Transaction on Microwave Theory and Tech.*, 43, 1, 87, 1995.

39. Mallavarpu, R., Teeter. D., and Snow, M., The importance of gate charge formulation in large-signal PHEMT modeling, *GaAs IC Symposium Technical Digest*, 87, 1998.

40. Horio, K. and Usarni, K., Analysis of kink-related backgating effect in GaAs MESFETs, *IEEE Electron Devices Letters*, 537, 16, 1995.

41. Curtice, W. R., Bennett, J. H., Suda, D., and Syrett, B. A., Modeling of current lag effects in GaAs IC's, *1998 IEEE MTT-S International Microwave Symposium Digest*, 2, 603, 1998.

42. K. Kunihiro and Y. Ohno, An equivalent circuit model for deep trap induced drain current transient behavior in HJFETs, *1994 GaAs IC Symposium Digest*, 267, 1994.

43. Chakrabarti, P., Shrestha, S. K., Srivastava, A., and Skxena, D., Switching characteristics of an optically controlled GaAs-MESFET, *IEEE Trans. on Microwave Theory and Techniques*, 42, 365, 1994.

44. Madjar, K., Paolella, A., and Herczfeld, P. R., Modeling the optical switching of MESFET's considering the external and internal photovoltaic effects, *IEEE Trans. on Microwave Theory and Tech.*, 42, 62, 1994.

45. Teyssier, J-P, Bouysse, P., Ouarch, A., Barataud, D., Peyretaillade, T., and Quere, R., 40-GHz/150-ns versatile pulsed measurement system for microwave transistor isothermal characterization, *IEEE Trans. on Microwave Theory and Tech.*, 46, 2043, 1998.

46. A. Platzker et al., Characterization of GaAs devices by a versatile pulsed I-V measurement system, *1990 IEEE MTT Symposium Digest*, 1137.

47. Anlholt and Swirhun, Experimental characterization of the temperature dependence of GaAs FET equivalent circuits, *IEEE Trans. on Electron Devices*, 39, 2029, 1992.

48. Grossman, P. C. and Oki, A., A large signal DC model for GaAs/GaAlAs heterojunction bipolar transistors, *Proc. IEEE BCTM*, 258, 1989.

49. Ye, F. Q., A BJT model with self-heating for WATAND computer simuation, M. S. Thesis, Youngstown State University, Youngstown, OH, 1990.

50. McAndrew, C. C., A complete and consistent electrical/thermal HBT model, *Proc. IEEE BCTM*, 200, 1992.

51. Corcoran, J., Poulton, K. and Knudsen, K., GaAs HBTs: an analog circuit design perspective, *Proc. IEEE BCTM*, 245, 1991.

52. Fox, R. M. and Lee, S-G, Predictive modeling of thermal effects in BJTs, *Proc. IEEE BCTM*, 89, 1991.

53. Fossum, J. G., Modeling issues for advanced bipolar device/circuit simulation, *Proc. 1989 IEEE BCTM*, 234, 1989.

54. McAndrew, C. C., Seitchik, J., Bowers, D., Dunn, M., Foisy, M., Getreu, I., Moinian, S., Parker, J., van Wijnen, P., and Wagner, L., VBIC95: An improved vertical IC bipolar transistor model, *Proc. 1995 BCTM*, 170, 1995.

55. de Graaff, H. C. and Kloosterman, W. J., New formulation of the current and charge relations in bipolar transistor modeling for CACD purposes, *IEEE Trans. on Electron Devices*, ED-32, 2415, 1986.

56. Perugupalli, P., Trivedi, M., Shenai, K., and Leong, S. K., Modeling and characterization of 80v LDMOSFET for RF communications, *1997 IEEE BCTM*, 92, 1997.

57. Ho, M. C., Green, K., Culbertson, R., Yang, J. Y., Ladwig, D., and Ehnis, P., A physical large signal Si model for RF circuit design, *1997 MTT-S International Microwave Symposium Digest*, 1997.

58. Miller, M., Dinh, T., and Shumate, E., A new empirical large signal model for silicon RF LDMOS FET's, *1997 IEEE MTT Symposium on Technologies for Wireless Applications Digest*, Vancouver, Canada, 19, 1997.

59. Curtice, W. R., Pla, J. A., Bridges, D., Liang, T., and Shumate, E., A new dynamic electro-thermal model for silicon RF LDMOS FET's, *1999 IEEE MTT-S International Microwave Symposium Digest*, 1999.

60. Heo, D., Chen, E., Gebara, E., Yoo, S., Lasker, J., and Anderson, T., Temperature dependent MOSFET RF large signal model incorporating self-heating effects, *1999 MTT-S International Microwave Symposium Digest*, 1999.

61. Golio, J. M., *Microwave MESFETs and HEMTs*, Artech House, Boston, 1991.

62. Demmler, M. and Tasker, P. J., A vector corrected on-wafer large-signal waveform system for novel characterization and onlinear modeling techniques for transistors, Presented at the workshop on New Direction in Nonlinear RF and Microwave Characterization, 1996 International Microwave Symposium, 1996.

63. Wei, C.-J., Lan, Y. E., Hwang, J. C. M., Ho, W.-J., and Higgins, J. A. Waveform-based modeling and characterization of microwave power heterojunction transistors, *IEEE Trans. on Microwave Theory and Techniques*, 43, 2898, 1995.

64. Watanabe, S, Takatuka, S., Takagi, K., Kukoda, H., and Oda, Y., Simulation and experimental results of source harmonic tuning on linearity of power GaAs FET, 1996 MTT-S International Microwave Symposium, 1996.

Index